T0365519

Introduction to Mathematical Analysis

Naokant Deo · Ryozi Sakai

Introduction to Mathematical Analysis

Naokant Deo
Department of Applied Mathematics
Delhi Technological University
New Delhi, India

Ryozi Sakai
Department of Mathematics
Meijo University
Nagoya, Japan

ISBN 978-981-97-6567-6 ISBN 978-981-97-6568-3 (eBook)
https://doi.org/10.1007/978-981-97-6568-3

Mathematics Subject Classification: 22E30, 26E40, 54E45, 54E50, 54E35, 26A42

This Springer imprint is published by the registered company Springer Nature Singapore Pte Ltd.
The registered company address is: 152 Beach Road, #21-01/04 Gateway East, Singapore 189721, Singapore

Preface

Real analysis has earned its place as a core subject in the undergraduate mathematics curriculum. The prime focus of this textbook is primarily on those students who are learning the topics of real analysis for the first time and are planning to pursue a professional career in advanced mathematics. It is often the case that core mathematics curricula are time-bound, so they do not include all the topics that one might like. This book includes important topics that may be skipped in required courses, but the mathematician who wants to excel in his/her profession will definitely seek to learn all the relevant topics through self-study.

This book is a straightforward and comprehensive presentation of the concepts and methodology of elementary real analysis, written as per the understanding and level of undergraduate students. We have tried to keep the focus on the core of the analysis.

This book differentiates itself from other books available in this area by incorporating and explaining each and every theorem as well as definitions through geometrical interpretations (pictures and graphs).

The theory of sequences and series forms the backbone of this book. Initial chapters look into the fundamentals of real variables, Peano axioms, countable and uncountable sets, the boundedness of sets, the convergence of the sequence, and the Cauchy sequence.

Later chapters deal with infinite series and alternating series, followed by a chapter on limit, continuity, and differentiability.

In Metric spaces, an attempt has been made to explain the theorems and definitions through several pictures. Limits of sequences and functions, topology of metric spaces, continuity of functions, and the Cauchy sequence have been thoroughly discussed in this chapter.

Riemann's integral and its properties, sequences, and series of functions have been explained up to undergraduate-level students in the sixth and seventh chapters, respectively.

The last chapter includes the Lebesgue theory of measure, measurable functions, the Lebesgue integral and its properties, differentiation, and integration. Further added features are the convergence of the sequence of functions, Dini theorem, Egorov theorem, Lusin theorem, and the Riemann-Stieltjes integral.

This book has the potential to serve as the building block for beginners in the field of Mathematics.

We are grateful to Prof. Vijay Gupta for encouraging this project and for making many suggestions about pursuing it, and to the students of the first author, Dr. Dhananjay Gopal, Neha, Navshakti Mishra, Lipi Choudhary, and Kapil Kumar, for helping with the readability. We would like to give a special thanks to Mr. Kagehiro Naito, a Japanese English teacher, for encouraging us to work even harder in our attempt to produce a good book. The typesetting is by *LaTeX*, and the figures are drawn with CorelDRAW. We plan to make corrections and other comments from readers. We plan to maintain a list of known corrections on the web page of the first author.

New Delhi, India — Naokant Deo
Nagoya, Japan — Ryozi Sakai

Acknowledgments

I dedicate this book to the memory of my father, the late S. L. Kurmi (1941–2020). He was a great source of inspiration and encouragement, guiding me to choose mathematics as my main subject from an early age. I am grateful to my mother, Sarojini Kurmi, for her continuous blessings. I am grateful for the unwavering support of my wife, Anjana, and my daughters, Bhumika (Vidhi) and Nidhi, throughout the entire process of writing this book.

Naokant Deo

Contents

1 The System of Real Numbers 1
- 1.1 Peano Postulates/Axioms 1
- 1.2 The Structure of the Real Number System 2
- 1.3 Density and Completeness 4
- 1.4 Fields 5
- 1.5 Ordered Fields 8
  - 1.5.1 The Absolute Value or Modulus 8
  - 1.5.2 Extended Real Numbers 9
- 1.6 Neighborhoods 10
- 1.7 Functions, Countable and Uncountable Sets 10
  - 1.7.1 Functions 10
  - 1.7.2 Direct and Inverse Images 10
  - 1.7.3 Composition of Functions 14
  - 1.7.4 Equivalence of Sets 14
  - 1.7.5 Countable Sets 15
  - 1.7.6 Uncountable Sets 19
  - 1.7.7 Bounded Above 21
  - 1.7.8 Bounded Below 21
- 1.8 Bounded Sets 22
  - 1.8.1 The Greatest Element (The Maximum Element) 22
  - 1.8.2 The Smallest Element (The Least Element) 22
  - 1.8.3 The Least Upper Bound (l.u.b.)/Supremum (Sup) 23
  - 1.8.4 Properties 23
  - 1.8.5 The Greatest Lower Bound (g.l.b.)/Infimum (Inf) 24
  - 1.8.6 Properties 24
  - 1.8.7 The Completeness Axiom 26
  - 1.8.8 Complete Ordered Fields 26
- 1.9 The Archimedean Property of Real Numbers 28
- 1.10 Denseness 30
- 1.11 Exercises 33

2 **Real Sequences** 35
2.1 Sequences 35
2.1.1 The Range of Sequences 36
2.1.2 Equality of Sequences 36
2.1.3 Constant Sequences 36
2.2 Bounded Sequences 37
2.3 Neighborhoods of a Point 37
2.4 Convergence of the Sequences (Limit of Sequences) 38
2.4.1 Working Rule to Prove That 40
2.4.2 Increasing Sequences 43
2.4.3 Decreasing Sequences 44
2.4.4 Strictly Increasing or Decreasing 44
2.4.5 Monotonic Sequences 44
2.4.6 Monotonic Sequences and Their Convergence 45
2.5 Operations of Convergent Sequences 51
2.6 Operations of Divergent Sequences 55
2.7 Nested Intervals 57
2.8 Subsequences 58
2.9 Cauchy Sequences 62
2.9.1 Cauchy's General Principle for Convergence 68
2.10 Limits Superior and Inferior 68
2.11 Exercises 75

3 **Infinite Series of Numbers** 79
3.1 Positive Terms Series 79
3.1.1 Convergence and Divergence of Series 80
3.1.2 Fundamental Properties 81
3.1.3 Comparison Tests 82
3.1.4 Comparison Tests (Limit Theorems) 84
3.1.5 p-Series Test 85
3.1.6 D'Alembert's Ratio Test 89
3.1.7 Cauchy's Root Test 91
3.1.8 Raabe's Test 93
3.1.9 The Integral Test (Cauchy-Maclaurin's Integral Test) 97
3.1.10 Logarithmic Test 100
3.2 Alternating Series 104
3.3 Absolute and Conditional Convergence 108
3.4 Rearrangement of Terms 110
3.5 Exercises 114

4 Limits, Continuity, and Differentiability 117
4.1 The Limit of a Function 117
4.1.1 Some Theorems on Limits 120
4.2 Continuity 121
4.2.1 Sums, Products, and Quotients of Continuous Functions 125
4.2.2 Some Theorems on Continuous Functions 125
4.2.3 Compositions of Continuous Functions 127
4.3 The Intermediate Value Theorem 128
4.4 Uniform Continuity 129
4.5 Types of Discontinuity 130
4.6 Differentiability 131
4.7 Interpretation of the Derivative 133
4.8 One-Sided Derivatives 135
4.9 The Mean-Value Theorem 136
4.9.1 The Geometric Interpretation of the Mean-Value Theorem 137
4.10 The Consequences of the Mean-Value Problem 139
4.11 Taylor's Theorem 140
4.11.1 Taylor Polynomials 140
4.12 The L'Hopital's Rule 142
4.13 Exercises 145

5 Metric Spaces 149
5.1 Definitions of Metric Spaces and Examples 149
5.1.1 The l^p-Spaces 151
5.1.2 Normed Linear Spaces 155
5.1.3 The Euclidean Space 156
5.1.4 The Pseudo-Metric (The Semi-Metric) 162
5.1.5 Quasi-metric Spaces 163
5.2 Limits of Sequences in (X, ρ) 163
5.3 Limits of Functions in (X, ρ) 164
5.3.1 The Diameter of a Set in Metric Spaces 166
5.3.2 The Distance Between a Point and a Set in Metric Spaces 166
5.3.3 The Distance Between Two Sets in Metric Spaces 167
5.3.4 Bounded Metric Spaces 168
5.4 Equivalent Metric Spaces 168
5.5 Product Metric Spaces 170

5.6 The Topology of Metric Spaces 171
5.6.1 Open and Closed Spheres; Balls 172
5.6.2 Neighborhoods 177
5.7 Continuity of Functions 178
5.8 Cluster Points 182
5.9 Open Sets 183
5.10 Closed Sets 187
5.11 The Closure of a Set 191
5.11.1 Interior Points 192
5.11.2 The Interior of a Set 193
5.11.3 Exterior Points and Exterior Sets 194
5.11.4 Frontier Points and Boundary Points 195
5.12 Subspaces of a Metric Space 196
5.12.1 Convergent Sequences 199
5.13 Cauchy Sequences 201
5.14 Complete Metric Spaces 205
5.15 Cantor Sets 206
5.15.1 Dense Sets 209
5.15.2 Separable Spaces 210
5.16 The Baire Category Theorem 211
5.17 Uniform Continuity 216
5.18 Homeomorphisms 218
5.19 The Banach Contraction Mapping Theorem 218
5.20 Compactness Arguments 221
5.20.1 The Finite Intersection Property (FIP) 227
5.20.2 Sequential Compactness 228
5.20.3 ε-Net and Totally Bounded 230
5.20.4 Separable Metric Spaces 231
5.21 Connectedness 235
5.21.1 Separated Sets 235
5.22 Connected and Disconnected Sets 236
5.23 Exercises 241

6 The Riemann Integral 243
6.1 Partitions 243
6.2 The Norm or Mesh of the Partition 244
6.3 Tagged Partitions 244
6.4 Refinement (Finer) 244
6.5 Riemann Sums 245
6.6 Upper and Lower Riemann Integrals 245
6.7 The Riemann Integral 246
6.8 Properties of the Riemann Integral 256
6.9 The Fundamental Theorem of Calculus 260
6.9.1 Mean-Value and Change-of-Variable Theorems 262
6.10 Exercises 264

7 Sequences and Series of Functions 267
7.1 Sequences of Functions 267
7.2 Series of Functions 267
7.3 Pointwise Convergence 268
7.4 Uniform Convergence 270
7.5 Comparing Uniform with Pointwise Convergence 271
7.6 Uniform Convergence and Continuity 277
7.7 Uniform Convergence and Integration 279
7.8 Uniform Convergence and Differentiation 281
7.9 The Weierstrass Approximation Theorem 283
7.10 Exercises 288

8 The Lebesgue Integral 289
8.1 Outer Measure and Measurable Sets 289
8.1.1 Algebra and σ-Algebra 289
8.1.2 Outer Measure and Measurable Sets 293
8.1.3 Fundamental Properties for the Measure 301
8.2 Lebesgue Integral on $\mathbb{R}$ 306
8.2.1 Measurable Functions 306
8.2.2 Simple Functions and Integrals 310
8.2.3 Fundamental Properties of the Lebesgue Integral 316
8.2.4 The Convergence of Sequences of Functions 321
8.2.5 The Riemann Integral and the Lebesgue Integral 331
8.2.6 Riemann-Stieltjes Integral 333
8.3 Exercises 339

Selected Answers 341

Bibliography 343

Symbols and Notations

$\mathbb{N}$	The set of natural numbers
$\mathbb{I}$	The set of integers
$\mathbb{Q}$	The set of rational numbers
$\mathbb{R}$	The set of real numbers
$\mathbb{R}_{ext}$	The set of extended real numbers
$\mathbb{C}$	The set of complex numbers
(a, b)	An open interval
$[a, b]$	A closed interval
(X, ρ)	Metric space equipped with metric ρ
$S(x_0, r)$ or $S_r(x_0)$	Open sphere (Open Ball)
$S[x_0, r]$ or $S_r[x_0]$	Closed sphere (Closed Ball)
$N_r(p)$ or $N(p, q)$	Neighborhood (neighborhood) of a point p
$\mathcal{P}$	Partition of interval $[a, b]$
$\mathcal{P}^*$	Refinement of $\mathcal{P}$
$L(\mathcal{P}, f)$	Lower sum
$U(\mathcal{P}, f)$	Upper sum
$\underline{\int_a^b} = \sup L(\mathcal{P}, f)$	Lower Riemann integral
$\overline{\int_a^b} fdx = \inf U(\mathcal{P}, f)$	Upper Riemann integral
$f \in \mathcal{R}[a, b]$	f is integrable in closed interval $[a, b]$
Λ	Index set
$(X, \lVert \cdot \rVert)$	Normed space
m^*	Outer measure
$\mathcal{L}$	Lebesgue integral

Chapter 1
The System of Real Numbers

"God created the integers and the rest is the work of man." (Leopold Kronecker, in an after-dinner speech at a conference, Berlin, 1886).

In this chapter, we go over the essential and foundational facts about the real number system. First, we will provide the structure of real numbers through certain axioms, and then we will show that between any two real numbers, there is an infinite number of rational and irrational numbers, which in turn are used to prove several interesting results.

The main concepts of Analysis are convergence, continuity, smoothness, differentiation and integration. Calculus and Real Analysis begin with natural numbers:

$$\mathbb{N} = \{1, 2, 3, \cdots\}.$$

Elements of $\mathbb{N}$ will also be called positive integers. We add and multiply natural numbers in the usual way.

An essential property of the natural numbers is the following induction principle, which expresses the idea that we can reach every natural number by counting upwards from one. We begin with the following axioms formulated by the Italian mathematician G. Peano (1858–1932) for the natural numbers $\mathbb{N}$.

1.1 Peano Postulates/Axioms

The properties of natural numbers are called Peano axioms:

(P1) $1 \in \mathbb{N}$, i.e., 1 is a natural number, and $\mathbb{N}$ **is a non-empty set.**
(P2) For each element $n \in \mathbb{N}$ there is a unique element $n_0 \in \mathbb{N}$ called the successor of n, i.e., **each element of $\mathbb{N}$ has a successor in $\mathbb{N}$.**
(P3) For each $n \in \mathbb{N}$, $n_0 \neq 1$, 1 **is not the successor of any element of** $\mathbb{N}$.

N. Deo and R. Sakai, *Introduction to Mathematical Analysis*,
https://doi.org/10.1007/978-981-97-6568-3_1

(P4) For every $m, n \in \mathbb{N}$ with $m \neq n$ and $m_0 \neq n_0$, distinct elements in $\mathbb{N}$ have distinct successors, i.e., **each successor is unique.**
(P5) If $A \subseteq \mathbb{N},\ 1 \in A$ and $p \in A \Longrightarrow p_0 \in A$, then $A = \mathbb{N}$.

These axioms completely define the set of natural numbers $\mathbb{N}$.

As an illustration of how induction can be used, we prove the following result for the sum of the first n squares, written in summation notation as

$$\sum_{k=1}^{n} k^2 = 1^2 + 2^2 + 3^2 + \cdots n^2.$$

Proposition 1.1 *For every $n \in \mathbb{N}$, we have*

$$\sum_{k=1}^{n} k^2 = \frac{1}{6} n(n+1)(2n+1).$$

Proof If $A \subseteq \mathbb{N},\ 1 \in A$ and $n \in A$, then

$$\begin{aligned}
\sum_{k=1}^{n+1} k^2 &= \sum_{k=1}^{n} k^2 + (n+1)^2 \\
&= \frac{1}{6} n(n+1)(2n+1) + (n+1)^2 \\
&= \frac{1}{6}(n+1)(2n^2 + 7n + 6) \\
&= \frac{1}{6}(n+1)(n+2)(2n+3).
\end{aligned}$$

It follows that Peano axioms $(P5)$ holds when n is replaced by $n + 1$. Thus $n \in \mathbb{N}$ implies that $(n + 1) \in \mathbb{N}$, so $A = \mathbb{N}$, and the proposition follows by induction. ∎

Equations for the sum of the first n cubes

$$\sum_{k=1}^{n} k^3 = \frac{1}{4} n^2 (n+1)^2,$$

and other powers can be similarly proved by induction.

1.2 The Structure of the Real Number System

When we add two natural numbers, we get a natural number. But the inverse operation (subtraction) is not always possible if the domain is a set of natural numbers

$$\mathbb{N} = \{1, 2, 3, \cdots\} = \mathbb{I}^+.$$

We note that $1_0 = 2,\ 2_0 = 3,\ 3_0 = 4, \cdots$, and so $n_0 - n = 1$ for any $n \in \mathbb{N}$, in other words,

$$n + 1 = n_0 = \min\{p;\ n < p,\ p \in \mathbb{N}\}.$$

Hereafter, we use these facts without notice. Let

$$a + x = b, \quad \forall a, b \in \mathbb{N}.$$

For $a = b$ or $a > b$, there is no solution in domain $\mathbb{N}$. This equation gives a solution within domain $\mathbb{N}$ for only $a < b$. We consider the equation $x + 3 = 2$ and its solution is not an element of domain $\mathbb{N}$.

So negative integers

$$\mathbb{I}^- = \{-1, -2, -3, \cdots\} \text{ and } a + (-a) = 0, \quad \forall a \in \mathbb{N}$$

are included in domain $\mathbb{I}$, which constitutes the system of integers

$$\mathbb{I} = \{\cdots, -3, -2, -1, 0, 1, 2, 3, \cdots\} = \mathbb{I}^- \cup \{0\} \cup \mathbb{I}^+.$$

The concept of fractions for positive and negative numbers and the number zero is called the system of rational numbers. For the equation

$$ax = b, \quad \forall a, b \in \mathbb{I} \text{ and } a \neq 0,$$

there is no solution in $\mathbb{I}$ for this equation.

The system of rational numbers $\mathbb{Q}$ is denoted by $\left\{\frac{p}{q};\ q \neq 0\ and\ p, q \in \mathbb{I}\right\}$.

The process of extracting the root of numbers; there is no rational number whose square is equal to 2, that is,

$$x^2 - 2 = 0, \quad 2 \in \mathbb{Q}.$$

In order to give answers to such questions, the system of rational numbers had been further extended by introducing the so-called *irrational* numbers like $\sqrt{2},\ \pi,\ e$, etc (Fig. 1.1).

The union or combination of *rational* and *irrational* numbers is the *Real Numbers*. Geometrically, the set of real numbers represented by a *real line*. The positive real numbers correspond to the point to the right of the origin, and the negative real

$1.5 = \frac{3}{2}$ > ***Ratio*** $\qquad \pi = 3.14159 \ldots = \frac{?}{?}$ ***(No Ratio)***

Rational number **Irrational number**

Fig. 1.1 Example of rational and irrational numbers

numbers correspond to the point to the left of the origin. The set of all real numbers is symbolically denoted by

$$\mathbb{R} = (-\infty, \infty).$$

Example 1.1 $\sqrt{2}$ is an irrational number.

Let us suppose that $\sqrt{2}$ is a rational number. Then it can be written as p/q, where p and q are integers that are not both even numbers. In other ways, we can say that p and q are relatively prime, that is, their greatest common divisor is 1.

Thus we have

$$\begin{aligned} &\frac{p}{q} = \sqrt{2} \quad (q \neq 0, \text{ no common factor in } p \text{ and } q) \\ \Longrightarrow\ & q\sqrt{2} = p \\ \Longrightarrow\ & 2q^2 = p^2. \end{aligned}$$

This shows that p^2 is even, therefore p is also even so that p^2 is divisible by 4, so that q^2 is even, which implies that q is even. Now if both p and q are even, it is a contradiction of our choice of p and q. Hence $\sqrt{2}$ is irrational.

Example 1.2 $\sqrt{8}$ is not a rational number.

We see

$$x^2 = 8 = 2^2 \times 2.$$

So we have

$$\left(\frac{x}{2}\right)^2 = 2,$$

that is

$$\frac{x}{2} = \sqrt{2}.$$

Consequently, $x = \sqrt{8}$ is not a rational number (see Example 1.1).

1.3 Density and Completeness

Theorem 1.1 *For $a < b,\ a, b \in \mathbb{R}$, there is a number $r \in \mathbb{Q}$ such that $a < r < b$.*

Proof We can find $n \in \mathbb{N}$ such that $(b-a)^{-1} < n$ by the Archimedes principle (see 1.9). Hence we see

$$\frac{1}{n} < b - a, \quad \text{that is,} \quad a < b - \frac{1}{n}. \tag{1.1}$$

Now, we can find the minimum number $p \in \mathbb{I}$ (the set of integers) satisfying $nb \leqslant p < nb + 1$, so we have

$$b - \frac{1}{n} \leqslant \frac{p-1}{n} < b.$$

Consequently, by (1.1) we have

$$a < r = \frac{m}{n} < b, \quad \text{where } m = p - 1.$$

■

From Theorem 1.1 we see that $\mathbb{Q}$ is dense in $\mathbb{R}$, that is, the closure of $\mathbb{Q}$ is equal to $\mathbb{R}$ (see 1.10). Furthermore, using an irrational number $\alpha \in \mathbb{R}$, we can separate $\mathbb{Q}$ as follows:

$$Q_1 = \{r < \alpha;\ r \in \mathbb{Q}\};\quad Q_2 = \{r > \alpha;\ r \in \mathbb{Q}\},\quad \text{then } Q_1 \cup Q_2 = \mathbb{Q}.$$

However, for $\mathbb{R}$ we cannot do so, that is, for any $\beta \in \mathbb{R}$ we have

$$R_1 = \{x < \beta;\ x \in \mathbb{R}\};\quad R_2 = \{x > \beta;\ x \in \mathbb{R}\},\quad \text{then } R_1 \cup R_2 \cup \{\beta\} = \mathbb{R},$$

and we see

$$R_1^* = \{x \leqslant \beta;\ x \in \mathbb{R}\};\quad R_2 = \{x > \beta;\ x \in \mathbb{R}\},\quad \text{then } R_1^* \cup R_2 = \mathbb{R}.$$

Similarly, we have

$$R_1 = \{x < \beta;\ x \in \mathbb{R}\};\quad R_2^* = \{x \geqslant \beta;\ x \in \mathbb{R}\},\quad \text{then } R_1 \cup R_2^* = \mathbb{R}.$$

Then we say that $\mathbb{R}$ is complete.

1.4 Fields

A field is a set $\mathbb{F}$ with two operations called addition and multiplication, which satisfy the following *field axioms* (A), (M) and (D):

(A) **Axioms for Addition (Abelian group for "+"):**
$(A1)$ If $x \in \mathbb{F}$ and $y \in \mathbb{F}$, then $x + y \in \mathbb{F}$; (by the Law of Closure)
$(A2)$ $x + y = y + x,\ \forall x, y \in \mathbb{F}$; (by Commutative Law)
$(A3)$ $(x + y) + z = x + (y + z),\ \forall x, y, z \in \mathbb{F}$; (by Associative Law)
$(A4)$ For any $x \in \mathbb{F}$, $\exists$ an element 0 in $\mathbb{F}$ such that

$$0 + x = x + 0 = x,\ \forall x \in \mathbb{F};\quad \text{(by Additive Identity)}$$

$(A5)$ For any $x \in \mathbb{F}$, $\exists$ an element $-x$ in $\mathbb{F}$ such that

$$x + (-x) = (-x) + x = 0, \quad \forall x \in \mathbb{F}. \quad \text{(by Additive Inverse)}$$

(M) **Axioms for Multiplication (Abelian group for "·"):**
$(M1)$ If $x \in \mathbb{F}$ and $y \in \mathbb{F}$, then $xy \in \mathbb{F}$; (by the Law of Closure)
$(M2)$ $xy = yx, \quad \forall x, y \in \mathbb{F}$; (by Commutative Law)
$(M3)$ $(xy)z = x(yz), \quad \forall x, y, z \in \mathbb{F}$; (by Associative Law)
$(M4)$ For any $x \in \mathbb{F}$, $\exists$ an element 1 in $\mathbb{F}$ such that

$$1 \cdot x = x \cdot 1 = x, \quad \forall x \in \mathbb{F}; \quad \text{(by Multiplicative Identity)}$$

$(M5)$ For any $x \in \mathbb{F}$ and $x \neq 0$, $\exists$ an element $1/x \in \mathbb{F}$ such that

$$x \cdot \frac{1}{x} = 1; \quad \text{(by Multiplicative Inverse).}$$

(D) **Distributive Law of multiplication over addition:**
For all $x, y, z \in \mathbb{F}$
$x \cdot (y + z) = x \cdot y + x \cdot z$, and $(y + z) \cdot x = y \cdot x + z \cdot x$.

Remark 1.1 The set $\mathbb{Q}$ of rational numbers and the real numbers $\mathbb{R}$ clearly hold field axioms, i.e., $(\mathbb{Q}, +, \cdot)$, $(\mathbb{R}, +, \cdot)$ are fields, but $(\mathbb{N}, +, \cdot)$ and $(\mathbb{I}, +, \cdot)$ are not fields.

Proposition 1.2 *The axioms for addition imply the following statements:*

(i) If $x + y = x + z$, *then* $y = z$; (by Cancellation Law).
(ii) If $x + y = x$, *then* $y = 0$; (by Uniqueness).
(iii) If $x + y = 0$, *then* $y = -x$.

Proof (i) Let $x + y = x + z$. Then we see

$$\begin{aligned} y &= 0 + y \\ &= (-x + x) + y && (\because -x + x = 0) \\ &= -x + (x + y) && \text{(by Associative Law)} \\ &= -x + (x + z) && \text{(by supposition)} \\ &= (-x + x) + z && \text{(by Associative Law)} \\ &= 0 + z && (\because -x + x = 0) \\ &= z. \end{aligned}$$

(ii) Put $z = 0$ in (i), then

$$x + y = x + 0 \Rightarrow y = 0 \quad \text{(by Cancellation Law).}$$

(iii) Put $z = -x$ in (i), then

$$x + y = x + (-x) \Rightarrow y = -x \quad \text{(by Cancellation Law).}$$

■

Proposition 1.3 *The axiom for multiplication implies the following statements:*

(i) If $x \neq 0$ and $x \cdot y = x \cdot z$, then $y = z$; (by Cancellation Law)
(ii) If $x \neq 0$ and $x \cdot y = x$, then $y = 1$; (by Uniqueness)
(iii) If $x \neq 0$ and $x \cdot y = 1$, then $y = 1/x$
(iv) If $x \neq 0$, then $1/(1/x) = x$.

Proof (i) If $x \cdot y = x \cdot z$, then we see

$$\begin{aligned} y &= 1 \cdot y \\ &= \left(\frac{1}{x} \cdot x\right) \cdot y & &\left(\because \frac{1}{x} \cdot x = 1\right) \\ &= \frac{1}{x}(x \cdot y) & &\text{(by Associative Law)} \\ &= \frac{1}{x}(x \cdot z) & &\text{(by supposition)} \\ &= \left(\frac{1}{x} \cdot x\right) \cdot z & &\text{(by Associative Law)} \\ &= 1 \cdot z & &\left(\because \frac{1}{x} \cdot x = 1\right) \\ &= z. \end{aligned}$$

(ii) Put $z = 1$ in (i), then

$$x \cdot y = x \cdot 1 \Rightarrow y = 1.$$

(iii) Put $z = \frac{1}{x}$ in (i), then

$$x \cdot y = x \cdot \frac{1}{x} \Rightarrow x \cdot y = 1 \Rightarrow y = \frac{1}{x}.$$

(iv) Since $(1/x) \cdot x = 1$, (iii) gives $x = \frac{1}{1/x}$.

■

1.5 Ordered Fields

A field $\mathbb{F}$ is an *ordered field* if it satisfies the following properties:

(O1) ***Trichotomy Law***; If $a, b \in \mathbb{F}$, one and only one of the followings is true:

$$a > b, \ a = b, \ b > a.$$

(O2) ***Transitivity***; $a > b$ and $b > c \Rightarrow a > c, \quad \forall a, b, c \in \mathbb{F}$.

(O3) ***Compatibility***; The order relation with the additive and multiplicative compositions:

$$a > b \Rightarrow a + c > b + c, \quad \forall a, b, c \in \mathbb{F},$$

and

$$a > b \text{ and } c > 0 \Rightarrow ac > bc, \quad \forall a, b, c \in \mathbb{F}.$$

1.5.1 *The Absolute Value or Modulus*

If x is a real number, then the modulus (absolute value) of x is denoted by $|x|$, and defined as follows:

$$|x| = \begin{cases} x, & \text{if } x \geqslant 0 \\ -x, & \text{if } x < 0. \end{cases}$$

Thus we always have $|x| \geqslant 0$, and by definition we see $|-x| = |x|$. Then we always have

$$-|x| \leqslant x \leqslant |x|.$$

Thus $|4| = 4, \ |-7| = 7$ and $|0| = 0$.

Let us give the main features of the absolute value as a proposition.

Theorem 1.2

$$|x| = \max(x, -x).$$

Proof We see

$$|x| = x \geqslant -x, \quad \text{if } x \geqslant 0,$$

and

$$|x| = -x > x, \quad \text{if } x < 0.$$

Hence $|x|$ is the greater of the two numbers x and $-x$, i.e.,

$$|x| = \max(x, -x).$$

■

Theorem 1.3

$$-|x| = \min(x, -x).$$

Proof We see

$$-|x| = -x < x, \qquad \text{if } x \geqslant 0,$$

and

$$-|x| = -(-x) = x < -x, \quad \text{if } x < 0,$$

that is, $-|x|$ is the smaller of the two numbers x and $-x$, i.e.,

$$-|x| = \min(x, -x).$$

■

1.5.2 Extended Real Numbers

The extended real number system consists of the real field $\mathbb{R}$ and two symbols, $+\infty$ and $-\infty$. This enlarged set is called the set of extended real numbers. Define it as

$$\mathbb{R}_{ext} = \mathbb{R} \cup \{-\infty, \infty\} = [-\infty, \infty].$$

The extended real number system does not form a field, but it is customary to make the following conventions:

(i) If x is real, then

$$x + \infty = +\infty, \quad x - \infty = -\infty, \quad \frac{x}{+\infty} = \frac{x}{-\infty} = 0.$$

(ii) If $x > 0$, then $x \cdot (+\infty) = +\infty, \quad x \cdot (-\infty) = -\infty$.
(iii) If $x < 0$, then $x \cdot (+\infty) = -\infty, \quad x \cdot (-\infty) = +\infty$.

Furthermore, we set

(i) $\infty + \infty = \infty, \quad -\infty - \infty = -\infty$, and
(ii) $\infty \cdot (\pm\infty) = \pm\infty, \quad -\infty \cdot (\pm\infty) = \mp\infty$.

The operation $0 \cdot \infty$ and $(\infty - \infty)$ are left undefined.

1.6 Neighborhoods

Before discussing the concept of an *open set*, we define the fundamental concept of the *neighborhood.*

Definition 1.1 Let $a \in \mathbb{R}$ and $\varepsilon > 0$. Then the interval $(a - \varepsilon, a + \varepsilon)$ is called the ε-neighborhood of a and denoted by $N_\varepsilon(a)$, where

$$N_\varepsilon(a) = \{x \in \mathbb{R}; |x - a| < \varepsilon\}.$$

Geometrically, $N_\varepsilon(a)$ is a set of all points that are within a distance of ε from a.

Notation 1.1 If the number a itself has been deleted or excluded from the set $\{x \in \mathbb{R}; |x - a| < \varepsilon\}$, i.e., an open interval $(a - \varepsilon, a) \cup (a, a + \varepsilon)$, then this set is called the *deleted neighborhood* of a.

1.7 Functions, Countable and Uncountable Sets

In this section we give a definition of the function concept.

1.7.1 Functions

Let A and B be two non-empty sets. Then we associate each element x of A, in some manner, with at least a unique element of B, which is denoted by $f(x)$. Then f is said to be a function from A to B (or a mapping of A to B). The set A is called the domain of f (we can also say f is defined on A), and the elements $f(x)$ are called the values of f. The set of all values of f; $\{f(x) : x \in A\}$ is called the range set of f.

1.7.2 Direct and Inverse Images

Let A and B be two non-empty sets, and let f be a mapping of A to B, i.e., $f : A \to B$. Let $E \subset A$, and $f(E)$ is defined to be the set of all elements $f(x)$, i.e.,

$$f(E) = \{f(x); x \in E\}.$$

$f(E)$ is the direct image of E under f, and $f(A)$ is the range of f, i.e.,

$$f(A) = \{y \in B; y = f(x), \text{ for some } x \in A\}.$$

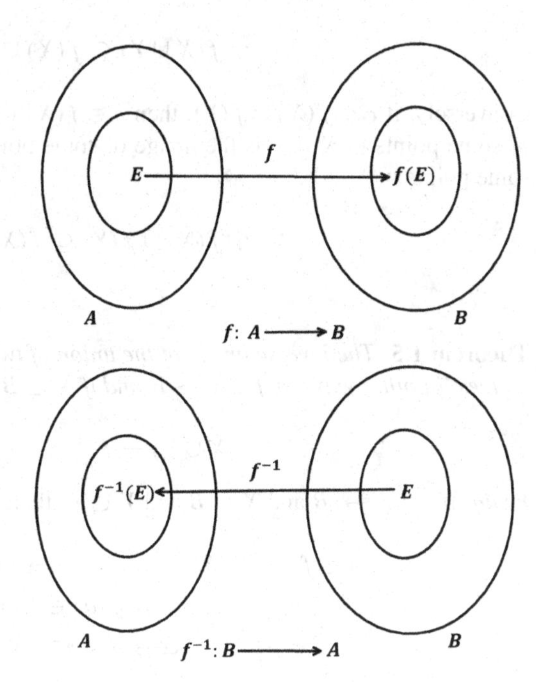

Fig. 1.2 Direct and inverse images of E under f

It is clear $f(A) \subset B$. Then we say that f maps A into B. If $f(A) = B$, i.e., if co-domain B of f itself is the range set of f, then we say that f maps A *onto* B. (Note that *onto mapping* is more specific than *into mapping*) (Fig. 1.2).

If $E \subset B$, $f^{-1}(E)$ denotes the set of all $x \in A$ such that $f(x) \in E$, i.e., $f^{-1}(E) = \{x \in A; f(x) \in E\}$, then we call $f^{-1}(E)$ the inverse image of E. If $y \in B$, then $f^{-1}(y)$ is the set of all $x \in A$ such that $f(x) = y$.

If for each $y \in B$, $f^{-1}(y)$ consists of at most one element of A, then f is said to be a $1-1$ (one-to-one) mapping from A to B. This can also be expressed as follows: f is a $1-1$ mapping from A to B provided that for $x_1, x_2 \in A$, we have $f(x_1) \neq f(x_2)$ whenever $x_1 \neq x_2$, or we see that $f(x_1) = f(x_2)$ implies $x_1 = x_2$.

From the definition, we see that f^{-1} is one-to-one and onto A, if $f : A \to B$ is one-to-one and onto B. Then we can say that f has a one-to-one correspondence between A and B.

Theorem 1.4 *The image of the union of two sets is the union of the images. In other ways, if $f : A \to B$ and if $X \subset A$ and $Y \subset A$, then*

$$f(X \cup Y) = f(X) \cup f(Y).$$

Proof If $b \in f(X \cup Y)$, then $b = f(a)$ for some $a \in X \cup Y$. Therefore $a \in X$ or $a \in Y$, so $f(a) \in f(X)$ or $f(a) \in f(Y)$. Thus $f(a) \in f(X) \cup f(Y)$.

$$\therefore\ f(X \cup Y) \subseteq f(X) \cup f(Y).$$

Conversely, if $c \in f(X) \cup f(Y)$, then $c \in f(X)$ or $c \in f(Y)$. Then c is the image of some points in X or c is the image of some points in Y. Thus c is the image of some points in $X \cup Y$.

$$\therefore\ f(X) \cup f(Y) \subseteq f(X \cup Y).$$

■

Theorem 1.5 *The inverse image of the union of two sets is the union of the inverse images. In other ways, if $f : A \to B$ and if $X \subset B$ and $Y \subset B$, then*

$$f^{-1}(X \cup Y) = f^{-1}(X) \cup f^{-1}(Y).$$

Proof If $f : A \to B$ and $X \subset B$ and $Y \subset B$, then we can write

$$\begin{aligned} a \in f^{-1}(X \cup Y) &\Longleftrightarrow f(a) \in X \cup Y \\ &\Longleftrightarrow f(a) \in X \text{ or } f(a) \in Y \\ &\Longleftrightarrow a \in f^{-1}(X) \text{ or } a \in f^{-1}(Y) \\ &\Longleftrightarrow a \in f^{-1}(X) \cup f^{-1}(Y). \end{aligned}$$

■

Theorem 1.6 *The inverse image of the intersection of two sets is the intersection of the inverse images. In other ways, if $f : A \to B$ and $X \subset B$ and $Y \subset B$, then*

$$f^{-1}(X \cap Y) = f^{-1}(X) \cap f^{-1}(Y).$$

Proof If $f : A \to B$ and $X \subset B$ and $Y \subset B$, then we can write

$$\begin{aligned} a \in f^{-1}(X \cap Y) &\Longleftrightarrow f(a) \in X \cap Y \\ &\Longleftrightarrow f(a) \in X \text{ and } f(a) \in Y \\ &\Longleftrightarrow a \in f^{-1}(X) \text{ and } a \in f^{-1}(Y) \\ &\Longleftrightarrow a \in f^{-1}(X) \cap f^{-1}(Y). \end{aligned}$$

■

Theorem 1.7 *It is necessarily true that the image of the intersection of two sets is equal to the intersection of images.*

In other ways, if $f : A \to B$ and $X \subset A$ and $Y \subset A$, then

$$f(X \cap Y) \subset f(X) \cap f(Y).$$

But the converse is not necessarily true.

Fig. 1.3 $f(X) \cap f(Y) \not\subset f(X \cap Y)$

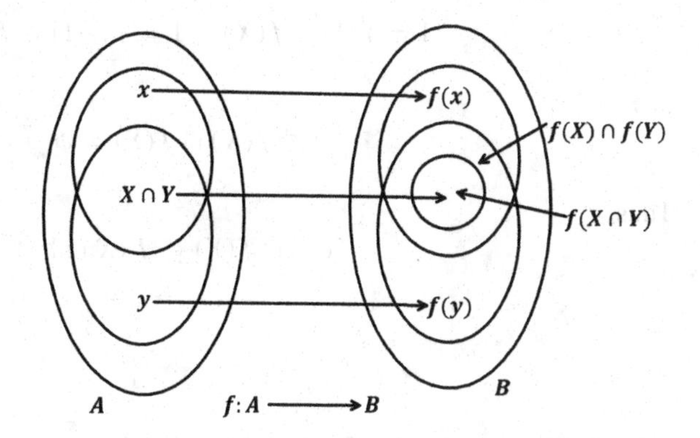

Proof Let us suppose that $b \in f(X \cap Y)$ such that $b = f(a)$ for some $a \in X \cap Y$. It means $a \in X$ and $a \in Y$. Therefore

$$a \in X \Longrightarrow b = f(a) \in f(X)$$
$$a \in Y \Longrightarrow b = f(a) \in f(Y).$$

Hence (Fig. 1.3)

$$b = f(a) \in f(X) \cap f(Y).$$

Thus

$$f(X \cap Y) \subset f(X) \cap f(Y).$$

But the converse is not necessarily true. Let $b \in f(X) \cap f(Y)$. Then $b \in f(X)$ and $b \in f(Y)$.

$$a_1 \in X \Longrightarrow f(a_1) = b$$
$$a_2 \in Y \Longrightarrow f(a_2) = b.$$

Therefore we get $f(a_1) = f(a_2)$, which is **only** true if the function f is *one-to-one*, i.e., $a_1 = a_2$.

If $f : A \to B$ is *onto* function and $X, Y \subseteq A$ and a point $b = f(a) \in f(X) \cap f(Y)$, then b is not necessarily a point of $f(X \cap Y)$.

Alternatively: Let $A = B = \mathbb{R} = (-\infty, \infty)$ and $f(x) = x^2, \ \forall x \in (-\infty, \infty)$. Let X be the set of all positive integers and Y be the set of non-positive numbers.

$$\therefore X \cap Y = \phi,$$

so that

$$f(X \cap Y) = \phi.$$

But we see

$$1 = f(1) \in f(X), \quad 1 = f(-1) \in f(Y),$$

that is,

$$1 \in f(X) \cap f(Y) \neq \phi.$$

Thus

$$f(X) \cap f(Y) \not\subset f(X \cap Y).$$

■

1.7.3 Composition of Functions

If $f : A \to B$ and $g : B \to C$, the composition of g by f is defined to be the set by

$$(g \circ f)(x) = g(f(x)), \quad \forall x \in A,$$

i.e., the image of x under $g \circ f$ is defined to be the image of $f(x)$ under g. The function $g \circ f$ is called the composition of f with g.

For example, if

$$f(x) = 1 + \cos x, \ \forall x \in (-\infty, \infty),$$
$$\text{and} \quad g(x) = x^2, \ \forall x \in [0, \infty),$$

then

$$(g \circ f)(x) = g(f(x)) = g(1 + \cos x) = 1 + 2\cos x + \cos^2 x, \ \forall x \in (-\infty, \infty).$$

1.7.4 Equivalence of Sets

Two sets A and B are said to be equivalent, if there exists a one-to-one correspondence between A and B. If A and B are equivalent, we denote this relation by the symbol $\sim$, i.e., $A \sim B$ means that A and B are equivalent.

Definition 1.2 Two sets A and B are called equivalent (written in $A \sim B$) if and only if there is a one-to-one and onto function f from A to B.

For example, every set is equivalent to itself. Two finite sets are equivalents if they have equal numbers of elements. Furthermore, if $A \sim B$, then $B \sim A$. For if f is a one-to-one correspondence between A and B, then f^{-1} will establish a one-to-one correspondence between B and A. If $A \sim B$ and $B \sim C$, then $A \sim C$. The concept of equivalence is applicable to both finite and infinite sets.

1.7.5 Countable Sets

Definition 1.3 A set A is said to be countable (or denumerable or enumerable) if A is equivalent to the set of natural numbers (positive integers) $\mathbb{N}$, i.e., if f is one-to-one and onto from $\mathbb{N}$ to A, symbolically,

$$f : \mathbb{N} \xrightarrow[onto]{1-1} A,$$

then the set A is called the countable set.

Since A is countable, the elements of A are the images of the positive integers which we can write as

$$A = \{f(1), f(2), f(3), \cdots\}.$$

Example 1.3 (i) The set of $3, 9, \cdots, 3^n, \cdots$ of powers of 3 is countable and defined as $n \leftrightarrow 3^n$.

(ii) The set of $2, 4, 8, \cdots, 2^n, \cdots$ of powers of 2 is countable and defined as $n \leftrightarrow 2^n$.

(iii) The set of all positive even (odd) numbers is countable and defined as $n \leftrightarrow 2n$ $(n \leftrightarrow (2n-1))$.

(iv) The set of all ordered pair of integers is countable.

(v) The set $\mathcal{P}$ of all polynomial functions with integer coefficients is countable.

Theorem 1.8 *If $A_1, A_2, A_3, \cdots$ are countable sets, then $\bigcup_{n=1}^{\infty} A_n$ is countable, i.e., the union of a countable collection of sets is also countable.*

Proof Since $A_1, A_2, A_3, \cdots$ are countable sets and every set A_n is arranged in a sequence $\{a_{nk}\}$, $k = 1, 2, 3, \cdots$, we may write as follows:

$$\begin{aligned}
A_1 &= \{a_{11}, a_{12}, a_{13}, \ldots a_{1n}, \ldots\} \\
A_2 &= \{a_{21}, a_{22}, a_{23}, \ldots a_{2n}, \ldots\} \\
A_3 &= \{a_{31}, a_{32}, a_{33}, \ldots a_{3n}, \ldots\} \\
\cdots &= \cdots\cdots\cdots\cdots\cdots\cdots\cdots\cdots \\
A_n &= \{a_{n1}, a_{n2}, a_{n3}, \ldots a_{nn}, \ldots\}.
\end{aligned}$$

Now we consider the elements in the above sets. There is only one element a_{11}, the sum of whose subscripts is 2. There are two elements a_{21}, a_{12}, the sum of whose subscripts is 3. There are three elements a_{13}, a_{22}, a_{31}, the sum of whose subscripts is 4, and so on. We observe that for any positive integer m (sum of subscripts) $\geqslant 2$ there are only $m - 1$ elements. Now we arrange (count) the elements of $\bigcup_{n=1}^{\infty} A_n$ according to the sum of the subscripts as follows:

$$a_{11}; a_{21}, a_{12}; a_{13}, a_{22}, a_{31}; \cdots,$$

where we have removed all the elements a_{nk} which have been counted already. Pictorially, the above process can be represented as follows:

$$
\begin{array}{ccccc}
a_{11}, & a_{12}, \rightarrow & a_{13}, & a_{14}, & \cdots \\
\downarrow & \nearrow \quad \swarrow & \nearrow & & \\
a_{21}, & a_{22}, & a_{23}, & a_{24}, & \cdots \\
a_{31}, & a_{32}, & a_{33}, & a_{34}, & \cdots \\
\downarrow & \nearrow & & & \\
a_{41}, & a_{42}, & a_{43}, & a_{44}, & \cdots \\
\cdots & \cdots & \cdots & \cdots & \cdots
\end{array}
$$

Therefore we have established a one-to-one and onto mapping between natural numbers $\mathbb{N}$ and $\bigcup_{n=1}^{\infty} A_n$, which proves that $\mathbb{N} \sim \bigcup_{n=1}^{\infty} A_n$. Thus $\bigcup_{n=1}^{\infty} A_n$ is countable. ∎

Corollary 1.1 *The set of integers is countable.*

Proof The set $\mathbb{I}$ can be written as a union of three countable sets as follows:

$$\mathbb{I} = \mathbb{N} \cup \{0\} \cup \{-n; n \in \mathbb{N}\}.$$

Thus $\mathbb{I}$ is countable. ∎

Corollary 1.2 *The set of all rational numbers is countable.*

Proof It is sufficient to show that the set of all positive rational numbers is countable, then the collection of positive and negative rational numbers and zero can be interspersed on a single list. Consider the set $\mathbb{Q}^+$ of all positive rational numbers, i.e., $\mathbb{Q}^+ = \left\{\frac{n}{m}; m, n \in \mathbb{N}\right\}$ where fractions n/m are written in the *lowest terms*. Arrange all ratios in an infinite matrix $\left[\frac{n}{m}\right]_{n,m=1}^{\infty}$:

$$
\begin{bmatrix}
1/1 & 1/2 & 1/3 & 1/4 & 1/5 & \ldots \\
2/1 & 2/2 & 2/3 & 2/4 & 2/5 & \ldots \\
3/1 & 3/2 & 3/3 & 3/4 & 3/5 & \ldots \\
4/1 & 4/2 & 4/3 & 4/4 & 3/5 & \ldots \\
5/1 & 5/2 & 5/3 & 5/4 & 5/5 & \ldots \\
\ldots & \ldots & \ldots & \ldots & \ldots & \ldots
\end{bmatrix}
$$

and define a map $f : \mathbb{N} \to \mathbb{Q}^+$ from the above sloping diagonals in succession. Now start at the left-hand corner and move along the northeasterly diagonals of this table, i.e., start with $1/1$, then move toward $2/1$, then $1/2$. Again proceed to $3/1$, $2/2$ and move along $2/2$ and $1/3$, and so on.

$$\begin{array}{cccccc}
1/1 & 1/2 & 1/3 & 1/4 & 1/5 & \cdots \\
2/1 & 2/2 & 2/3 & 2/4 & 2/5 & \cdots \\
3/1 & 3/2 & 3/3 & 3/4 & 3/5 & \cdots \\
4/1 & 4/2 & 4/3 & 4/4 & 4/5 & \cdots \\
5/1 & 5/2 & 5/3 & 5/4 & 5/5 & \cdots \\
\cdots & \cdots & \cdots & \cdots & \cdots & \cdots
\end{array}$$

When you hit a repetition $(2/2 = 1/1)$, skip it.

$$\begin{array}{llll}
f(1) = 1/1; & & & \\
f(2) = 2/1; & f(3) = 1/2; & & \\
f(4) = 3/1; & f(5) = 1/3; & & (2/2 = f(1) \text{ skipped}); \\
f(6) = 4/1; & f(7) = 3/2; & f(8) = 2/3; \ f(9) = 1/4; & \\
f(10) = 5/1; & f(11) = 1/5; & (4/2 = f(2),\ 3/3 = f(1), & 2/4 = f(3) \text{ skipped}); \\
\ldots & \ldots & \ldots &
\end{array}$$

Clearly f is a one-to-one and onto mapping, thus $\mathbb{N} \sim \mathbb{Q}^+$. Therefore the set of all rational numbers is countable. ■

Remark 1.2 In Corollary 1.2, which terms do we skip?

Let $l, m, n, r \in \mathbb{N}$, and let $n + 1 - s = ml,\ s = rl$. Then

$$\frac{n+1-s}{s}; \text{ irreducible } \iff \frac{n+1}{s}; \text{ irreducible.}$$

In fact, when $n + 1 - s = ml$ and $s = rl$, we see

$$\frac{n+1-s}{s} = \frac{ml}{rl} = \frac{m}{r}, \quad \frac{n+1}{s} = \frac{(m+r)l}{rl} = \frac{m+r}{r}.$$

Conversely, let $n + 1 = m'l'$ and $s = r'l'$, where $l', m', r' \in \mathbb{N}$ we see

$$\frac{n+1-s}{s} = \frac{(m'-r')l'}{r'l'} = \frac{m'-r'}{r'}, \quad \frac{n+1}{s} = \frac{m'l'}{r'l'} = \frac{m'}{r'}.$$

Now, we consider two cases:

(i) Let $n + 1 = p$ be a prime number. Then we have

$$\frac{n+1-s}{s} = \frac{p}{s} - 1,$$

and so we see that every fractional number,

$$\frac{n+1-s}{s}, \quad s = 1, 2, \cdots, n,$$

is irreducible.

(ii) On the other hand, we let

$$n+1 = \prod_{i=1}^{m} p_i^{k_i}, \; 2 \leqslant p_1 < p_2 < \cdots < p_m; \text{ prime numbers},$$

where $n+1$ is a composite number. Then we see that for $s \neq 1,\ n$ the rational numbers

$$\frac{n+1-s}{s} = \frac{n+1}{s} - 1, \quad s = 2, \cdots, n-1,$$

are skipped except the irreducible rational numbers, that is, we leave only rational numbers, where s does not have common factors.

Alternate Proof

Let $\mathbb{Q}^+$ be the set of all positive rational numbers. Then we arrange $S_n,\ n = 1, 2, \cdots$ as follows:

$$\begin{aligned}
S_1 &= \{1/1,\ 1/2,\ 1/3,\ 1/4,\ 1/5, \cdots, 1/n, \cdots\} \\
S_2 &= \{2/1,\ 2/2,\ 2/3,\ 2/4,\ 2/5, \cdots, 2/n, \cdots\} \\
S_3 &= \{3/1,\ 3/2,\ 3/3,\ 3/4,\ 3/5, \cdots, 3/n, \cdots\} \\
S_4 &= \{4/1,\ 4/2,\ 4/3,\ 4/4,\ 4/5, \cdots, 4/n, \cdots\} \\
S_5 &= \{5/1,\ 5/2,\ 5/3,\ 5/4,\ 5/5, \cdots, 5/n, \cdots\} \\
\cdots &= \cdots\cdots\cdots\cdots\cdots\cdots\cdots\cdots\cdots\cdots\cdots\cdots \\
S_n &= \{n/1,\ n/2,\ n/3,\ n/4,\ n/5, \cdots, n/n, \cdots\} \\
\cdots &= \cdots\cdots\cdots\cdots\cdots\cdots\cdots\cdots\cdots\cdots\cdots\cdots .
\end{aligned}$$

Since all the sets $S_1, S_2, S_3, \cdots$ are countable, from Theorem 1.8, the union of S_n is also countable. Thus $\mathbb{Q}^+ = \bigcup_{n=1}^{\infty} S_n$ is countable.

Theorem 1.9 *If* $\mathbb{N}$ *is a set of natural numbers, then the set* $\mathbb{N} \times \mathbb{N}$ *is a countable set.*

Proof We consider that the following collection of sets is countable.

$$\begin{aligned}
A_1 &= \{(1, 1), (1, 2), (1, 3), \cdots (1, n), \cdots\} \\
A_2 &= \{(2, 1), (2, 2), (2, 3), \cdots (2, n), \cdots\} \\
\cdots &= \cdots\cdots\cdots\cdots\cdots\cdots\cdots\cdots\cdots\cdots
\end{aligned}$$

$$A_n = \{(n,1), (n,2), (n,3), \cdots (n,n), \cdots\}$$
$$\cdots = \cdots\cdots\cdots\cdots\cdots\cdots\cdots\cdots\cdots\cdots .$$

Let the function $f : \mathbb{N} \to A_n$ be defined by $f(n,m) = m$. Clearly, it is *one-to-one* and *onto*. Thus A_n is equivalent to $\mathbb{N}$, i.e., $A_n \sim \mathbb{N}$, therefore A_n is countable. Now consider

$$\mathbb{N} \times \mathbb{N} = \cup \{A_n : n \in \mathbb{N}\}.$$

By Theorem 1.8 "the union of a countable collection of sets is also countable," therefore $\mathbb{N} \times \mathbb{N}$ itself is countable. ■

1.7.6 Uncountable Sets

"A set which is not countable is called uncountable."

We will give an example of an infinite set that is not countable. We will assume that every real number x can be written in decimal form.

$$x = a.a_1a_2a_3\cdots = a + \frac{a_1}{10} + \frac{a_2}{10^2} + \frac{a_3}{10^3} + \cdots,$$

where $0 \le a_i \le 9$ for each $i \in \mathbb{N}$. This can be written as decimals in two ways, either with an infinite run of zeros or an infinite run of nines. For example,

$$x = \frac{1}{2} = \frac{5}{10} = 0.5000000\cdots = 0.4999999\cdots$$

shows that the fact that two decimals are distinct does not necessarily mean that they represent distinct real numbers.

Theorem 1.10 *The set* $[0, 1]$ *is uncountable.*

Proof By contradiction, we prove $[0, 1]$ is uncountable. Let us suppose that $[0, 1]$ is a countable set, i.e., all the real numbers lying on the segment $[0, 1]$ can be written in the form of an infinite (uniquely) decimal expansion (non-terminating decimal), and it can be arranged in the form of a sequence as follows:

$$[0,1] = \{x_1, x_2, x_3, \cdots, x_n, \cdots\}.$$

For example, instead of $0.19999\cdots$ we would use $0.20000\cdots$, therefore we can express x_i as follows:

$$\begin{aligned} x_1 &= 0.a_{11}a_{12}a_{13}\cdots\cdots\cdots \\ x_2 &= 0.a_{21}a_{22}a_{23}\cdots\cdots\cdots \\ x_3 &= 0.a_{31}a_{32}a_{33}\cdots\cdots\cdots \\ \cdots &= \cdots\cdots\cdots\cdots\cdots\cdots\cdots \\ x_n &= 0.a_{n1}a_{n2}a_{n3}\cdots a_{nn}\cdots \\ \cdots &= \cdots\cdots\cdots\cdots\cdots\cdots\cdots, \end{aligned}$$

where each a_{ij} is one of the numbers $0, 1, 2, \cdots, 9$. Now we will construct the real number

$$y = 0.b_1b_2b_3\cdots b_n\cdots\cdots.$$

We choose each integer b_n $(n = 1, 2, \cdots)$ from 1 to 9 such that $b_1 \neq a_{11}$. Then choose another integer b_2 such that $b_2 \neq a_{22}$. Proceeding like this, for each n let b_n be any integer such that $b_n \neq a_{nn}$.

Since $b_1 \neq a_{11}$, it follows that $y \neq x_1$ and $b_2 \neq a_{22}$, $y \neq x_2$. Similarly, since $b_n \neq a_{nn}$, it follows that $y \neq x_n$ for each n. It means the decimal expansion of y is different from the decimal expansion of $x_1, x_2, x_3, \cdots, x_n, \cdots$ listed above. Therefore the decimal expansion of y is unique.

Since $y \in [0, 1]$, this is a contradiction to the assumption that every element in $[0, 1]$ can be listed as $x_1, x_2, x_3, \cdots, x_n, \cdots$. Thus the assumption that $[0, 1]$ is countable leads to a contradiction. Thus $[0, 1]$ is not countable. ■

Corollary 1.3 *The set of all real numbers $\mathbb{R}$ is uncountable.*

Suppose that $\mathbb{R}$ is a countable set and $[0, 1]$ is an infinite subset of $\mathbb{R}$. Since every infinite subset of a countable set is countable, $[0, 1]$ is countable. This is contradicting the Theorem 1.10. Hence the set $\mathbb{R}$ cannot be countable.

Corollary 1.4 *The set of all irrational numbers is uncountable.*

Suppose that S is a set of irrational numbers and countable. We know that the set $\mathbb{Q}$ of rational numbers is countable. Therefore $\mathbb{R} = \mathbb{Q} \cup S$ is also countable, but $\mathbb{R}$ is uncountable. Hence the set $S = \mathbb{R} - \mathbb{Q}$ is uncountable.

Some other examples of uncountable sets equivalent to $[0, 1]$ are given here:

(i) The set of points on the real line.
(ii) The set of points in any open interval (a, b).
(iii) The set of all points in the plane or in space.
(iv) The set of all points on a sphere or inside a sphere.
(v) The set of all lines in the plane.
(vi) The set of all continuous real functions of one or several variables.

1.7.7 Bounded Above

A set $S\,(\subset \mathbb{R})$ of real numbers is said to be bounded above if there exists a real number K such that

$$x \leqslant K, \quad \forall x \in S.$$

The real number K is called an upper bound of S.

Example 1.4

$$\mathbb{I}^{-} = \{-1, -2, -3, \cdots\}, \; i.e., \quad x \leqslant -1, \quad \forall x \in \mathbb{I}^{-}.$$

Therefore, -1 is the upper bound of S.

1.7.8 Bounded Below

A set $S\,(\subset \mathbb{R})$ of real numbers is said to be bounded below if there exists a real number k such that

$$k \leqslant x, \quad \forall x \in S.$$

The real number k is called a lower bound of S.

Example 1.5

$$\mathbb{N} = \{1, 2, 3, \cdots\}, \; i.e., \quad 1 \leqslant x, \quad \forall x \in \mathbb{N}.$$

Hence, 1 is the lower bound of $\mathbb{N}$.

Example 1.6

$$\mathbb{I}^{-} = \{-1, -2, -3, \cdots\} \implies \text{No lower bound exists, that is, not bounded below.}$$

Remark 1.3 (i) If a set S is bounded below, then it has infinite lower bounds. (ii) If a set S is bounded above, then it has infinite upper bounds (Fig. 1.4).

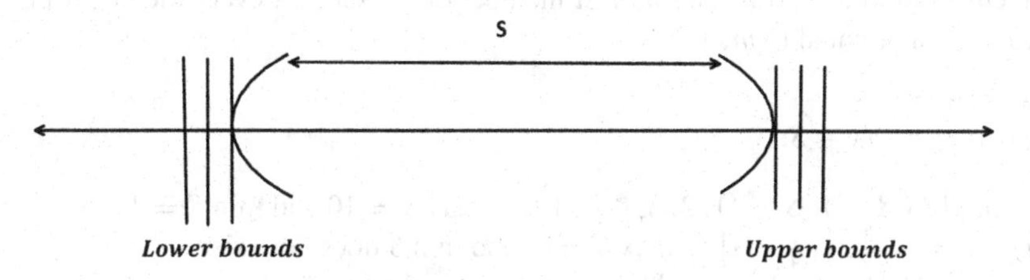

Fig. 1.4 Infinite lower and upper bounds of a set S

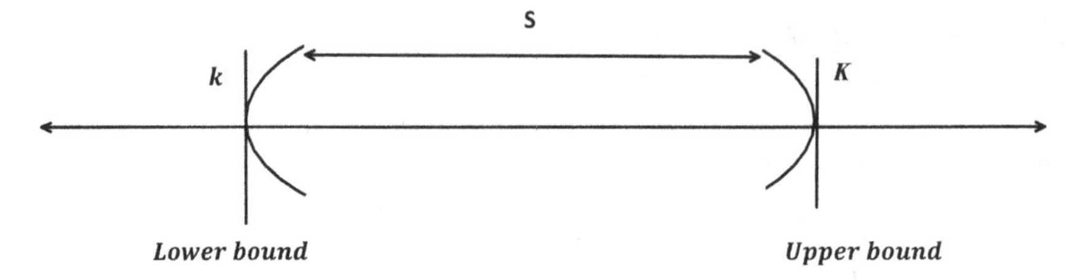

Fig. 1.5 Bounded set S

1.8 Bounded Sets

A set is said to be bounded if it is bounded above as well as bounded below. So, a set S is bounded if there exist $k,\ K \in \mathbb{R}$ such that $k \leqslant x \leqslant K, \quad \forall x \in S$ (Fig. 1.5).

Example 1.7 (i) $S = \left\{1, \frac{1}{2}, \frac{1}{3}, \cdots\right\}$, *i.e.*, $0 < x \leqslant 1, \ \forall x \in S \implies$ bounded.
(ii) $S = \{1, 2, 3, \cdots, 9\}$, *i.e.*, $1 \leqslant x \leqslant 9, \ \forall x \in S \implies$ bounded.
(iii) $\mathbb{N} = \{1, 2, 3, \cdots\}$, *i.e.*, $1 \leqslant x < \infty, \ \forall x \in \mathbb{N} \implies$ Not bounded, it is only bounded below.
(iv) $\mathbb{I}^- = \{-1, -2, -3, \cdots\}$, *i.e.*, $-\infty < x \leqslant -1, \ \forall x \in \mathbb{I}^- \implies$ Not bounded, it is only bounded above.

1.8.1 The Greatest Element (The Maximum Element)

A member M of a set S is called the *greatest* member of the set S, if every member of S is less than or equal to M, i.e.,

(i) $M \in S$,
(ii) $x \leqslant M, \ \forall x \in S$.

1.8.2 The Smallest Element (The Least Element)

A member m of a set S is the *smallest* member of the set S, if every member of S is greater than or equal to m, i.e.;

(i) $m \in S$,
(ii) $m \leqslant x, \ \forall x \in S$.

Example 1.8 (i) $S = \{1, 2, 3, \cdots, 10\}$; $\max S = 10$ and $\min S = 1$.
(ii) $S = \left\{1, \frac{1}{2}, \frac{1}{3}, \frac{1}{4}, \cdots\right\}$; $\max S = 1$, and $\min S$ does not exist.
(iii) $\mathbb{N} = \{1, 2, 3, \cdots\}$; $\max \mathbb{N}$ does not exist and $\min \mathbb{N} = 1$.

1.8.3 The Least Upper Bound (l.u.b.)/Supremum (Sup)

The smallest member in the set of all upper bounds of a set S is called the *least upper bound (l.u.b.) or the supremum (sup)* of S.

1.8.4 Properties

$Sup\ S = M$ is equivalent to hold the next (i) and (ii) (Fig. 1.6);

(i) M is an upper bound of S, i.e., $x \leqslant M, \ \forall x \in S$.
(ii) No number less than M can be an upper bound of S,
i.e., $\forall\ \varepsilon > 0\ \exists\ y \in S$ such that $(M - \varepsilon) < y \leqslant M$.

Necessity: For $\varepsilon > 0$, $(M - \varepsilon)$ is not an upper bound of S
$\Longrightarrow \exists\ y \in S$ such that $(M - \varepsilon) < y \leqslant M$ (by property (i)).

Sufficiency: By (i), M is an upper bound of S, so by (ii), we see that M is the least upper bound.

Example 1.9 (i) $S = \{1, 2, 3, \cdots, 10\}$; $\sup S = 10 \in S$.
(ii) $S = \left\{1, \frac{1}{2}, \frac{1}{3}, \frac{1}{4}, \cdots\right\}$; $\sup S = 1 \in S$.
(iii) $B = [a, b)$; $\sup B = b \notin B$.

Remark 1.4 Let $\alpha = \sup E$ exist. Then α may or may not be a member of E.

Example 1.10 (i) Let E_1 be the set of all $r \in \mathbb{Q}$ and $r < 0$. Let E_2 be the set of all $r \in \mathbb{Q}$ and $r \leqslant 0$. Then

$$\sup E_1 = \sup E_2 = 0 \ \text{ but } \ 0 \notin E_1,\ 0 \in E_2.$$

(ii) $A = (a, b) = \{x \in \mathbb{R}; a < x < b\}$; $\sup A = b \notin A$.
(iii) $B = [a, b] = \{x \in \mathbb{R}; a \leqslant x \leqslant b\}$; $\sup B = b \in B$.

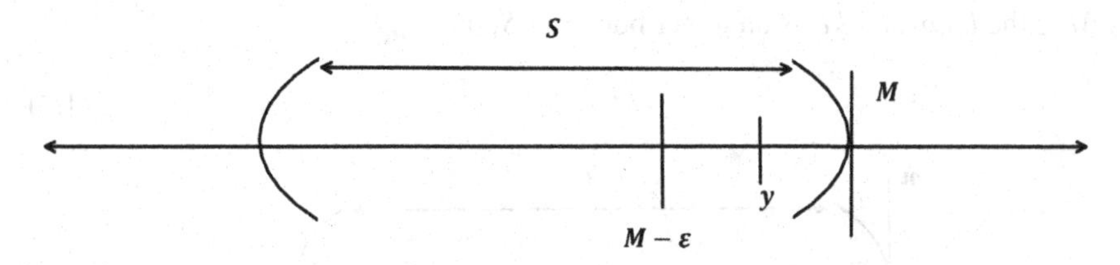

Fig. 1.6 sup S = M

1.8.5 The Greatest Lower Bound (g.l.b.)/Infimum (Inf)

The greatest member in the set of all lower bounds of a set S is called the greatest lower bound (g.l.b.) or the infimum (inf) of S.

1.8.6 Properties

Inf $S = m$ is equivalent to hold the next (i) and (ii) (Fig. 1.7):

(i) m is a lower bound of the set S, i.e., $m \leqslant x, \quad \forall x \in S$.
(ii) No number greater than m can be a lower bound of S,
i.e., $\forall\, \varepsilon > 0 \; \exists\, z \in S$ such that

$$m \leqslant z < (m + \varepsilon).$$

Necessity: For $\varepsilon > 0$, $(m + \varepsilon)$ is not a lower bound of S
$\Longrightarrow$ $\exists z \in S$ such that $m \leqslant z < (m + \varepsilon)$ (by property (i)).

Sufficiency: By (i), m is a lower bound of S, so by (ii), we see that m is the greatest lower bound.

Example 1.11 (i) $\mathbb{N} = \{1, 2, 3, \cdots, 10\}$; $\inf \mathbb{N} = 1$.
(ii) $S = \left\{1, \frac{1}{2}, \frac{1}{3}, \frac{1}{4}, \cdots\right\}$; $\inf S = 0 \notin S$.
(iii) $A = (a, b) = \{x \in \mathbb{R} : a < x < b\}$; $\inf A = a \notin A$.
(iv) $B = [a, b] = \{x \in \mathbb{R} : a \leqslant x \leqslant b\}$; $\inf B = a \in B$.

Theorem 1.11 *Sup (Inf) of any bounded set is unique, i.e., the set A cannot have more than one* sup (inf).

Proof Suppose that S is a non-empty bounded set, and then show that sup of S is unique.

Let M and M' be two sups of the set S. Therefore, both M and M' are upper bounds of S (by the definition of sup).

If M is the *l.u.b.* and M' is an upper bound of S, then

$$M \leqslant M'. \tag{1.2}$$

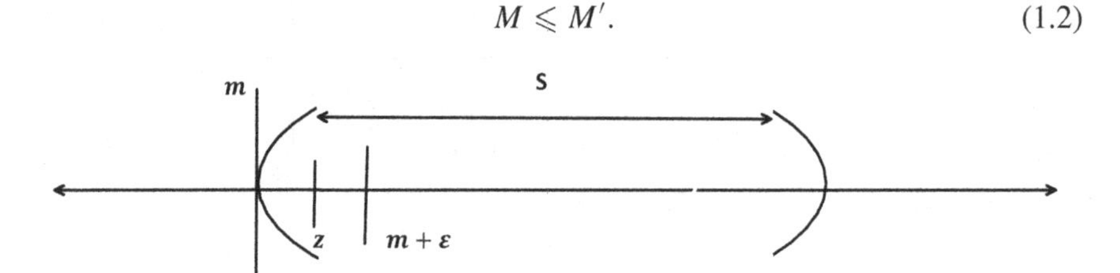

Fig. 1.7 inf S = m

If M' is the *l.u.b.* and M is an upper bound of S, then

$$M' \leqslant M. \tag{1.3}$$

From (1.2) and (1.3) we have $M = M'$.
Hence, the sup of the set S is unique. ■

Example 1.12 We show that the greatest member of a set, if it exists, is the sup of the set. Let M be the greatest member of the set S. Then we prove that

$$\sup S = M.$$

Since M is the greatest member of S,

$$\max S = M \implies x \leqslant M, \ \forall\, x \in S,$$

$$i.e., \ M \text{ is an upper bound of the set } S. \tag{1.4}$$

Let $\varepsilon > 0$. Then there exists $y \in S$ such that

$$M - \varepsilon < y = M \in S \ \text{ (see } (ii) \text{ in Sec. 1.8.4 } Properties). \tag{1.5}$$

From (1.4) and (1.5), we have

$$\sup S = M.$$

Example 1.13 We show that the lowest member of a set, if it exists, is the inf of the set. Let m be the smallest member of the set S. Then we prove that

$$\inf S = m.$$

Since m is the smallest member of S,

$$\min S = m \implies m \leqslant x, \ \forall\, x \in S.$$

Therefore, m is a lower bound of the set S. (1.6)

Let $\varepsilon > 0$. Then there exists $y \in S$ such that

$$m = y < m + \varepsilon \ \text{ (see } (ii) \text{ in Sec. 1.8.6 } Properties). \tag{1.7}$$

From (1.6) and (1.7), we have

$$\inf S = m.$$

1.8.7 The Completeness Axiom

(i) Every non-empty set of real numbers, which is bounded above, has the *supremum (l.u.b.)* in $\mathbb{R}$.
(ii) Every non-empty set of real numbers, which is bounded below, has the *infimum (g.l.b.)* in $\mathbb{R}$.

1.8.8 Complete Ordered Fields

A set of real numbers $\mathbb{R}$ is called the *complete ordered field* if it satisfies the following axioms:

(i) Field axioms.
(ii) Ordered field axioms.
(iii) Completeness axioms.

Theorem 1.12 *The set of rational numbers is not a completely ordered field, that is, there exists a bounded above subset of $\mathbb{Q}$ which does not have a* sup *in $\mathbb{Q}$.*

Proof Let S ($\subseteq \mathbb{Q}$) be the set of all those positive rational numbers, whose square is less than 2, i.e.,

$$S = \left\{x; x \in \mathbb{Q},\ x > 0 \wedge x^2 < 2\right\}.$$

Since

$$1^2 = 1 < 2 \Longrightarrow 1 \in S,$$

S is a non-empty set. Clearly, $\sqrt{2}$ is an upper bound of S. Hence S is bounded above.

Now, we show that there is not any rational number which can be the sup of S.

Let, if possible, the rational number K be its $l.u.b.$ (sup). Clearly, K is positive, then by law of *trichotomy*, one and only one of cases as follows is true.

$$(i)\ K^2 < 2, \qquad (ii)\ K^2 = 2, \qquad (iii)\ K^2 > 2.$$

(i) Let $K^2 < 2$. Let us take the positive rational number

$$y = \frac{4+3K}{3+2K}.$$

Then

$$K - y = K - \frac{4+3K}{3+2K} = \frac{2\left(K^2-2\right)}{3+2K} < 0 \quad \left(\because K^2 < 2\right)$$
$$\Longrightarrow y > K. \tag{1.8}$$

Now,

$$
\begin{aligned}
&2-y^2 = 2-\left(\frac{4+3K}{3+2K}\right)^2 = \frac{2-K^2}{(3+2K)^2} > 0 \quad (\because K^2 < 2)\\
&\implies y^2 < 2\\
&\implies y \in S. \qquad (1.9)
\end{aligned}
$$

Thus, by (1.9) and (1.8) we have $y \in S$ and $K < y$, so that K cannot be an upper bound of S.

Hence, there is a contradiction.

(ii) Let $K^2 = 2$. Then we know that $x \in \mathbb{Q}$ and $x^2 \neq 2$

$$K^2 \neq 2.$$

Thus, this case is not possible.

(iii) Let $K^2 > 2$. For $y = \frac{4+3K}{3+2K}$,

$$
\begin{aligned}
&K - y = K - \left(\frac{4+3K}{3+2K}\right) = \frac{2\left(K^2-2\right)}{3+2K} > 0 \quad (\because K^2 > 2)\\
&\implies K > y. \qquad (1.10)
\end{aligned}
$$

Now

$$
\begin{aligned}
y^2 - 2 &= \left(\frac{4+3K}{3+2K}\right)^2 - 2\\
&= \frac{16+24K+9K^2-2\left(9+12K+4K^2\right)}{(3+2K)^2} \quad (\because K^2 > 2)\\
&= \frac{K^2-2}{(3+2K)^2} > 0,
\end{aligned}
$$

so that $y^2 > 2 \implies y \notin S$.

Therefore, y is an upper bound of S. But we see $K > y$ by (1.10), and so we have a contradict.

Thus, none of the three possible cases hold. Hence, our supposition that a rational number K is sup (l.u.b.) of S is wrong. ∎

1.9 The Archimedean Property of Real Numbers

An ordered field F is said to have the *Archimedean property* if for any $x, y \in F$ with $x > 0$, there exists a multiple of x which exceeds y:

$$nx > y, \quad \text{for some } n \in \mathbb{N}.$$

Theorem 1.13 (The Archimedean Theorem)

"If $x, y \in \mathbb{R}$ and $x > 0$, then there is a positive integer n such that $nx > y$," that is, "the field of real numbers satisfies the Archimedean *property."*

Proof (i) If $y \leqslant 0$, then the theorem is true.

(ii) We need to show that the following:
If $x, y \in \mathbb{R}$ and $x, y > 0$, then $\exists\, n \in \mathbb{N}$ such that $nx > y$.
Suppose this is false, then $nx \leqslant y, \quad \forall\, n \in \mathbb{N}$.
Now, let

$$S = \{nx : n \in \mathbb{N}\}.$$

Thus the set S is bounded above and y will be an upper bound of S. By the *Complete Axiom,*

$$\begin{aligned}
&\sup S = M, \ \textit{i.e.}, \ M \text{ is a } \textit{l.u.b.} \text{ of } S \\
&\Longrightarrow M \text{ is an upper bound of } S, \ \textit{i.e.}, \ nx \leqslant M, \qquad (\forall x \in S \ \& \ \forall\, n \in \mathbb{N}) \\
&\Longrightarrow (n+1)x \leqslant M \qquad (\forall n \in \mathbb{N} \Longrightarrow n+1 \in \mathbb{N}) \\
&\Longrightarrow nx \leqslant M - x \\
&\Longrightarrow (M - x) \text{ is an upper bound of } S, \text{ and} \\
&\Longrightarrow (M - x) < M \quad (\because x > 0).
\end{aligned}$$

Hence, we have the upper bound $(M - x)$ less than the supremum M, which is a contradiction. ■

The next theorem states that every real number is located between consecutive integers n and $n + 1$.

Corollary 1.5 *If c is a real number, then there exists an integer n such that*

$$n \leqslant c < n + 1.$$

Proof If $c = 0$, then $0 \leqslant c < 1$. Now let us think $c > 0$. By the *Archimedean property* with $x = 1$ and $y = c$ we show that there is an integer M with $M > c$. Then there exists

$$n + 1 = \min\{c < M : M \in \mathbb{I}\},$$

such that $c < n + 1$. Then we see

$$n \leqslant c < n + 1.$$

Next, let $c < 0$. Then from the above considerations there exists $m \in \mathbb{I}$ such that

$$m \leqslant -c < m + 1,$$

that is,

$$-(m + 1) < c \leqslant -m.$$

If $c < -m$, then we have $n = -(m + 1) \leqslant c < n + 1$. And if $c = -m$, then we see

$$n = -m = c < n + 1.$$

■

Theorem 1.14 *Let $x \in \mathbb{R}$. Then there exists $K \in \mathbb{I}$ such that*

$$x - 1 \leqslant K < x.$$

Proof Let $x \in \mathbb{R}$. If $x = 1$, then we have

$$x - 1 = 0 = K < x.$$

Let $x > 1$. Then by the Archimedean property there exists $n \in \mathbb{N}$ such that

$$x - 1 < n \cdot 1.$$

Now we let

$$k = \min\{n;\ x - 1 < n,\ n \in \mathbb{N}\}.$$

Then we see

$$x - 1 < k \leqslant x.$$

If $k < x$, then we have

$$x - 1 \leqslant K = k < x,$$

and if $k = x$, then we see

$$x - 1 = k - 1 = K < x.$$

Let $x < 1$. Then by the above considerations there exists $k \in \mathbb{N}$ such that

$$0 < 1 - x \leqslant k < 2 - x,$$

that is,

$$x - 2 < -k \leqslant x - 1.$$

If $-k < x - 1$, then we have

$$x - 1 \leqslant -k + 1 = K < x,$$

and if $-k = x - 1$, then we have

$$x - 1 \leqslant -k = K < x.$$

■

1.10 Denseness

Definition 1.4 A set E of real numbers is said to be *dense* in $\mathbb{R}$ if between any two real numbers there lies a member of E.

The next theorem states that every open interval (x, y) with $x < y$ contains a rational number. In other words, it says that the set of rational numbers is *dense* in $\mathbb{R}$.

Theorem 1.15 (The Rational Density Theorem)
"Between any two real numbers there exists at least one rational number," that is, "if $x, y \in \mathbb{R}$ and $x < y$, then there exists a rational number $p \in \mathbb{Q}$ such that $x < p < y$."

Proof Let $x, y \in \mathbb{R}$, where $x < y$. Then we prove that there is a rational number p between x and y.
Since $x < y$, we have $y - x > 0$.

By the *Archimedean property*, there exists $n \in \mathbb{N}$ such that

$$\begin{aligned} & n(y - x) > 1 \\ \Longrightarrow\ & ny - nx > 1 \\ \Longrightarrow\ & ny > 1 + nx. \end{aligned} \tag{1.11}$$

Again, we apply the *Archimedean property* to obtain two positive integers m_1 and m_2 such that

$$m_1 \cdot 1 > nx \quad \text{and} \quad m_2 \cdot 1 > -nx,$$

then

$$-m_2 < nx < m_1.$$

Hence, there is an integer m ($with$ $-m_2 \leqslant m \leqslant m_1$) such that

$$m - 1 \leqslant nx < m \implies nx + 1 \geqslant m \text{ and } m > nx. \tag{1.12}$$

We repeatedly apply from (1.11) and (1.12), we obtain

$$\begin{aligned}
& nx < m \leqslant nx + 1 < ny \\
& \implies nx < m < ny \\
& \implies x < \frac{m}{n} < y \\
& \implies x < p < y \\
& \implies p = \frac{m}{n}, \; i.e., \text{ the rational number.}
\end{aligned}$$

■

Theorem 1.16 (The Irrational Density Theorem) *"If $x, y \in \mathbb{R}$ and $x < y$, then there exists a number $\beta \in (\mathbb{R} - \mathbb{Q})$ such that $x < \beta < y$."*

Proof Let $x, y \in \mathbb{R}$ and $0 < x < y$. Then there exists a rational number p (by the Rational Density Theorem) such that

$$0 < \frac{x}{\alpha} < p < \frac{y}{\alpha},$$

where α is an irrational number,

$$\begin{aligned}
& \implies x < \alpha p < y \\
& \implies x < \beta < y,
\end{aligned}$$

where $\beta = \alpha p$ is an irrational number as we know that the product of rational numbers and irrational numbers is an irrational number.

Alternate Proof 1.
Let x and y be distinct real numbers with $x < y$. Now $\sqrt{2}$, the positive square root of 2, is irrational and $x - \sqrt{2} < y - \sqrt{2}$. By the *Rational Density Theorem* there is a rational number p such that

$$x - \sqrt{2} < p < y - \sqrt{2} \implies x < p + \sqrt{2} < y,$$

since $p + \sqrt{2}$ is an irrational number.

Alternate Proof 2.
Let x and y be distinct real numbers with $x < y$. Now $\sqrt{2}$, the positive square root of 2, is irrational and $x\sqrt{2} < y\sqrt{2}$. By the *Rational Density Theorem* there is a rational number p ($\neq 0$) such that

$$x\sqrt{2} < p < y\sqrt{2} \implies x < \frac{p}{\sqrt{2}} < y.$$

Here $p/\sqrt{2} = p\sqrt{2}/2$ is an irrational number. ■

Example 1.14 If $p \neq 0$ is a rational number and x is an irrational number, then we prove that $p + x$ and px are irrational numbers.

Proof We will prove this theorem by contradiction. Let us assume that $p + x$ (where $p \neq 0$ is a rational number and x is an irrational number) is a rational number. Therefore

$$p + x = \frac{m}{n}, \quad (m, n \text{ are integers with } n \neq 0)$$
$$\implies x = \frac{m}{n} - p.$$

Since p is rational, $p = j/k$, where j, k are integers with $k \neq 0$ so that

$$x = \frac{m}{n} - \frac{j}{k} \implies x = \frac{mk - nj}{nk}.$$

Since both km and jn are integers, x is also a rational number. This leads to a contradiction. Thus, $p + x$ is an irrational number.

Similarly, let us suppose that px (where $p \neq 0$ is a rational number and x is an irrational number) is a rational number. Therefore

$$px = \frac{m}{n}, \quad (m, n \text{ are integers with } n \neq 0).$$

Since p is rational, $p = j/k$, where j, k are integers with $k \neq 0$ so that

$$\frac{j}{k} \cdot x = \frac{m}{n} \implies x = \frac{k}{j} \cdot \frac{m}{n} = \frac{km}{jn}.$$

Since both km and jn are integers, x is also a rational number which leads to a contradiction. Hence, px is an irrational number. ■

1.11 Exercises

1. Show that $\sqrt{5}$ is an irrational number.
2. Show that $\mathbb{R}$ is a field. Is $\mathbb{R}$ an ordered field?
3. Let $A = \{1, 2, 3, 4\}$ and $B = \{a, b, c\}$. Determine the number of one-to-one, onto and total functions from the set A to B.
4. Let $f(x) = x^2$ and $g(x) = \sin x$. Find fog and gof.
5. Prove that any infinite set contains a countable set.
6. Prove that a subset of a countable set is countable.
7. Which of the following sets are countable?
   (i) $[0, 1] \cap \mathbb{Q}$ (ii) $(0, 1)$ (iii) $\left\{\frac{n}{n+1}; n \in \mathbb{N}\right\}$

   (iv) $\left\{\frac{1}{n}; n \in \mathbb{I}\right\}$ (v) $[0, 1] \cap \mathbb{Q}^c$.
8. Prove that the intersection of two countable sets is countable.
9. Let $p \in \mathbb{R}$, if x is any real number in the neighborhood $N_\epsilon(p)$ for every $\epsilon > 0$, then show that $x = p$.
10. Let A and B be two sets and $B \subseteq A$. Show that

    (i) if A is a countable set, B is also a countable set.
    (ii) if B is an uncountable set, A is also an uncountable set.

11. Find $l.u.b.$ and $g.l.b.$ for $A = \left\{1 + \frac{(-1)^n}{n+1}; n \in \mathbb{N}\right\}$.
12. Is $A = \left\{x \in \mathbb{Q}; 0 < x^3 < 27\right\}$ bounded above in $\mathbb{Q}$?
13. Give an example of a countable bounded subset of A of $\mathbb{R}$ whose $l.u.b.$ and $g.l.b.$ are in A^c.
14. Show that an infinite subset of a countable set is countable.
15. Give an example of a set which has a supremum but does not have a maximum.
16. Define $A = \left\{x \in \mathbb{R} \mid x^2 < x\right\}$. Prove that $\sup A = 1$.
17. If A is a non-empty bounded subset of $\mathbb{R}$ and $l.u.b.A = g.l.b.A$, what can be said about A?
18. Show by example that the product of two irrational numbers may be rational or irrational.
19. Show that irrational numbers are dense in $\mathbb{R}$.
20. Show that the Archimedean property is a consequence of the assertion that every interval (a, b) contains a rational number.
21. Prove that the set of all sequences of natural numbers is uncountable.

    (i) $|x + y| \leq |x| + |y|$ for all $x, y \in \mathbb{R}$
    (ii) $|x + y| < |x| + |y|$ iff $xy < 0$

    (iii) $|x + y| = |x| + |y|$ iff $xy \geq 0$

    (iv) $\frac{|x+y|}{1+|x+y|} \leq \frac{|x|}{1+|x|} + \frac{|y|}{1+|y|}$ for all $x, y \in \mathbb{R}$.

22. Let S be a non-empty subset of $\mathbb{R}$ that is bounded. Prove that

$$inf S = -sup\{-s; s \in S\} \text{ and } sup S = -inf\{-s; s \in S\}.$$

23. If $f(x) = \frac{1}{1-x}, x \neq 1$, and $g(x) = f(f(x))$, $h(x) = f(g(x))$, then find $f(x)g(x)h(x)$.
24. Let $f : \mathbb{Z} \to \mathbb{N}$ by $f(x) = |x| + 1$. Is f one-to-one and onto? If yes, then what is the inverse of f?

Chapter 2
Real Sequences

In this chapter, we discuss sequences. We say what it means for a sequence to converge, and define the limit of a convergent sequence.

In mathematics, informally speaking, a sequence is an ordered list of objects (or events). Like a set, it contains members (also called elements, or terms). The number of ordered elements (possibly infinite) is called the length of the sequence.

2.1 Sequences

A function, whose domain is the set $\mathbb{N}$ of natural numbers and whose range is a subset of `real(complex)` numbers, is called a `real(complex)` sequence.

$$s : \mathbb{N} \longrightarrow \mathbb{R},$$

and

$$s : \mathbb{N} \longrightarrow \mathbb{C},$$

where $\mathbb{C}$ is the set of complex numbers.

Notation 2.1 *Since the domain of a sequence is always a set of natural numbers* $\mathbb{N}$, *the sequence may be denoted as follows: (Fig. 2.1)*

$$\{s_n\}_{n=1}^{\infty} \quad or \quad \{s_n; n \in \mathbb{N}\} \quad or \quad \{s_1, s_2, s_3, \cdots\} \quad or \quad \{s_n\} \quad or \quad \langle s_n \rangle .$$

The values $s_1, s_2, s_3, \cdots$ *are called the first term, the second term, the third term,* $\cdots$ *of the sequences. For example,*

(i) $\{s_n\} = \left\{\frac{1}{n}\right\} = \left\{1, \frac{1}{2}, \frac{1}{3}, \cdots\right\}$, $\forall n \in \mathbb{N}$. *Here all the elements are distinct.*

N. Deo and R. Sakai, *Introduction to Mathematical Analysis*,
https://doi.org/10.1007/978-981-97-6568-3_2

(*ii*) $\{s_n\} = \{(-1)^n\} = \{-1, 1, -1, 1, \cdots\}$, $\forall n \in \mathbb{N}$. *Here we have only two distinct elements, i.e.,* -1 *and* 1.

2.1.1 The Range of Sequences

The range or the range set of a sequence is the set consisting of all distinct elements of a sequence without repetition and without regard to the position of a term. For example,

(*i*) $\{1 + (-1)^n\}$ Range=$\{0, 2\}$.
(*ii*) $\{1 + (-1)^{n+1}\}$ Range=$\{0, 2\}$.
(*iii*) $\{(-1)^n\}$ Range=$\{-1, 1\}$.
(*iv*) $\{n^2\}$ Range=$\{1^2, 2^2, 3^2, \cdots\}$.

2.1.2 Equality of Sequences

Any two sequences $\{s_n\}$ and $\{t_n\}$ are equal if $s_n = t_n, \ \forall\, n \in \mathbb{N}$.

2.1.3 Constant Sequences

Any function $s : \mathbb{N} \longrightarrow \mathbb{R}$ is called a constant sequence if

$$s_n = k, \ \ \forall\, n \in \mathbb{N},$$

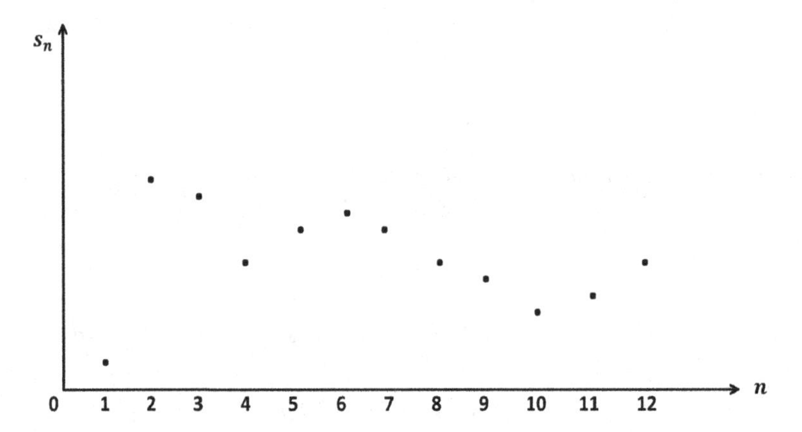

Fig. 2.1 Graphic interpretation of sequence s_n

i.e., $s_n = k$ for all n, and we get the constant sequence $\{k, k, k, \cdots\}$.
For example, $\{s_n\} = \left\{(-1)^{2n}\right\}$, where $s_n = 1, \ \forall\, n \in \mathbb{N}$.
$\{x_n\} = \{x\}, \ \ \forall\, n \in \mathbb{N}$ (Fig. 2.2).

2.2 Bounded Sequences

A sequence $\{s_n\}$ is said to be bounded, if there exists a positive real number λ such that

$$|s_n| \leqslant \lambda, \ \forall n \in \mathbb{N}.$$

In other words, a sequence $\{s_n\}$ is said to be bounded, when it is bounded both above and below by K and k which are respectively the upper and lower bounds of the sequence such that

$$k \leqslant s_n \leqslant K, \ \ \forall n \in \mathbb{N}.$$

Remark 2.1 "Evidently a sequence is bounded if and only if its range is bounded."

2.3 Neighborhoods of a Point

A set $\mathbf{S} \subset \mathbb{R}$ is called the neighborhood of a point $x \in \mathbf{S}$, if there exists an open interval I containing x and contained in $\mathbf{S}$, i.e.,

$$x \in I \subset \mathbf{S}.$$

In other words, a set $\mathbf{S} \subset \mathbb{R}$ is called the neighborhood of a point $x \in \mathbb{R}$, if there exists $\varepsilon > 0$ such that

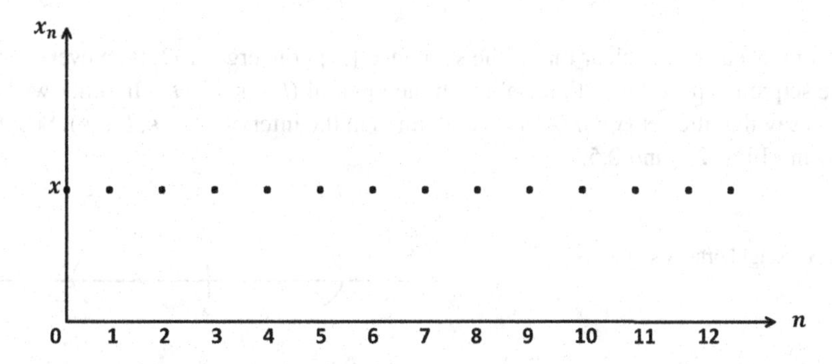

Fig. 2.2 Constant sequence

$$x \in (x - \varepsilon, x + \varepsilon) \subset \mathbf{S}.$$

Alternatively, a set **S** of real numbers is a neighborhood of a real number x if and only if **S** contains an interval of positive length centered at x, i.e., if and only if there is $\varepsilon > 0$ such that $(x - \varepsilon, x + \varepsilon) \subset \mathbf{S}$ (Fig. 2.3).

Remark 2.2 (i) The open interval (a, b) is the *neighborhood* of each point $x \in (a, b)$.

(ii) The closed interval $[a, b]$ is the *neighborhood* of each point of (a, b) but not at the end points a and b.

(iii) The set $\mathbb{R}$ of real numbers is the *neighborhood* of each of its points.

(iv) The sets $\mathbb{N}, \mathbb{I}, \mathbb{Q}$ are not the neighborhoods of any point in these sets as no interval can be contained in $\mathbb{N}, \mathbb{I}, \mathbb{Q}$, i.e., for all $x \in \mathbb{R},\ (x - \varepsilon, x + \varepsilon) \not\subset \mathbb{N}/\mathbb{I}/\mathbb{Q},\ \forall \varepsilon > 0$.

2.4 Convergence of the Sequences (Limit of Sequences)

A real sequence $\{s_n\}$ is said to converge to a real number l (or a real number l is said to be the *limit* of a sequence $\{s_n\}$) if and only if the following criterion is satisfied.

For every $\varepsilon > 0$ **there exists a positive integer** m **(that usually depends on** ε**)** such that

$$|s_n - l| < \varepsilon, \ \forall n \geqslant m.$$

That is, for every $\varepsilon > 0$ there exists a positive integer $m = m(\varepsilon)$ such that all the terms in the sequence after the m-th between $l - \varepsilon$ and $l + \varepsilon$, i.e.,

$$s_n \in (l - \varepsilon, l + \varepsilon)\,, \ \forall n \geqslant m.$$

Symbolically,

$$s_n \to l \text{ as } n \to \infty, \quad i.e., \quad \lim_{n \to \infty} s_n = l.$$

Geometrically, it is clear that if the sequence $\{s_n\}$ converges to l, then every term of the sequence past the m-th term lies in the interval $(l - \varepsilon, l + \varepsilon)$. In other words, we can say that the set $\{s_n; n \geqslant m\}$ is contained in the interval $(l - \varepsilon, l + \varepsilon)$. See the following Figs. 2.4 and 2.5.

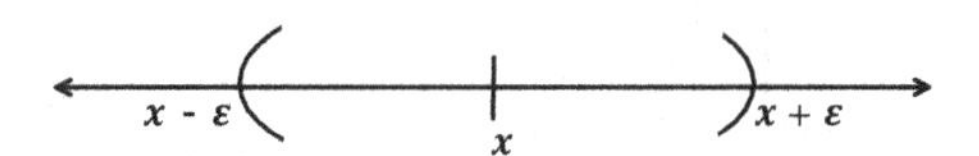

Fig. 2.3 Neighborhoods of a point x

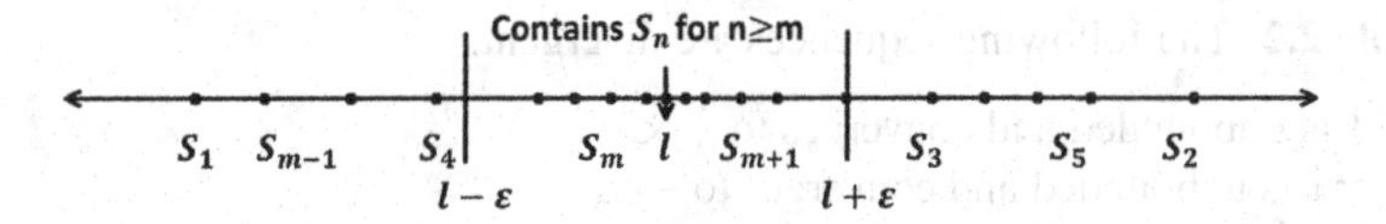

Fig. 2.4 Tail of a convergent sequence

If a sequence has a limit, we say that the sequence is **convergent**, and if it has no limit, we say that the sequence is **divergent**, i.e., a sequence that does not converge is said to diverge.

Example 2.1 Consider the sequence $\{x_n\} = \left\{3 + \frac{(-1)^n}{n}\right\}$. By plotting the function $y = x_n$ in a two-dimensional coordinate system, we observe in Fig. 2.6 that the horizontal line $y = 3$ is an asymptote. Thus we say that the sequence $\{x_n\}$ converges to the number 3 as its limit.

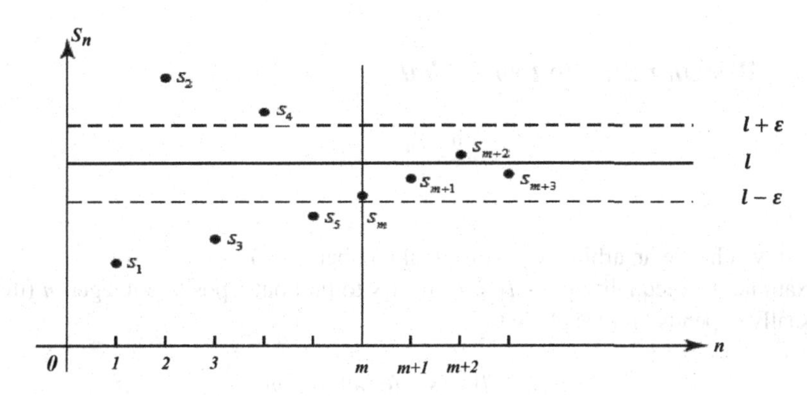

Fig. 2.5 Graph of a convergence sequence

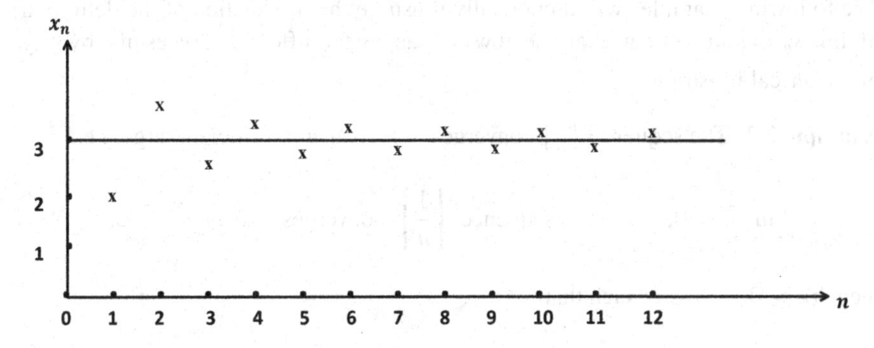

Fig. 2.6 Figure for example 2.1

Example 2.2 The following sequences are divergent:

(i) $\{n^2\}$ is unbounded and converges to $+\infty$.
(ii) $\{-n^2\}$ is unbounded and converges to $-\infty$.
(iii) $\{(-1)^n\}$ is bounded and oscillatory.
(iv) $\{(-2)^n\}$ is unbounded and oscillatory.

Remark 2.3 (i) When we say that a sequence has a limit, it means that the limit is finite and definitely a real number.
(ii) $|s_n - l| < \varepsilon, \ \ \forall n \geqslant m$

$$\Longrightarrow l - \varepsilon < s_n < l + \varepsilon, \ n = m, m+1, m+2, \cdots .$$

$\Longrightarrow$ all but finitely many terms of the sequence $\{s_n\}$ lie to the left of $(l + \varepsilon)$ and to the right of $(l - \varepsilon)$.

$$\Longrightarrow s_n \in (l - \varepsilon, l + \varepsilon), \quad n = m, m+1, m+2, \cdots .$$

2.4.1 Working Rule to Prove That

$$\lim_{n\to\infty} s_n = l.$$

- First we choose an arbitrary positive real number $\varepsilon > 0$.
- Examine the inequality $|s_n - l| < \varepsilon$. We try to find out a positive integer m (that usually depends on ε) such that

$$|s_n - l| < \varepsilon, \quad \text{for all } n \geqslant m.$$

The following examples will numerically illustrate the implication of the definition of $\lim\limits_{n\to\infty} s_n = l$ and we can examine how m changes for different choices of ε by way of numerical illustration.

Example 2.3 The sequence $\left\{\frac{1}{n}\right\}$ converges to zero as n is sufficient large, i.e.,

$$\lim_{n\to\infty} \frac{1}{n} = 0, \ \ i.e., \text{ the sequence } \left\{\frac{1}{n}\right\} \text{ converges to 0 as } n \to \infty,$$

then $\forall \varepsilon > 0 \ \exists \ n \in \mathbb{N}$ such that

$$\left|\frac{1}{n} - 0\right| < \varepsilon, \quad \forall n \geqslant m.$$

Let $\varepsilon = 0.1$. Then

$$\left|\frac{1}{n}\right| < 0.1 \Longrightarrow n \geqslant 11.$$

Now let $\varepsilon = 0.01$. Then

$$\left|\frac{1}{n}\right| < 0.01 \Longrightarrow n \geqslant 101.$$

Example 2.4 The sequence $\{s_n\} = \left\{\frac{n+1}{2n-1}\right\}$ converges to $1/2$ as n is sufficient large.

$$\lim_{n\to\infty} s_n = \lim_{n\to\infty} \frac{n+1}{2n-1} = \frac{1}{2}.$$

Once we obtain the value of $l = 1/2$, we may examine as follows:

$$\forall\, \varepsilon > 0\; \exists\, m \in \mathbb{N} \quad \text{such that}$$

$$\left|\frac{n+1}{2n-1} - \frac{1}{2}\right| = \left|\frac{3}{2\,(2n-1)}\right| < \varepsilon, \quad \forall n \geqslant m,$$

where

$$n \geqslant \frac{3+2\varepsilon}{4\varepsilon} + 1 = m\ (say).$$

Now we choose $\varepsilon = 1$, and we have

$$n \geqslant \frac{3+2}{4} + 1 = \frac{9}{4} = 2.25.$$

Similarly,

$$\text{for } \varepsilon = 0.1, \text{ we have } \; n \geqslant \frac{3+0.2}{0.4} + 1 = 9.$$

$$\text{For } \varepsilon = 0.01, \text{ we have } \; n \geqslant \frac{3+0.02}{0.04} + 1 = 76.5, \quad \text{and so on.}$$

Now we obtain integers m as follows:

$$n \geqslant m = 3, \quad \text{we have} \quad \left|\frac{n+1}{2n-1} - \frac{1}{2}\right| < 1,$$

$$n \geqslant m = 9, \quad \text{we have} \quad \left|\frac{n+1}{2n-1} - \frac{1}{2}\right| < 0.1,$$

$$n \geqslant m = 77, \quad \text{we have} \quad \left|\frac{n+1}{2n-1} - \frac{1}{2}\right| < 0.01,$$

$$n \geqslant m = 751, \quad \text{we have} \quad \left|\frac{n+1}{2n-1} - \frac{1}{2}\right| < 0.001.$$

Theorem 2.1 *"Every convergent sequence is bounded."*

Proof Let a sequence $\{s_n\}$ converge to the limit l. Then

$$\forall \varepsilon > 0 \, \exists \, m \in \mathbb{N} \quad \text{such that}$$
$$|s_n - l| < \varepsilon, \quad \forall n \geqslant m.$$
$$\text{Then we write that} \quad l - \varepsilon < s_n < l + \varepsilon, \quad \forall n \geqslant m.$$

Let $g = \min\{l - \varepsilon, s_1, s_2, \cdots, s_{m-1}\}$ and $G = \max\{l + \varepsilon, s_1, s_2, \cdots, s_{m-1}\}$.

Then we get

$$g \leqslant s_n \leqslant G, \quad \forall n \in \mathbb{N}.$$

Hence, the sequence $\{s_n\}$ is a bounded sequence.

Alternate Method: The given sequence $\{s_n\}$ of real numbers is convergent, therefore s_n converges to l as n approaches infinity, i.e.,

$$\lim_{n \to \infty} s_n = l.$$

Then, given $\varepsilon = 1$, there exists $m \in \mathbb{N}$ such that

$$|s_n - l| < 1, \quad \forall n \geqslant m.$$

For all $n \geqslant m$ we have

$$|s_n| = |l + (s_n - l)| \leqslant |l| + |s_n - l| < |l| + 1.$$

Let $M = \max\{|s_1|, |s_2|, \cdots, |s_{m-1}|, \ |l| + 1\}$. Then we have

$$|s_n| < M, \quad n \in \mathbb{N},$$

which shows that $\{s_n\}$ is bounded. ■

Remark 2.4 The converse of the above theorem may not be true.

For example, $\{s_n\} = (-1)^n, \ \forall n \in \mathbb{N}$ is bounded but not convergent.

Range $= \{-1, 1\}$ is bounded but $\langle -1, 1, -1, 1, -1, 1, \cdots \rangle$ is an oscillating sequence.

Theorem 2.2 (Uniqueness of Limits)

"A sequence cannot converge to more than one limit."
In other words, "the limit of a convergent sequence is unique."

Proof Let us assume that a sequence $\{s_n\}$ has two distinct limits l and l', i.e.,

$$\lim_{n\to\infty} s_n = l \text{ and } \lim_{n\to\infty} s_n = l' \ (l \neq l').$$

Let $\varepsilon = |l - l'| > 0$. Since $\lim_{n\to\infty} s_n = l$, for a given $\varepsilon > 0$ there exists $m_1 \in \mathbb{N}$ such that

$$|s_n - l| < \frac{\varepsilon}{2}, \quad \forall n \geqslant m_1. \tag{2.1}$$

Similarly, since $\lim_{n\to\infty} s_n = l'$, for a given $\varepsilon > 0$ there exists $m_2 \in \mathbb{N}$ such that

$$|s_n - l'| < \frac{\varepsilon}{2}, \quad \forall n \geqslant m_2. \tag{2.2}$$

Now, using the triangle inequality and from (2.1) and (2.2), for $n \geqslant \max(m_1, m_2)$

$$0 \leqslant |l - l'| = |l - s_n + s_n - l'| \leqslant |l - s_n| + |s_n - l'| < \frac{\varepsilon}{2} + \frac{\varepsilon}{2} = \varepsilon,$$

i.e.,

$$|l - l'| < |l - l'|.$$

This is in *contradiction to our assumption*, therefore the *limit is unique* or we can say this way

$$0 \leqslant |l - l'| \leqslant |l - s_n| + |s_n - l'| < \frac{\varepsilon}{2} + \frac{\varepsilon}{2} = \varepsilon.$$

This implies $l = l'$. ∎

2.4.2 Increasing Sequences

A sequence $\{s_n\}$ is said to be an *increasing sequence* if

$$s_{n+1} \geqslant s_n, \ \forall\, n \geqslant 1.$$

"A non-decreasing sequence $\{s_n\}$ is always bounded below by s_1."

2.4.3 Decreasing Sequences

A sequence $\{s_n\}$ is said to be a *decreasing sequence* if

$$s_{n+1} \leqslant s_n, \ \forall\, n \geqslant 1.$$

"A non-increasing sequence $\{s_n\}$ is always bounded above by s_1".

2.4.4 Strictly Increasing or Decreasing

A sequence $\{s_n\}$ is called *strictly increasing or decreasing* according to

$$s_{n+1} > s_n \ or \ s_{n+1} < s_n, \ \forall n \geqslant 1.$$

2.4.5 Monotonic Sequences

A sequence $\{s_n\}$ is said to be an *monotonic sequence* if it is either increasing or decreasing (Figs. 2.7 and 2.8):

$\{s_n\}$ is positive and monotonically increasing if $s_{n+1} - s_n \geqslant 0$

$$or \ \frac{s_{n+1}}{s_n} \geqslant 1, \ \forall n \geqslant 1.$$

$\{s_n\}$ is positive and monotonically decreasing if $s_n - s_{n+1} \geqslant 0$

$$or \ \frac{s_n}{s_{n+1}} \geqslant 1, \ \forall n \geqslant 1.$$

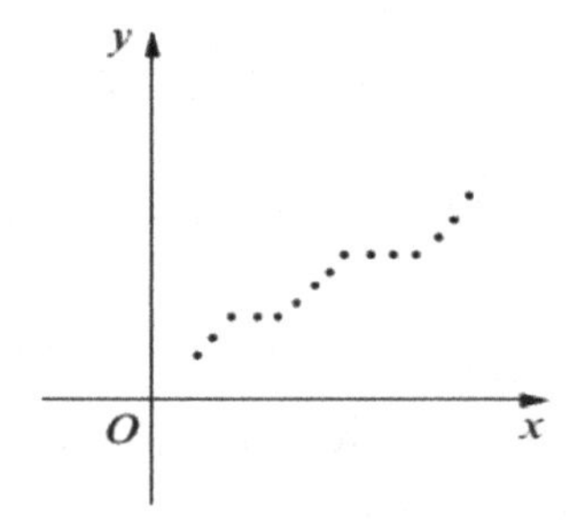

Fig. 2.7 An increasing sequence s_n

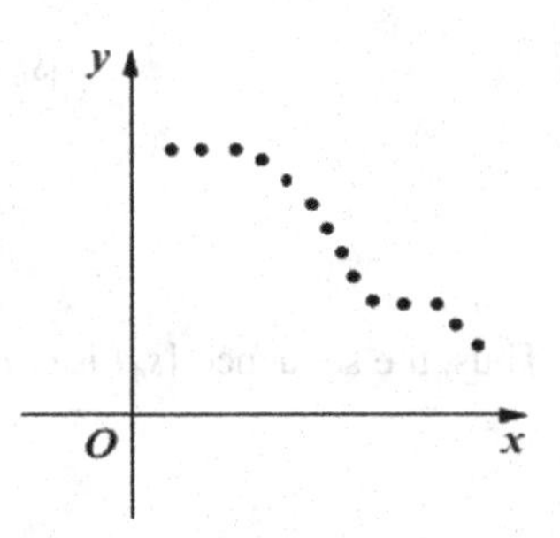

Fig. 2.8 A decreasing sequence S_n

2.4.6 Monotonic Sequences and Their Convergence

Theorem 2.3 *A monotonic increasing sequence, which is bounded above, converges to its least upper bound.*

Proof Suppose that $\{s_n\}$ is a monotonic increasing and bounded above sequence. Then the set

$$A = \{s_1, s_2, s_3, \cdots\}$$

is a non-empty subset of $\mathbb{R}$, which is bounded above. By the *Completeness Axiom* of a set of real numbers, every bounded above set has a *l.u.b.*. Let K be a *l.u.b.* for the set A. Then (Fig. 2.9)

$$K = l.u.b. \{s_1, s_2, s_3, \cdots\}, \; i.e., \; \underset{n}{l.u.b.}\, s_n = K.$$

Now we have to show that

$$s_n \to K, \text{ as } n \to \infty.$$

For each $\varepsilon > 0$, there exists a $n_0 \in \mathbb{N}$ such that

$$K - \varepsilon < s_{n_0} \leqslant K,$$

i.e., the number $K - \varepsilon$ is not an upper bound for A but, on the other hand, K is an upper bound for A.

Since $\{s_n\}$ is monotonic increasing and K is the *l.u.b.*, we get

$$K - \varepsilon < s_{n_0} \leqslant s_n \leqslant K.$$

Hence, it can be written as

$$K - \varepsilon < s_{n_0} \leqslant s_n < K + \varepsilon, \quad \text{for all } n \geqslant n_0.$$

Therefore,

$$|s_n - K| < \varepsilon, \quad \text{for all } n \geqslant n_0,$$

i.e.,

$$\lim_{n\to\infty} s_n = K.$$

Thus, the sequence $\{s_n\}$ is convergent and

$$\lim_{n\to\infty} s_n = \underset{n}{l.u.b.}\, s_n. \qquad \blacksquare$$

Example 2.5 The sequence $\left\{\frac{2n-1}{n}\right\}$ is monotonic increasing and bounded above, since

$$\lim_{n\to\infty} \frac{2n-1}{n} = 2 = \sup_n \left\{\frac{2n-1}{n}\right\}.$$

Thus, the sequence $\left\{\frac{2n-1}{n}\right\}$ is convergent.

Corollary 2.1 *A monotonic decreasing (non-increasing) sequence, which is bounded below, converges to its greatest lower bound.*

Example 2.6 The sequence $\left\{\frac{1}{n}\right\}$ is monotonic decreasing and bounded below, since

$$\lim_{n\to\infty} \frac{1}{n} = 0 = \inf_n \left\{\frac{1}{n}\right\}.$$

Thus, the sequence $\left\{\frac{1}{n}\right\}$ is convergent.

Corollary 2.2 ***The fundamental theorem of a monotonic sequence:***
"a monotonic sequence converges if and only if it is bounded."

Proof From Theorem 2.1, we know that "every convergent sequence is bounded." Conversely, let $\{s_n\}$ be a monotonic (increasing or decreasing) bounded sequence.

(i) When the sequence $\{s_n\}$ is bounded above and monotonic increasing, let $M > 0$. Then

$$s_n \le M, \quad \forall n \in \mathbb{N}.$$

By the *Completeness Axiom*, the supremum K exists, i.e.,

$$K = \sup\{s_n; n \in \mathbb{N}\} \; in \; \mathbb{R}.$$

If a given $\varepsilon > 0$, there exists a $n_0 \in \mathbb{N}$ such that

$$K - \varepsilon < s_{n_0} \leqslant K,$$

i.e., the number $K - \varepsilon$ is not an upper bound of the set $\{s_n; n \in \mathbb{N}\}$. Since the sequence $\{s_n\}$ is a monotonic increasing, it implies that

$$s_{n_0} \leqslant s_n, \quad \forall n \geqslant n_0.$$

Hence, we have

$$|s_n - K| < \varepsilon, \quad \forall n \geqslant n_0.$$

Thus, the sequence $\{s_n\}$ converges to K.

(ii) When the sequence $\{u_n\}$ is bounded and monotonic decreasing, let

$$\{s_n\} = \{-u_n\}, \quad \forall n \in \mathbb{N}$$

be a bounded monotonic increasing sequence. From Theorem 2.3, we obtain

$$\lim_{n\to\infty} s_n = \sup\{-u_n; n \in \mathbb{N}\}.$$

Now, $\lim\limits_{n\to\infty} s_n = -\lim\limits_{n\to\infty} u_n \Longrightarrow \sup\{-u_n; n \in \mathbb{N}\} = -\inf\{u_n; n \in \mathbb{N}\}$. Thus,

$$\lim_{n\to\infty} u_n = -\lim_{n\to\infty} s_n = \inf\{u_n; n \in \mathbb{N}\}.$$

■

Example 2.7 (The number e) Show that the sequence $\left\{\left(1+\frac{1}{n}\right)^n\right\}$ converges.

Let $s_n = \left(1+\frac{1}{n}\right)^n$. Expanding s_n by the Binomial Theorem, we get

$$\begin{aligned}
s_n &= 1 + 1 + \frac{n(n-1)}{2!}\frac{1}{n^2} + \frac{n(n-1)(n-2)}{3!}\frac{1}{n^3} + \cdots + \frac{1}{n^n} \\
&= 1 + 1 + \frac{1}{2!}\left(1-\frac{1}{n}\right) + \frac{1}{3!}\left(1-\frac{1}{n}\right)\left(1-\frac{2}{n}\right) + \cdots \\
&\quad + \frac{1}{n!}\left(1-\frac{1}{n}\right)\left(1-\frac{2}{n}\right)\cdots\left(1-\frac{(n-1)}{n}\right) \\
&< 1 + \frac{1}{1!} + \frac{1}{2!} + \frac{1}{3!} + \cdots + \frac{1}{n!} < 3.
\end{aligned}$$

Therefore, $\{s_n\}$ is bounded above. Now

$$\begin{aligned}
s_{n+1} &= 1 + 1 + \frac{1}{2!}\left(1-\frac{1}{n+1}\right) + \frac{1}{3!}\left(1-\frac{1}{n+1}\right)\left(1-\frac{2}{n+1}\right) + \cdots \\
&\quad + \frac{1}{(n+1)!}\left(1-\frac{1}{n+1}\right)\left(1-\frac{2}{n+1}\right)\cdots\left(1-\frac{n}{n+1}\right) \\
&> 1 + 1 + \frac{1}{2!}\left(1-\frac{1}{n}\right) + \frac{1}{3!}\left(1-\frac{1}{n}\right)\left(1-\frac{2}{n}\right) + \cdots
\end{aligned}$$

$$+\frac{1}{n!}\left(1-\frac{1}{n}\right)\left(1-\frac{2}{n}\right)\cdots\left(1-\frac{n-1}{n}\right)=s_n.$$

Hence, $s_{n+1} > s_n$. Therefore, $\{s_n\}$ is a monotonic increasing sequence and bounded consequently, and by the fundamental theorem of a monotone sequence the given sequence converges to e, i.e.,

$$\lim_{n\to\infty}\left(1+\frac{1}{n}\right)^n = e,$$

and $2 < e < 3$. Thus the sequence $\{s_n\}$ is convergent and bounded.

Theorem 2.4 *A non-decreasing sequence, which is not bounded above, diverges to infinity.*

Proof Suppose that $\{s_n\}$ is a non-decreasing sequence and not bounded above. To prove this theorem, we have to show that for a given $K > 0$ there exists $n_0 \in \mathbb{N}$ such that

$$s_n \geqslant s_{n_0} > K, \quad \text{for all } n \geqslant n_0.$$

Since the sequence $\{s_n\}$ is unbounded above, and furthermore monotonic increasing $K > 0$ cannot be an upper bound of $\{s_1, s_2, s_3, \cdots\}$. Then there exists $n_0 \in \mathbb{N}$ such that

$$s_n > K, \quad \text{for all } n \geqslant n_0.$$

Hence,

$$\lim_{n\to\infty} s_n = +\infty.$$

■

In a similar manner, we can prove the following theorem.

Theorem 2.5 *A monotonic decreasing sequence, which is not bounded below, diverges to minus infinity.*

Example 2.8 Let $k, m \in \mathbb{N}, k \geqslant 1, m > 1$ and $\alpha, s \in \mathbb{R},\ \alpha > 1, s \geqslant 1$.

(i) The sequence $\{s_n\}$ satisfying

$$s_{n+1} = \sqrt{m + (m-1)s_n}, \quad s_1 = \sqrt{m}$$

converges to m.

Since $s_1 = \sqrt{m}\ (< m)$, we see

$$s_2 = \sqrt{m + (m-1)\sqrt{m}} < \sqrt{m + (m-1)m} = m.$$

Now we suppose $s_n < m$, then we have

$$s_{n+1} = \sqrt{m + (m-1)s_n} < \sqrt{m + (m-1)m} = m.$$

Hence we have

$$s_n < m, \quad n = 1, 2, \cdots .$$

On the other hand, we see

$$s_n < s_{n+1}, \quad n = 1, 2, \cdots .$$

In fact, we can easily see $s_1 < s_2$. And if we suppose $s_n < s_{n+1}$ for some $n \geqslant 1$, then from

$$s_{n+1} = \sqrt{m + (m-1)s_n},$$

and

$$s_{n+2} = \sqrt{m + (m-1)s_{n+1}},$$

we see

$$s_{n+1} < s_{n+2}.$$

Consequently, we see that the sequence $\{s_n\}$ converges to some $s \leqslant m$. The value s satisfies

$$s = \sqrt{m + (m-1)s},$$

that is,

$$(s-m)(s+1) = 0.$$

Thus, we have $s = m$.

(ii) The sequence $\{s_n\}$ satisfying

$$s_{n+1} = \sqrt{m^{1/k} + (m^{1/k} - 1)s_n}, \quad s_1 = \sqrt{m^{1/k}}$$

converges to $m^{1/k}$.

Since $s_1 = \sqrt{m^{1/k}}$ $(< m)$, we see

$$s_2 = \sqrt{m^{1/k} + (m^{1/k} - 1)\sqrt{m^{1/k}}} < \sqrt{m^{1/k} + (m^{1/k} - 1)m^{1/k}} = m^{1/k}.$$

Now we suppose $s_n < m^{1/k}$, then we have

$$s_{n+1} = \sqrt{m^{1/k} + (m^{1/k} - 1)s_n} \leqslant \sqrt{m^{1/k} + (m^{1/k} - 1)m^{1/k}} = m^{1/k}.$$

Hence we have

$$s_n < m^{1/k}, \quad n = 1, 2, \cdots .$$

On the other hand, we see

$$s_n < s_{n+1}, \quad n = 1, 2, \cdots .$$

In fact, we can easily see $s_1 < s_2$. And if we suppose $s_n < s_{n+1}$ for some $n \geqslant 1$, then from

$$s_{n+1} = \sqrt{m^{1/k} + (m^{1/k} - 1)s_n},$$

and

$$s_{n+2} = \sqrt{m^{1/k} + (m^{1/k} - 1)s_{n+1}},$$

we see

$$s_{n+1} < s_{n+2}.$$

Consequently, we see that the sequence $\{s_n\}$ converges to some $s \leqslant m^{1/k}$. The value s satisfies

$$s = \sqrt{m^{1/k} + (m^{1/k} - 1)s},$$

that is,

$$(s - m^{1/k})(s + 1) = 0.$$

Thus, we have $s = m^{1/k}$.

(iii) Let $s > 1$ and $\alpha > 0$. The sequence $\{s_n\}$ satisfying

$$s_{n+1} = \{s^{1/\alpha} + (s^{1/\alpha} - 1)(s_n)^{1/\alpha}\}^{\alpha/2}, \quad s_1 = \sqrt{s}$$

converges to s.

Since $s_1 = \sqrt{s}$ $(< s)$, we see

$$s_2 = \{s^{1/\alpha} + (s^{1/\alpha} - 1)(\sqrt{s})^{1/\alpha}\}^{\alpha/2} < \{s^{1/\alpha} + (s^{1/\alpha} - 1)s^{1/\alpha}\}^{\alpha/2} = s.$$

Now we suppose $s_n < s$, then we have

$$s_{n+1} = \{s^{1/\alpha} + (s^{1/\alpha} - 1)(s_n)^{1/\alpha}\}^{\alpha/2} < \{s^{1/\alpha} + (s^{1/\alpha} - 1)s^{1/\alpha}\}^{\alpha/2} = s.$$

Thus, we have

$$s_n < s, \quad n = 1, 2, \cdots .$$

On the other hand, we see

$$s_n < s_{n+1}, \quad n = 1, 2, \cdots .$$

In fact, we can easily see $s_1 < s_2$. And if we suppose $s_n < s_{n+1}$ for some $n \geqslant 1$, then from

$$s_{n+1} = \{s^{1/\alpha} + (s^{1/\alpha} - 1)(s_n)^{1/\alpha}\}^{\alpha/2},$$

and

$$s_{n+2} = \{s^{1/\alpha} + (s^{1/\alpha} - 1)(s_{n+1})^{1/\alpha}\}^{\alpha/2},$$

we see

$$s_{n+1} < s_{n+2}.$$

Consequently, we see that the sequence $\{s_n\}$ converges to some $t \leqslant s$. The value t satisfies

$$t = \{s^{1/\alpha} + (s^{1/\alpha} - 1)t^{1/\alpha}\}^{\alpha/2},$$

that is,

$$(t^{1/\alpha} - s^{1/\alpha})(t^{1/\alpha} + 1) = 0.$$

Thus, we have $t = s$.

2.5 Operations of Convergent Sequences

The sum of sequences $\{s_n\}$ and $\{t_n\}$ is defined to be the sequence $\{s_n + t_n\}$. We have the following useful consequences of the definition of convergence that show how limits team up with the basic algebraic operations.

Theorem 2.6 (Algebra of limits for convergent sequences) *Suppose that* $\{s_n\}$ *and* $\{t_n\}$ *are two sequences such that* $\lim_{n\to\infty} s_n = L$ *and* $\lim_{n\to\infty} t_n = M$*, where* $L, M \in \mathbb{R}$*. Then*

(i) $\lim_{n\to\infty} (s_n + t_n) = L + M.$ *(by Linearity rule for sequences)*

(ii) $\lim_{n\to\infty} cs_n = cL, \quad c \in \mathbb{R}.$

(iii) $\lim_{n\to\infty} s_n^2 = L^2.$

(iv) $\lim_{n\to\infty} (s_n - t_n) = L - M.$

(v) $\lim_{n\to\infty} s_n \cdot t_n = LM.$ *(by Product rule for sequences)*

(vi) $\lim_{n\to\infty} \frac{1}{t_n} = \frac{1}{M}, \ M \neq 0.$ *(by Reciprocal rule)*

(vii) $\lim_{n\to\infty} \frac{s_n}{t_n} = \frac{L}{M}, \ M \neq 0.$ *(by Quotient rule for sequences)*

Proof (*i*) Since $\lim_{n\to\infty} s_n = L$, for a given $\varepsilon > 0$, there exists $n_1 \in \mathbb{N}$ such that

$$|s_n - L| < \frac{\varepsilon}{2}, \quad \text{for all } n \geqslant n_1.$$

Since $\lim\limits_{n\to\infty} t_n = M$, for a given $\varepsilon > 0$, there exists $n_2 \in \mathbb{N}$ such that

$$|t_n - M| < \frac{\varepsilon}{2}, \quad \text{for all } n \geqslant n_2.$$

Let $n_0 = \max(n_1, n_2)$. Then

$$\begin{aligned}|(s_n + t_n) - (L + M)| &= |s_n - L + t_n - M| \\ &\leqslant |s_n - L| + |t_n - M| < \frac{\varepsilon}{2} + \frac{\varepsilon}{2} = \varepsilon, \quad \text{for all } n \geqslant n_0.\end{aligned}$$

Thus, $\lim\limits_{n\to\infty} (s_n + t_n) = L + M$.

(ii) If $c = 0$, then the theorem is obvious. So we only have to prove the theorem for $c \neq 0$. For this we have to show that for a given $\varepsilon > 0$ there exists $n_0 \in \mathbb{N}$ such that

$$|cs_n - cL| < \varepsilon, \quad \text{for all } n \geqslant n_0.$$

Now, since $\lim\limits_{n\to\infty} s_n = L$, there exists $n_0 \in \mathbb{N}$ such that

$$|s_n - L| < \frac{\varepsilon}{|c|}, \quad \text{for all } n \geqslant n_0.$$

But then

$$|cs_n - cL| = |c|\,|s_n - L| < |c|\,\frac{\varepsilon}{|c|} = \varepsilon, \quad \text{for all } n \geqslant n_0.$$

Hence, we get $\lim\limits_{n\to\infty} cs_n = cL, \quad c \in \mathbb{R}$.

(iii) Given $\lim\limits_{n\to\infty} s_n = L$, we have to prove $\lim\limits_{n\to\infty} s_n^2 = L^2$, i.e., for a given $\varepsilon > 0$ we must find $n_0 \in \mathbb{N}$ such that

$$\left|s_n^2 - L^2\right| < \varepsilon, \quad \text{for all } n \geqslant n_0$$

or

$$|s_n - L| \cdot |s_n + L| < \varepsilon, \quad \text{for all } n \geqslant n_0. \tag{2.3}$$

We know that every convergent sequence is bounded. Thus for some $K > 0$

$$|s_n| \leqslant K, \quad \text{for all } n \in \mathbb{N}.$$

Therefore,

$$|s_n + L| \leqslant |s_n| + |L| \leqslant K + |L|, \quad \text{for all } n \in \mathbb{N}. \tag{2.4}$$

Since $\lim_{n\to\infty} s_n = L$, there exists $n_0 \in \mathbb{N}$ such that

$$|s_n - L| < \frac{\varepsilon}{K + |L|}, \quad \text{for all } n \geqslant n_0. \tag{2.5}$$

From (2.4) and (2.5), we obtain (2.3):

$$|s_n - L| \cdot |s_n + L| < \frac{\varepsilon}{K + |L|} \cdot (K + |L|) = \varepsilon, \quad \text{for all } n \geqslant n_0.$$

The proof is complete.

(iv) Since $\lim_{n\to\infty} ct_n = cM$, now taking $c = -1$, and from (ii) we obtain

$$\lim_{n\to\infty} (-t_n) = -M.$$

Thus, we have

$$\begin{aligned}\lim_{n\to\infty} (s_n - t_n) &= \lim_{n\to\infty} [s_n + (-t_n)] = \lim_{n\to\infty} s_n + \lim_{n\to\infty} (-t_n) = L + (-M) \\ &= L - M, \quad (\text{from } (i) \text{ and } (ii)).\end{aligned}$$

(v) Since every convergent sequence must be bounded, there exists $K > 0$ such that

$$|s_n| \leqslant K \text{ (say)}, \quad \text{for all } n \in \mathbb{N}.$$

Let $s_n \to L$ as $n \to \infty$ and let $\varepsilon > 0$ be given. Then there exists $n_1 \in \mathbb{N}$ such that

$$|s_n - L| < \frac{\varepsilon}{2(1 + |M|)}, \quad \text{for all } n \geqslant n_1.$$

(We remark that we cannot use $\varepsilon/2\,|M|$ instead of $\varepsilon/[2\,(1 + |M|)]$ because M can be zero.)

By the hypothesis $t_n \to M$ as $n \to \infty$, let $\varepsilon > 0$ be given. Then there exists $n_2 \in \mathbb{N}$ such that

$$|t_n - M| < \frac{\varepsilon}{2K}, \quad \text{for all } n \geqslant n_2.$$

Finally, for $n_0 = \max\{n_1, n_2\}$, we have

$$\begin{aligned}|s_n t_n - LM| &\leqslant |s_n|\,|t_n - M| + |M|\,|s_n - L| \\ &\leqslant K\frac{\varepsilon}{2K} + |M|\,\frac{\varepsilon}{2(1 + |M|)} < \varepsilon, \quad \text{for all } n \geqslant n_0.\end{aligned}$$

The Product rule clearly follows.

Alternate Proof of (v). Without use of ε, we can show (v).

From (i) we have

$$s_n + t_n \to L + M \;\; as \;\; n \to \infty.$$

And from (iii), we have

$$(s_n + t_n)^2 \to (L + M)^2 \;\; as \;\; n \to \infty. \tag{2.6}$$

Again from (iv), we have

$$s_n - t_n \to L - M \;\; as \;\; n \to \infty.$$

And from (iii), we have

$$(s_n - t_n)^2 \to (L - M)^2 \;\; as \;\; n \to \infty. \tag{2.7}$$

Using (2.6) and (2.7), we get

$$(s_n + t_n)^2 - (s_n - t_n)^2 \to (L + M)^2 - (L - M)^2 = 4LM \;\; as \;\; n \to \infty. \tag{2.8}$$

Finally, with the help of (2.8), we obtain

$$s_n t_n = \frac{1}{4}\left[(s_n + t_n)^2 - (s_n - t_n)^2\right] \to \frac{1}{4}(4LM) = LM \;\; as \;\; n \to \infty.$$

(vi) First, we put $\varepsilon = \frac{|M|}{2} > 0$ (note $\lim_{n\to\infty} t_n = M \neq 0$). There exists $n_1 \in \mathbb{N}$ such that

$$|t_n - M| < \frac{|M|}{2}, \quad \text{for all } n \geqslant n_1.$$

Therefore

$$|t_n| = |M + (t_n - M)| \geqslant ||M| - |t_n - M|| > \frac{|M|}{2}, \quad \text{for all } n \in \mathbb{N}.$$

If we take ε to be $\frac{\varepsilon}{2}|M|^2$, there exists $n_2 \in \mathbb{N}$ such that

$$|t_n - M| < \frac{\varepsilon}{2}|M|^2, \quad \text{for all } n \geqslant n_2,$$

and let $n_0 = \max\{n_1, n_2\}$. We have

$$\left|\frac{1}{t_n} - \frac{1}{M}\right| = \left|\frac{M - t_n}{t_n M}\right| = \frac{|t_n - M|}{|t_n|\,|M|} < \frac{2}{|M|^2}\frac{\varepsilon}{2}|M|^2 = \varepsilon, \quad \text{for all } n \geqslant n_0.$$

Hence

$$\lim_{n\to\infty} \frac{1}{t_n} = \frac{1}{M}.$$

(vii) From (iv) and (v) we get $\lim_{n\to\infty} s_n \cdot \frac{1}{t_n} = L \cdot \frac{1}{M}$. This proves (vii).

■

Theorem 2.7 (The Sandwich Theorem) *Suppose that* $\{s_n\}$ *and* $\{t_n\}$ *are two sequences of real numbers and*

$$\lim_{n\to\infty} s_n = L \text{ and } \lim_{n\to\infty} t_n = L.$$

If $s_n \leqslant u_n \leqslant t_n$ $(n = 1, 2, \cdots)$, *then*

$$\lim_{n\to\infty} u_n = L.$$

Proof Since $s_n \to L$ as $n \to \infty$, we know that for a given $\varepsilon > 0$ there exists $n_1 \in \mathbb{N}$ such that

$$|s_n - L| < \varepsilon, \quad \text{for all } n \geqslant n_1 \quad \Longrightarrow \quad L - \varepsilon < s_n < L + \varepsilon. \tag{2.9}$$

Similarly, since $t_n \to L$ as $n \to \infty$, there exists $n_2 \in \mathbb{N}$ such that

$$|t_n - L| < \varepsilon, \quad \text{for all } n \geqslant n_2 \quad \Longrightarrow \quad L - \varepsilon < t_n < L + \varepsilon, \tag{2.10}$$

and the given condition

$$s_n \leqslant u_n \leqslant t_n \ (n = 1, 2, \cdots). \tag{2.11}$$

Let $n_0 = \max\{n_1, n_2\}$. From (2.9) to (2.11), we have

$$L - \varepsilon < s_n \leqslant u_n \leqslant t_n < L + \varepsilon.$$

Hence,

$$L - \varepsilon < u_n < L + \varepsilon \quad \Longrightarrow \quad |u_n - L| < \varepsilon, \quad \text{for all } n > n_0.$$

Thus,

$$\lim_{n\to\infty} u_n = L.$$

■

2.6 Operations of Divergent Sequences

Theorem 2.8 *If* $\{s_n\}$ *and* $\{t_n\}$ *are sequences of real numbers that diverge to infinity, then the sum* $\{s_n + t_n\}$ *and the product* $\{s_n \cdot t_n\}$ *diverge to infinity.*

Proof Since $\{s_n\}$ diverges to infinity, for a given $M > 0$ there exists $n_1 \in \mathbb{N}$ such that

$$s_n > M, \quad \text{for all } \ n \geqslant n_1,$$

and choose $n_2 \in \mathbb{N}$ such that

$$t_n > 1, \quad \text{for all } \ n \geqslant n_2.$$

Since $s_n \to \infty$ and $t_n \to \infty$ as $n \to \infty$, for $n_0 = \max(n_1, n_2)$ we have

$$s_n + t_n > M + 1 > M, \quad \text{for all } \ n \geqslant n_0,$$

and

$$s_n \cdot t_n > M \cdot 1 = M, \quad \text{for all } \ n \geqslant n_0.$$

Since M is any arbitrary positive number, this shows that

$$\{s_n + t_n\} \to \infty \quad \text{and} \quad \{s_n \cdot t_n\} \to \infty.$$

■

Theorem 2.9 *If $\{s_n\}$ and $\{t_n\}$ are sequences of real numbers, and if $\{s_n\}$ diverges to infinity and $\{t_n\}$ is bounded, then the sum $\{s_n + t_n\}$ diverges to infinity.*

Proof Since $\{t_n\}$ is bounded, there exists a constant $K > 0$ such that

$$|t_n| \leqslant K, \quad \text{for all } n \in \mathbb{N}.$$

Since the sequence $\{s_n\}$ diverges to infinity, for a given $M > 0$ there exists $n_1 \in \mathbb{N}$ such that

$$s_n > M + K, \quad \text{for all } n \geqslant n_1.$$

Then, for all $n \geqslant n_1$, we have

$$s_n + t_n \geqslant s_n - |t_n| > (M + K) - K = M,$$

i.e.,

$$s_n + t_n > M, \quad \text{for all } n \geqslant n_1,$$

which shows that $\{s_n + t_n\} \to \infty$ as $n \to \infty$. ■

Corollary 2.3 *If $\{s_n\}$ diverges to infinity and if $\{t_n\}$ converges, then the sum $\{s_n + t_n\}$ diverges to infinity.*

Proof Since every convergent sequence is bounded (see Theorem 2.1), from Theorem 2.9 the proof follows. ■

2.7 Nested Intervals

A sequence of intervals $\{I_n\}$ is called a *nest* if

(*i*) $I_1 \supset I_2 \supset I_3 \supset I_4 \supset \cdots$,
(*ii*) the length of $I_n \to 0$ as $n \to \infty$.

Theorem 2.10 (Cantor's Theorem with Nested Intervals) *Let $\{I_n\}$ be a sequence of non-empty closed bounded intervals such that*

$$I_n = [a_n, b_n], \quad I_{n+1} \subset I_n, \quad \text{for all } n \geqslant 1, \quad \text{and} \quad (b_n - a_n) \to 0 \text{ as } n \to \infty.$$

Then $\bigcap_{n=1}^{\infty} I_n$ contains one and only one point.

Proof Since $I_{n+1} \subset I_n$, $\forall n \geqslant 1$ (Fig. 2.10),

$$a_1 \leqslant a_2 \leqslant a_3 \leqslant \cdots \leqslant a_{n-1} \leqslant a_n \leqslant b_n \leqslant b_{n-1} \leqslant \cdots \leqslant b_3 \leqslant b_2 \leqslant b_1.$$

Note that the sequence $\{a_n\}$ is monotonic increasing and bounded above by b_1, therefore from Corollary 2.2, we have

$$\lim_{n\to\infty} a_n = a \ (say),$$

and the sequence $\{b_n\}$ is monotonic decreasing and bounded below by a_1, therefore from Corollary 2.2, we have

$$\lim_{n\to\infty} b_n = b \ (say),$$

i.e., as $n \to \infty$ we have $\{a_n\} \to a$ and $\{b_n\} \to b$. Since the length of $I_n = b_n - a_n \to 0$ as $n \to \infty$, it follows that $a = b$. Hence, we see

$$a_n \leqslant a = b \leqslant b_n, \ \forall n \in \mathbb{N}$$

and so

$$a \in \bigcap_{n=1}^{\infty} I_n.$$

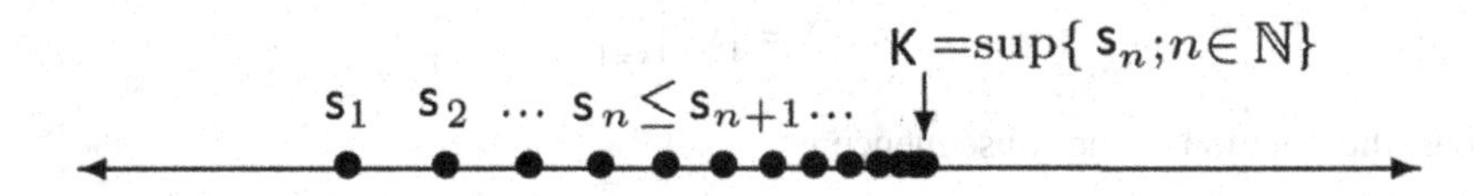

Fig. 2.9 l.u.b.(sup) of a sequence s_n = K

Clearly, no c ($\neq a$) can lie in $\bigcap_{n=1}^{\infty} I_n$ by our hypothesis $|c-a|$, which is greater than the length of I_n for n sufficiently large. Thus $\bigcap_{n=1}^{\infty} I_n$ contains only a, and does not contain other points. ■

2.8 Subsequences

Suppose that

$$\{x_1, x_2, \cdots, x_m, \cdots\} \tag{2.12}$$

is a sequence of real numbers. Then the sequences $\{x_1, x_3, x_5, x_7, \cdots\}$ and $\{x_2, x_5, x_8, x_{11}, \cdots\}$ are examples of subsequences of (2.12). More generally, suppose that $\{n_1, n_2, \cdots, n_k, \cdots\}$ is an increasing sequence of positive integers. Then we say that

$$\left\{x_{n_1}, x_{n_2}, \cdots, x_{n_m}, \cdots\right\}$$

is a subsequence of (2.12). The choice $n_1 = 1, n_2 = 3, n_3 = 5, n_4 = 7, \cdots$ is an example of a subsequence of (2.12) above. To avoid double subscripts, which are cumbersome, we will frequently write $y_1 = x_{n_1}, y_2 = x_{n_2}, \cdots, y_n = x_{n_n}, \cdots$, in which case $y_1, y_2, \cdots, y_n, \cdots$ is a subsequence of (2.12).

A subsequence of $\{x_n\}_{n=1}^{\infty}$ is a sequence $\{y_k\}_{k=1}^{\infty}$ defined by $y_k = x_{n_k}$, where $n_1 < n_2 < \cdots$ is an increasing sequence of indices. In other words, the terms of a subsequence are members of the original sequence. A subsequence of $\{s_n\}_{n=1}^{\infty}$ is usually written as $\left\{s_{n_k}\right\}_{k=1}^{\infty}$.

A sub-mapping in the sequence which preserves the order is called a subsequence.

Example 2.9 Let $S = \{s_n\}_{n=1}^{\infty}$ be a sequence of real numbers and $N = \{n_k\}_{k=1}^{\infty}$ be a subsequence of the sequence of positive integers. Then the composite function $(S \circ N)$ is called a subsequence of S.
For $k \in \mathbb{N}$, we have $N(k) = n_k$. Therefore

$$(S \circ N)(k) = S(N(k)) = S(n_k) = s_{n_k}.$$

Hence

$$(S \circ N) = \left\{s_{n_k}\right\}_{k=1}^{\infty}.$$

Thus, the notation of the subsequence is

$$\left\{s_{n_1}, s_{n_2}, s_{n_3}, \cdots\right\}.$$

Example 2.10 Let a sequence

$$f : \mathbb{N} \xrightarrow[into]{1-1} \mathbb{R}$$

be defined by

$$f(n) = x^n, \ \forall n \in \mathbb{N}, \ i.e., \ \left\{x^1, x^2, x^3, \cdots, x^n, \cdots\right\}$$

and a subsequence (sub-mapping)

$$g : \mathbb{N} \xrightarrow[into]{1-1} \mathbb{R}$$

be defined by

$$g(n) = x^{n^2}, \ \ \forall \, n \in \mathbb{N}, \ i.e., \ \left\{x^1, x^4, x^9, \cdots, x^{n^2}, \cdots\right\}.$$

Example 2.11 The sequences

$$\{2n - 1\} = \{1, 3, 5, \cdots\} \quad \text{and} \quad \{2n\} = \{2, 4, 6, \cdots\}$$

are subsequences of the sequence

$$\{n\} = \{1, 2, 3, \cdots\}.$$

Theorem 2.11 *A sequence $\{x_n\}$ converges if and only if every subsequence $\left\{x_{n_k}\right\}$ of $\{x_n\}$ converges to the same limit.*

Proof Suppose that a sequence $\{x_n\}$ converges to a real number l, then show that the subsequence $\left\{x_{n_k}\right\}$ also converges to l because of

$$\lim_{n\to\infty} x_n = l.$$

Let $\varepsilon > 0$. There exists $n_0 \in \mathbb{N}$ such that

$$|x_n - l| < \varepsilon, \quad \text{for all } n \geqslant n_0.$$

We observe that $\{n_k\}$ is a strictly increasing sequence of natural numbers. By the method of induction we can show that

$$n_k \geqslant k, \quad \text{for all } k \in \mathbb{N}.$$

In fact, $n_1 \geqslant 1$ and $n_k \geqslant k \implies n_{k+1} > n_k \geqslant k$. Thus $n_{k+1} \geqslant k + 1$.

Now $k \geqslant n_0$ implies $n_k \geqslant k \geqslant n_0$, which implies

$$\left|x_{n_k} - l\right| < \varepsilon.$$

Thus, the subsequence $\{x_{n_k}\}$ converges to the same limit l.

The converse of this theorem is trivially true: if every subsequence of the sequence $\{x_n\}$ converges to the same number, say l, then because any sequence is a subsequence of the sequence, the sequence $\{x_n\}$ must converge to l. ■

Corollary 2.4 *If a sequence $\{x_n\}$ has two subsequences that converge to different limits, then $\{x_n\}$ does not converge.*

Example 2.12 (i) Consider the sequence $\{(-1)^n\}$. This sequence has two subsequences $\{(-1)^{2n}\}$ and $\{(-1)^{2n+1}\}$. The first subsequence $\{(-1)^{2n}\}$ converges to 1, and the second subsequence converges to -1. Since these two subsequences converge to two different limits, the given sequence $\{x_n\}$ does not converge.

(ii) The sequence

$$\left\{\frac{1+(-1)^n}{2}\right\} = \{0, 1, 0, 1, 0, 1, 0, 1, \cdots\}$$

is divergent. It has two subsequences, the first one $\{0, 0, 0, \cdots\}$ converges to 0, and the second one $\{1, 1, 1, \cdots\}$ converges to 1.

Theorem 2.12 (The Bolzano-Weierstrass Theorem) *Each bounded sequence has a convergent subsequence.*

Proof Since $\{s_n\}$ is a bounded sequence of real numbers, there exists an interval $[a_1, b_1]$ such that $a_1 \leqslant s_n \leqslant b_1, \ \forall n \in \mathbb{N}$, and

$$a_1 = \inf s_n \ \text{ and } \ b_1 = \sup s_n.$$

Now bisect the interval $[a_1, b_1]$, then either $\left[a_1, \frac{a_1+b_1}{2}\right]$ or $\left[\frac{a_1+b_1}{2}, b_1\right]$ contains infinite numbers of terms of the sequence $\{s_n\}$. If $\left[a_1, \frac{a_1+b_1}{2}\right]$ contains infinitely many terms of $\{s_n\}$, let $[a_2, b_2] = \left[a_1, \frac{a_1+b_1}{2}\right]$, otherwise take $[a_2, b_2] = \left[\frac{a_1+b_1}{2}, b_1\right]$ (Fig. 2.11).

Then either $\left[a_2, \frac{a_2+b_2}{2}\right]$ or $\left[\frac{a_2+b_2}{2}, b_2\right]$ contains infinite numbers of terms of the sequence $\{s_n\}$. If $\left[a_2, \frac{a_2+b_2}{2}\right]$ contains infinitely many terms of $\{s_n\}$, let $[a_3, b_3] = \left[a_2, \frac{a_2+b_2}{2}\right]$, otherwise take $[a_3, b_3] = \left[\frac{a_2+b_2}{2}, b_2\right]$. By continuing this process we obtain a sequence of closed intervals $\{[a_n, b_n]\}$ with the following properties:

(i) $b_n - a_n = \frac{b_1-a_1}{2^{n-1}}$.
(ii) $[a_n, b_n]$ contains infinitely many points of $\{s_n\}$.
(iii) $[a_n, b_n] \subset \left[a_{n-1}, b_{n-1}\right] \subset \cdots \subset [a_1, b_1]$.

Fig. 2.10 Nested intervals

$a_1 \quad a_2 \quad a_3 \qquad\qquad b_3 \quad b_2 \quad b_1$

By the *nested intervals Theorem* 2.10 we get a unique point s such that

$$s = \bigcap_{n=1}^{\infty} [a_n, b_n] \quad \text{and} \quad b_n - a_n = \frac{b_1 - a_1}{2^{n-1}} \to 0 \text{ as } n \to \infty.$$

Now we construct a subsequence of $\{s_n\}$, which will converge to s.

Since $[a_1, b_1]$ contains s_n for infinitely many values of n, choose one of the $s_n \in [a_1, b_1]$ and call it s_{n_1}. Choose $s_{n_2} \in [a_2, b_2]$ such that $n_1 < n_2$, and so on, $s_{n_k} \in [a_k, b_k]$, $s_{n_{k+1}} \in \left[a_{k+1}, b_{k+1}\right]$ such that $n_k < n_{k+1}$, and so on. By continuing this process (by induction), we obtain a sequence $\left\{s_{n_k}\right\}$.

The sequence $\left\{s_{n_k}\right\}$ is a subsequence of the sequence $\{s_n\}$ and now using the nested intervals Theorem 2.10, there exists a unique common point s in $[a_k, b_k]$ for all $k \in \mathbb{N}$. Moreover, since both s_{n_k} and s belong to $[a_k, b_k]$, we have

$$\left|s_{n_k} - s\right| \leqslant \frac{b_1 - a_1}{2^{k-1}}.$$

This proves that the subsequence $\left\{s_{n_k}\right\}$ converges to s. ■

Alternate Proof.
Let $\{s_n\}$ be a bounded sequence of real numbers. Then show that the sequence $\{s_n\}$ has a limit point.

Let $S = \{s_n; n \in N\}$ be the range set of the sequence $\{s_n\}$. Since the sequence $\{s_n\}$ is bounded, the range set S is also bounded. Now there are two possibilities: (i) S is finite and (ii) S is infinite.

(*i*) S **is finite**: If S is finite, then there exists at least one member $\xi \in S$ such that $\{s_n\} = \xi$ for an infinite number of values of n. For $\varepsilon > 0$,

$$s_{n_k} (= \xi) \in (\xi - \varepsilon, \xi + \varepsilon), \ \forall k \geqslant k(\varepsilon).$$

Thus, ξ is the limit point of the sequence $\left\{s_{n_k}\right\}$.

(*ii*) S **is infinite**: Since S is an infinite and bounded set, by the *Bolzano-Weierstrass Theorem (for sets)*, there exists at least one limit point of S.
Let η be a limit point of S. Therefore every neighborhood $(\eta - \varepsilon, \eta + \varepsilon)$ of η contains an infinite number of members of S, $(\varepsilon > 0)$.
$\Longrightarrow$ Every neighborhood $(\eta - \varepsilon, \eta + \varepsilon)$, $\varepsilon > 0$ of η contains an infinite number of members of S, that is, $s_{n_k} \in (\eta - \varepsilon, \eta + \varepsilon)$, $\forall k \geqslant k(\varepsilon)$. Thus η is the limit point of the sequence $\{s_{n_k}\}$.

■

2.9 Cauchy Sequences

We recall the definition of a convergent sequence $\{s_n\}$. A sequence converges to a limit l if for every $\varepsilon > 0$ there is a positive integer N such that

$$|s_n - l| < \varepsilon, \text{ whenever } n > N. \tag{2.13}$$

Suppose that we are given a sequence and wish to examine the possibility of convergence. Usually, the number l is not given, so that the condition (2.13) above cannot be verified directly. For this reason, it is important to have a technique for deciding convergence that does not employ the limit l of the sequence. Such a criterion, presented below, was given first by Cauchy.

Definition 2.1 An infinite sequence $\{s_n\}$ is called a Cauchy sequence if and only if for each $\varepsilon > 0$, there is a positive integer $N(\varepsilon)$ such that

$$|s_n - s_m| < \varepsilon, \quad \text{for all } m, n \geqslant N. \tag{2.14}$$

Roughly, a sequence $\{s_n\}$ is a Cauchy sequence if s_m and s_n are close together when m and n are large.

Example 2.13 Suppose $\{s_n\}$ is a sequence of real numbers, where

$$s_n = \sum_{k=1}^{n} \frac{1}{k^2}, \quad \forall n \in \mathbb{N}.$$

Let $\varepsilon > 0$ and choose a positive integer N such that $\frac{1}{N} < \varepsilon$. If $m, n \geqslant N$ with $n > m$, then for $n > m + 1$

$$|s_n - s_m| = \left| \sum_{k=1}^{n} \frac{1}{k^2} - \sum_{k=1}^{m} \frac{1}{k^2} \right| = \sum_{k=m+1}^{n} \frac{1}{k^2} \leqslant \int_m^n \frac{1}{x^2} dx < \frac{1}{m}.$$

Since $\frac{1}{m} \to 0$ as $m \to \infty$, we have

$$\frac{1}{m} < \varepsilon, \quad m \geqslant N.$$

Thus,

$$|s_n - s_m| < \varepsilon, \quad m, n \geqslant N.$$

Hence, the sequence $\{s_n\}$ is a Cauchy sequence.

Theorem 2.13 *Every Cauchy sequence is bounded.*

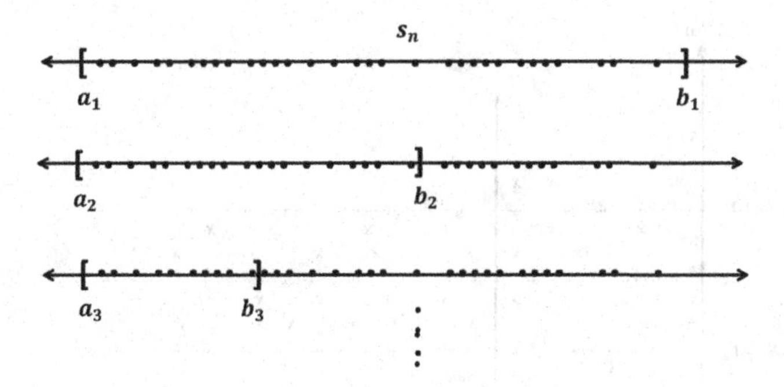

Fig. 2.11 Bisection of intervals

Proof Let $\{s_n\}$ be a Cauchy sequence of real numbers. Then for any $\varepsilon > 0$ there exists a positive integer $N(\varepsilon)$ such that

$$|s_n - s_m| < \varepsilon, \quad \text{for all } n, m \geqslant N.$$

In particular, choosing $m = N + 1$, we find that

$$|s_n - s_{N+1}| < \varepsilon, \quad \text{for all } n \geqslant N.$$

Therefore,

$$|s_n| = |s_n - s_{N+1} + s_{N+1}| \leqslant |s_n - s_{N+1}| + |s_{N+1}| < \varepsilon + |s_{N+1}|, \quad \text{for all } n \geqslant N.$$

Let

$$M = \max\{|s_1|, |s_2|, \cdots |s_N|, \varepsilon + |s_{N+1}|\}.$$

Thus,

$$|s_n| \leqslant M, \quad \text{for all } n \in \mathbb{N}.$$

Alternate Method: Suppose that $\{s_n\}$ is a Cauchy sequence of real numbers. Then taking $\varepsilon = 1$ there exists a positive integer $N(1)$ such that (Fig. 2.12)

$$|s_n - s_m| < 1, \text{ for all } m, n \geqslant N.$$

Then

$$\begin{aligned} n \geqslant N &\Longrightarrow |s_n - s_N| < 1, \\ &\Longrightarrow -1 < s_n - s_N < 1 \\ &\Longrightarrow s_N - 1 < s_n < s_N + 1 \end{aligned}$$

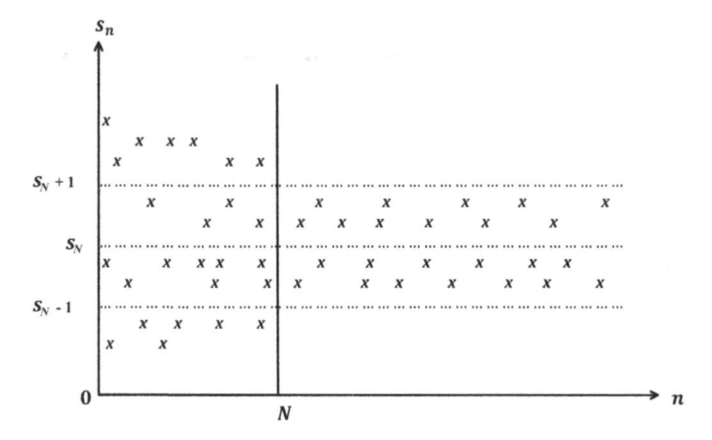

Fig. 2.12 Graphic interpretation for bounded Cauchy sequence

Let $g = \min\{s_1, s_2, \cdots, s_N, s_N - 1\}$ and $G = \max\{s_1, s_2, \cdots, s_N, s_N + 1\}$. Then we get

$$g \leqslant s_n \leqslant G, \quad \forall n \in \mathbb{N}.$$

Hence, the Cauchy sequence $\{s_n\}$ is a bounded sequence. ■

Theorem 2.14 *Every convergent sequence of real numbers is a Cauchy sequence.*

Proof Suppose $\{s_n\}$ is a convergent sequence such that

$$\lim_{n\to\infty} s_n = l.$$

Choose $\varepsilon/2 > 0$, then there is a positive integer N such that

$$|s_n - l| < \frac{\varepsilon}{2}, \quad \text{for all } n > N.$$

For all $m > N$,

$$|s_m - l| < \frac{\varepsilon}{2}.$$

Thus,

$$\begin{aligned} |s_n - s_m| &= |s_n - l + l - s_m| \\ &\leqslant |s_n - l| + |l - s_m| = |s_n - l| + |s_m - l| < \frac{\varepsilon}{2} + \frac{\varepsilon}{2} = \varepsilon. \end{aligned}$$

Hence, $\{s_n\}$ is a Cauchy sequence. ■

Theorem 2.15 (Cauchy's Completeness Principle) *Every Cauchy sequence of real numbers is convergent.*

Proof Let $\{s_n\}$ be a Cauchy sequence. And every Cauchy sequence is bounded by Theorem 2.13. Therefore, by the *Bolzano-Weierstrass Theorem* 2.12, it has a convergent subsequence $\{s_{n_k}\}$.

Suppose that $s_{n_k} \to l$ as $k \to \infty$. Then for any $\varepsilon > 0$ there exists a positive integer N_1 such that

$$\left|s_{n_k} - l\right| < \frac{\varepsilon}{2}, \quad \text{for all } n_k > N_1.$$

Since $\{s_n\}$ is a Cauchy sequence of real numbers, for any $\varepsilon > 0$ there exists a positive integer N_2 such that

$$|s_n - s_m| < \varepsilon/2, \quad \text{for all } n, m \geqslant N_2.$$

In particular, we have

$$|s_n - s_{n_k}| < \varepsilon/2, \quad \text{for all } n, n_k \geqslant \max(N_1, N_2).$$

Hence

$$|s_n - l| = \left|s_n - s_{n_k} + s_{n_k} - l\right| < \left|s_n - s_{n_k}\right| + \left|s_{n_k} - l\right| < \frac{\varepsilon}{2} + \frac{\varepsilon}{2} = \varepsilon,$$

for $n \geqslant \max(N_1, N_2)$. Thus, $\lim_{n\to\infty} s_n = l$. ■

Example 2.14 The sequence $\sum \frac{1}{n!}$ converges, then the sequence $\left\{\sum_{k=1}^{n} \frac{1}{k!}\right\}$ is also convergent.

For all $n \in \mathbb{N}$, let

$$s_n = \sum_{k=1}^{n} \frac{1}{k!}.$$

Therefore, for $n > m$ and $2^{n-1} \leqslant n!$, $\forall n \in N$, we have

$$\begin{aligned} |s_n - s_m| &= \left|\sum_{k=1}^{n} \frac{1}{k!} - \sum_{k=1}^{m} \frac{1}{k!}\right| = \sum_{k=m+1}^{n} \frac{1}{k!} \\ &\leqslant \sum_{k=m}^{n-1} \frac{1}{2^k} = \frac{1}{2^m} \sum_{k=1}^{n-m} \frac{1}{2^{k-1}} < \frac{1}{2^{m-1}}. \end{aligned}$$

Let $\varepsilon > 0$. Since $\frac{1}{2^m} \to 0$ as $m \to \infty$, there exists a positive real number N such that

$$\frac{1}{2^m} < \varepsilon, \quad m \geqslant N.$$

Thus,

$$|s_n - s_m| < \varepsilon, \quad m, n \geqslant N.$$

Hence, the sequence $\{s_n\}$ is a Cauchy sequence and we know that every Cauchy sequence is convergent by Theorem 2.15.

Remark 2.5 We notice that

$$\sum_{n=1}^{\infty} \frac{1}{n!} = e - 1. \tag{2.15}$$

Using Theorem 4.18, below, for $0 \leqslant x$ there exists $\xi \in (0, x)$ such that

$$e^x = 1 + \frac{1}{1!}x + \frac{1}{2!}x^2 + \cdots + \frac{1}{n!}x^n + \frac{e^\xi}{(n+1)!}x^{n+1}.$$

Let $\varepsilon > 0$. Then there exists a positive number $N = N(x, \varepsilon)$ such that

$$\frac{e^\xi}{(n+1)!}x^{n+1} < \varepsilon, \quad n \geqslant N. \tag{2.16}$$

Then we conclude that for each x

$$e^x = 1 + \frac{1}{1!}x + \frac{1}{2!}x^2 + \cdots + \frac{1}{n!}x^n + \cdots .$$

If $x \in [0, M]$, where $M > 0$ is a constant, then the series is uniformly convergent. We show (2.16). Let $x \in [0, M]$, and let $2M < m < n, \ m, n \in \mathbb{N}$. Then we see

$$\begin{aligned} \frac{e^\xi}{(n+1)!}x^{n+1} &< \frac{e^M}{(n+1)!}M^{n+1} = e^M \frac{M \cdot M \cdots M}{(n+1) \cdot n \cdots m \cdots 2 \cdot 1} \\ &< e^M \left(\frac{1}{2}\right)^{n+1-m} \frac{M^m}{m!} = e^M \left(\frac{1}{2}\right)^{-m} \frac{M^m}{m!} \left(\frac{1}{2}\right)^{n+1} \to 0 \text{ as } n \to \infty. \end{aligned}$$

Thus, we have (2.16), so we conclude (2.15).

Example 2.15 The sequence, whose n-th term is

(i) $s_n = 1 + \frac{1}{2} + \frac{1}{3} + \cdots + \frac{1}{n}$, is divergent.
(ii) $s_n = 1 + \frac{1}{2} + \frac{1}{3} + \cdots + \frac{1}{n} - \log n$, is convergent.

We can show (i), (ii) as follows:

(i) Given

$$s_n = 1 + \frac{1}{2} + \frac{1}{3} + \cdots + \frac{1}{n}, \quad \text{for any } n \geqslant 1,$$

then we have

$$s_{2n} - s_n = \frac{1}{n+1} + \frac{1}{n+2} + \cdots + \frac{1}{2n}.$$

Therefore,

$$\frac{1}{n+n} + \frac{1}{n+n} + \cdots + \frac{1}{2n} \leqslant s_{2n} - s_n$$
$$\Rightarrow \frac{1}{2} \leqslant s_{2n} - s_n.$$

This clearly implies that $\{s_n\}$ fails to satisfy Cauchy's condition. Therefore it is divergent.

(ii) Applying the integral definition of the logarithm function,

$$\log x = \int_1^x \frac{1}{t} dt.$$

If $0 < a < b$, then we have

$$\frac{b-a}{b} \leqslant \int_a^b \frac{1}{t} dt \leqslant \frac{b-a}{a}.$$

Since

$$\log n = \int_1^n \frac{1}{t} dt = \sum_{k=1}^{n-1} \int_k^{k+1} \frac{1}{t} dt,$$

we have

$$\log n \leqslant \sum_{k=1}^{n-1} \frac{k+1-k}{k} = 1 + \frac{1}{2} + \frac{1}{3} + \cdots + \frac{1}{n-1}.$$

Thus,

$$s_n = 1 + \frac{1}{2} + \frac{1}{3} + \cdots + \frac{1}{n} - \log n$$
$$= 1 + \frac{1}{2} + \frac{1}{3} + \cdots + \frac{1}{n-1} - \log n + \frac{1}{n} \geqslant \frac{1}{n} > 0.$$

On the other hand, we have

$$s_{n+1} - s_n = \frac{1}{n+1} - \log(n+1) + \log n = \frac{1}{n+1} - \int_n^{n+1} \frac{1}{t} dt < 0.$$

Two inequalities above imply that $\{s_n\}$ is decreasing and bounded below by 0. Therefore, by Corollary 2.2 the sequence $\{s_n\}$ is convergent.

2.9.1 Cauchy's General Principle for Convergence

A sequence of real numbers is convergent if and only if it is a Cauchy sequence.

Theorem 2.16 (Cauchy's criterion) *For a sequence $\{s_n\}$ of real numbers, the following conditions are equivalent:*

(i) the sequence $\{s_n\}$ is convergent;
(ii) for every $\varepsilon > 0$, there is an index N such that $|s_n - s_m| < \varepsilon$, whenever $m, n \geqslant N$; in symbols,

$$\forall \varepsilon > 0 \; \exists \, N > 0, \; \textit{for all } m, n \geqslant N \Rightarrow |s_n - s_m| < \varepsilon.$$

Proof The proofs of Necessary and Sufficient Conditions are given in Theorem 2.14 and Theorem 2.15, respectively. ■

2.10 Limits Superior and Inferior

The limit inferior (also called infimum limit, lim inf, inferior limit, lower limit, or inner limit) and limit superior (also called supremum limit, lim sup, superior limit, upper limit, or outer limit) of a sequence can be thought of as limiting (i.e., eventual and extreme) bounds on the sequence. The limit inferior and limit superior of a function can be thought of in a similar fashion (see limit of a function). The limit inferior and limit superior of a set are the infimum and supremum of the set's limit points, respectively. In general, when there are multiple objects around a place in which a sequence, a function, or a set accumulates, the inferior and superior limits extract the smallest and largest of them; the type of the object and the measure of the size are context-dependent, but the notion of extreme limits is invariant.

If the terms in the sequence are real numbers, the limit superior and limit inferior always exist, as real numbers or $\pm\infty$.

Let us suppose a sequence $\{s_n\}$ is bounded above, i.e.,

$$s_n \leqslant M, \quad \forall n \in \mathbb{N}.$$

Then, for a fixed $N (\in \mathbb{N})$ and $n > N$ the set $\{s_n, s_{n+1}, s_{n+2}, \cdots\}$ is clearly bounded above. Let

$$M_n = l.u.b. \{s_n, s_{n+1}, s_{n+2}, \cdots\},$$

and

$$M_{n+1} = l.u.b.\{s_{n+1}, s_{n+2}, s_{n+3}, \cdots\}.$$

Since $\{s_{n+1}, s_{n+2}, s_{n+3}, \cdots\} \subset \{s_n, s_{n+1}, s_{n+2}, \cdots\} \Rightarrow M_n \geqslant M_{n+1}$. Thus the sequence $\{M_n\}_{n=1}^{\infty}$ is *non-increasing*, so the sequence either converges or diverges to minus infinity.

Definition 2.2 Let $\{s_n\}$ be a sequence of real numbers that is bounded above, and let $M_n = l.u.b.\{s_n, s_{n+1}, s_{n+2}, \cdots\}$.

(i) If $\{M_n\}$ converges, we define

$$\overline{\lim_{n\to\infty}}\, s_n = \limsup_{n\to\infty} s_n = \lim_{N\to\infty} \sup \{s_n : n \geqslant N\}$$

to be $\lim_{n\to\infty} M_n$,

$$i.e., \quad \overline{\lim_{n\to\infty}}\, s_n = \limsup_{n\to\infty} s_n = \lim_{N\to\infty}\left(\sup_{n\geqslant N} s_n\right) = \inf_N \left(\sup_{n\geqslant N} s_n\right).$$

(ii) If $\{M_n\}$ diverges to minus infinity, we write

$$\overline{\lim_{n\to\infty}}\, s_n = \limsup_{n\to\infty} s_n = \lim_{N\to\infty} \sup \{s_n : n \geqslant N\} = -\infty.$$

Definition 2.3 Let $\{s_n\}$ be a sequence of real numbers that is not bounded above. Thus,

$$\overline{\lim_{n\to\infty}}\, s_n = \limsup_{n\to\infty} s_n = \lim_{N\to\infty} \sup \{s_n ; n \geqslant N\} = +\infty.$$

Example 2.16 (i) Let $s_n = (-1)^n$, $n \in \mathbb{N}$. Then $\{s_n\}$ is bounded above. In this case $M_n = 1$ for every $n \in \mathbb{N}$, and $\lim_{n\to\infty} M_n = 1$. Hence $\limsup_{n\to\infty} (-1)^n = 1$.

(ii) Let us consider the sequence $\{1, -1, 1, -2, 1, -3, 1, -4, \cdots\}$. Then we see $M_n = 1$ for every $n \in \mathbb{N}$, and so the limit superior of this sequence is 1.

(iii) If $s_n = -n$, $n \in \mathbb{N}$, then $M_n = l.u.b.\{-n, -n-1, -n-2, \cdots\} = -n$. Hence, $M_n \to -\infty$ as $n \to \infty$ and therefore $\limsup_{n\to\infty} (-n) = -\infty$.

Now we define the *limit inferior*. Let us suppose that a sequence $\{s_n\}$ of real numbers is bounded below. Then the set $\{s_n, s_{n+1}, s_{n+2}, \cdots\}$ has a $g.l.b.$ Let

$$m_n = g.l.b.\{s_n, s_{n+1}, s_{n+2}, \cdots\},$$

and

$$m_{n+1} = g.l.b.\{s_{n+1}, s_{n+2}, s_{n+3}, \cdots\}.$$

Since $\{s_n, s_{n+1}, s_{n+2}, \cdots\} \supset \{s_{n+1}, s_{n+2}, s_{n+3}, \cdots\} \Rightarrow m_n \leqslant m_{n+1}$. Thus the sequence $\{m_n\}$ is a *non-decreasing* sequence, so the sequence either converges or diverges to infinity.

Definition 2.4 Let $\{s_n\}$ be a sequence of real numbers that is bounded below, and let $m_n = g.l.b.\{s_n, s_{n+1}, s_{n+2}, \cdots\}$.

(i) If $\{m_n\}$ converges, we define

$$\underline{\lim}_{n\to\infty} s_n = \liminf_{n\to\infty} s_n = \liminf_{N\to\infty} \{s_n : n \geqslant N\} = \lim_{n\to\infty} m_n,$$

$$i.e., \quad \underline{\lim}_{n\to\infty} s_n = \liminf_{n\to\infty} s_n = \lim_{N\to\infty}\left(\inf_{n\geqslant N} s_n\right) = \sup_N \left(\inf_{n\geqslant N} s_n\right).$$

(ii) If $\{m_n\}$ diverges to infinity, we write

$$\underline{\lim}_{n\to\infty} s_n = \liminf_{n\to\infty} s_n = \liminf_{N\to\infty} \{s_n : n \geqslant N\} = \infty.$$

Definition 2.5 If $\{s_n\}$ is a sequence of real numbers that is not bounded below, then

$$\underline{\lim}_{n\to\infty} s_n = \liminf_{n\to\infty} s_n = \liminf_{N\to\infty} \{s_n ; n \geqslant N\} = -\infty.$$

Example 2.17 (i) Let $s_n = (-1)^n$, $n \in \mathbb{N}$. Then $\{s_n\}$ is bounded below. In this case $m_n = -1$ for every $n \in \mathbb{N}$, and $\lim_{n\to\infty} m_n = -1$. Hence $\liminf_{n\to\infty} (-1)^n = -1$.

(ii) Let $\{s_n\} = \{-1, -2, -3, \cdots\}$. Then $m_n = \inf\{-n, -n-1, -n-2, \cdots\} = -n$, and so that $\underline{\lim}_{n\to\infty} s_n = \lim_{n\to\infty} m_n = -\infty$.

(iii) Let $\{s_n\} = \{1, 2, 3, \cdots\}$. Then $m_n = \inf\{n, n+1, n+2, \cdots\} = n$, and so that $\underline{\lim}_{n\to\infty} s_n = \lim_{n\to\infty} m_n = \infty$.

Now we are giving alternative definitions for lim sup and lim inf, which are more useful for all practical purposes.

Definition 2.6 Let $\{s_n\}$ be a sequence of real numbers:

(a) Let $\{s_n\}$ be *bounded above*, and let

$$M_n = \sup\{s_n, s_{n+1}, \cdots\}.$$

(i) If $\{M_n\}$ converges, then $\limsup_{n\to\infty} s_n = \lim_{n\to\infty} M_n$.

(ii) If $\{M_n\}$ diverges to $-\infty$, then $\limsup_{n\to\infty} s_n = -\infty$.

(b) Let $\{s_n\}$ be *bounded below*, and let

$$m_n = \inf\{s_n, s_{n+1}, \cdots\}.$$

(i) If $\{m_n\}$ converges, then $\liminf_{n\to\infty} s_n = \lim_{n\to\infty} m_n$.

(ii) If $\{m_n\}$ diverges to ∞, then $\liminf_{n\to\infty} s_n = \infty$.

(c) Let $\{m_n\}$ be unbounded. Then

(i) If the sequence is not bounded above, then $\limsup_{n\to\infty} s_n = \infty$.

(ii) If the sequence is not bounded below, then $\liminf_{n\to\infty} s_n = -\infty$.

Theorem 2.17 *Let $\{s_n\}$ be a sequence of real numbers. Then*

$$\liminf_{n\to\infty} s_n \leqslant \limsup_{n\to\infty} s_n.$$

Proof Let $M_n = l.u.b.\{s_n, s_{n+1}, s_{n+2}, \cdots\} = \sup\{s_n, s_{n+1}, s_{n+2}, \cdots\}$, and

$$m_n = g.l.b.\{s_n, s_{n+1}, s_{n+2}, \cdots\} = \inf\{s_n, s_{n+1}, s_{n+2}, \cdots\}.$$

If a sequence $\{s_n\}$ is bounded above as well as bounded below, then for $n \geqslant 1$, M_n and m_n are defined and sequences $\{M_n\}$ and $\{m_n\}$ are also bounded.

We observe that the sequence $\{M_n\}$ is monotonic decreasing, and the sequence $\{m_n\}$ is monotonic increasing respectively. Therefore from Theorem 2.3 and Corollary 2.1 we have the convergent sequences $\{M_n\}$ and $\{m_n\}$. It is clear that

$$m_n \leqslant M_n, \quad n = 1, 2, 3, \cdots.$$

Then we see

$$m_j \leqslant M_k, \quad \forall j, k \geqslant 1.$$

In fact, if $j \leqslant k$, then we see

$$m_j \leqslant m_k \leqslant M_k,$$

on the other hand, if $j > k$, then we see

$$m_j \leqslant M_j \leqslant M_k.$$

Therefore, for any $k \geqslant 1$ we have

$$\liminf_{j\to\infty} s_j \leqslant M_k,$$

and so we have

$$\liminf_{j\to\infty} s_j \leqslant \limsup_{k\to\infty} s_k.$$

Consequently we have the result. ■

Theorem 2.18 *Let $\{s_n\}$ be a convergent sequence in $\mathbb{R}$ if and only if*

$$\underline{\lim}_{n\to\infty} s_n = \liminf_{n\to\infty} s_n = L = \limsup_{n\to\infty} s_n = \overline{\lim}_{n\to\infty} s_n.$$

Proof Let $\lim_{n\to\infty} s_n = L$. Then $\{s_n\}$ is a convergent sequence of real numbers. Now given $\varepsilon > 0$, then $N \in \mathbb{N}$ such that

$$|s_n - L| < \varepsilon \Longleftrightarrow L - \varepsilon < s_n < L + \varepsilon, \quad \text{for all } n \geqslant N. \tag{2.17}$$

Since $L + \varepsilon$ is an upper bound for $\{s_n, s_{n+1}, s_{n+2}, \cdots\}$ for all $n \geqslant N$.

$$M_n = l.u.b. \{s_n, s_{n+1}, s_{n+2}, \cdots\} < L + \varepsilon, \quad n \geqslant N.$$

Similarly, if we consider $m_n = g.l.b. \{s_n, s_{n+1}, s_{n+2}, \cdots\}$, we have

$$L - \varepsilon < m_n = g.l.b. \{s_n, s_{n+1}, s_{n+2}, \cdots\}, \quad n \geqslant N.$$

Hence we have

$$L - \varepsilon < m_n \leqslant M_n \leqslant L + \varepsilon, \quad n \geqslant N.$$

Therefore we have (as we see in the proof of Theorem 2.17)

$$L - \varepsilon < \liminf_{n\to\infty} s_n \leqslant \limsup_{n\to\infty} s_n \leqslant L + \varepsilon.$$

Consequently, we have the result.

Converse: If $\liminf_{n\to\infty} s_n = \limsup_{n\to\infty} s_n = L \in \mathbb{R}$, then we will prove that $\{s_n\}$ is a convergent sequence. Since

$$L = \limsup_{n\to\infty} s_n = l.u.b. \{s_n, s_{n+1}, s_{n+2}, \cdots\},$$

for all $\varepsilon > 0$, there exists a positive integer N_1 such that

$$|l.u.b. \{s_n, s_{n+1}, s_{n+2}, \cdots\} - L| < \varepsilon, \quad \text{for all } n \geqslant N_1.$$

Therefore,

$$s_n < L + \varepsilon, \quad \text{for all } n \geqslant N_1. \tag{2.18}$$

Similarly,

$$L = \liminf_{n\to\infty} s_n = g.l.b. \{s_n, s_{n+1}, s_{n+2}, \cdots\}.$$

For all $\varepsilon > 0$ there exists a positive integer N_2 such that

$$|g.l.b. \{s_n, s_{n+1}, s_{n+2}, \cdots\} - L| < \varepsilon, \quad \text{for all } n \geqslant N_2.$$

Again from (2.17), we have

$$L - \varepsilon < s_n, \quad \text{for all } n \geqslant N_2. \tag{2.19}$$

Let $N = \max\{N_1, N_2\}$, from (2.18) and (2.19). Then we have

$$L - \varepsilon < s_n < L + \varepsilon, \quad \text{for all } n \geqslant N.$$

Hence

$$|s_n - L| < \varepsilon, \quad \text{for all } n \geqslant N.$$

Thus,

$$\lim_{n\to\infty} s_n = L.$$

■

Theorem 2.19 *Let $\{s_n\}$ and $\{t_n\}$ be any two bounded sequences in $\mathbb{R}$. Then*

(i) $\limsup\limits_{n\to\infty} (s_n + t_n) \leqslant \limsup\limits_{n\to\infty} s_n + \limsup\limits_{n\to\infty} t_n$.

(ii) $\liminf\limits_{n\to\infty} (s_n + t_n) \geqslant \liminf\limits_{n\to\infty} s_n + \liminf\limits_{n\to\infty} t_n$.

Proof Since $\{s_n\}$ and $\{t_n\}$ are two bounded sequences, if we get

$$M_n = \sup\{s_n, s_{n+1}, s_{n+2}, \cdots\} \quad \text{and} \quad M'_n = \sup\{t_n, t_{n+1}, t_{n+2}, \cdots\},$$

then

$$s_k \leqslant M_n \quad \text{and} \quad t_k \leqslant M'_n, \quad \text{for all } k \geqslant n.$$

Thus,

$$\begin{aligned}&\limsup_{n\to\infty}\{s_n + t_n, s_{n+1} + t_{n+1}, s_{n+2} + t_{n+2}, \cdots\}\\ &\qquad \leqslant \lim_{n\to\infty} \left(M_n + M'_n\right) = \lim_{n\to\infty} M_n + \lim_{n\to\infty} M'_n.\end{aligned}$$

Similarly, we have (ii). ■

Theorem 2.20 *Let $\{s_n\}$ and $\{t_n\}$ be any two bounded sequences in $\mathbb{R}$. If $s_n \leqslant t_n$, $\forall n \in \mathbb{N}$, then*

(i) $\liminf\limits_{n\to\infty} s_n \leqslant \liminf\limits_{n\to\infty} t_n$.

(ii) $\limsup\limits_{n\to\infty} s_n \leqslant \limsup\limits_{n\to\infty} t_n$.

Proof Since for all $n \in \mathbb{N}$, $s_n \leqslant t_n$, it is clear that

$$\sup\{s_n, s_{n+1}, s_{n+2}, \cdots\} \leqslant \sup\{t_n, t_{n+1}, t_{n+2}, \cdots\},$$

and

$$\inf\{s_n, s_{n+1}, s_{n+2}, \cdots\} \leqslant \inf\{t_n, t_{n+1}, t_{n+2}, \cdots\}.$$

Taking the limits as $n \to \infty$ on both sides,

$$\limsup_{n\to\infty}\{s_n, s_{n+1}, s_{n+2}, \cdots\} \leqslant \limsup_{n\to\infty}\{t_n, t_{n+1}, t_{n+2}, \cdots\},$$

and

$$\liminf_{n\to\infty}\{s_n, s_{n+1}, s_{n+2}, \cdots\} \leqslant \liminf_{n\to\infty}\{t_n, t_{n+1}, t_{n+2}, \cdots\}.$$

Thus,

$$\limsup_{n\to\infty} s_n \leqslant \limsup_{n\to\infty} t_n,$$

and

$$\liminf_{n\to\infty} s_n \leqslant \liminf_{n\to\infty} t_n.$$

■

Example 2.18 We find lim sup, lim inf and lim of the sequence

$$\{s_n\} = \left\{\frac{n}{n+1}\right\}.$$

For $n \geqslant 1$,

$$M_n = \sup\{s_n, s_{n+1}, s_{n+2}, \cdots\} \quad \text{and} \quad m_n = \inf\{s_n, s_{n+1}, s_{n+2}, \cdots\}.$$

Then

$$M_1 = \sup\left\{\frac{1}{2}, \frac{2}{3}, \cdots\right\} = 1$$
$$M_2 = \sup\left\{\frac{2}{3}, \frac{3}{4}, \cdots\right\} = 1$$
$$\vdots$$
$$M_n = \sup\left\{\frac{n}{n+1}, \frac{n+1}{n+2}, \cdots\right\} = 1.$$

Hence, $M_n = 1$, for all $n \geqslant 1$. Therefore, we see

$$\limsup_{n\to\infty} s_n = \lim_{n\to\infty} M_n = 1. \tag{2.20}$$

Now

$$m_1 = \inf\left\{\frac{1}{2}, \frac{2}{3}, \cdots\right\} = \frac{1}{2}$$

$$m_2 = \inf\left\{\frac{2}{3}, \frac{3}{4}, \cdots\right\} = \frac{2}{3}$$

$$\vdots$$

$$m_n = \inf\left\{\frac{n}{n+1}, \frac{n+1}{n+2}, \cdots\right\} = \frac{n}{n+1}.$$

Hence,

$$m_n = \frac{n}{n+1}, \quad \forall n \geqslant 1.$$

Therefore, we see

$$\liminf_{n\to\infty} s_n = \lim_{n\to\infty} m_n = 1. \tag{2.21}$$

From (2.20) and (2.21) we have

$$\lim_{n\to\infty} s_n = 1.$$

2.11 Exercises

1. Choose the correct option:

   (i) The limit of the sequence $\left\{s_n = \frac{n+(-1)^n}{n}\right\}$ is
   (a) 0 (b) 1 (c) ∞ (d) $-\infty$

   (ii) The limit of the sequence $\left\{s_n = \frac{(10/11)^n}{(9/10)^n+(11/12)^n}\right\}$ is
   (a) 0 (b) 1 (c) ∞ (d) $-\infty$

   (iii) The limit of the sequence $\left\{s_n = \frac{1-n^3}{70-4n^2}\right\}$ is
   (a) 0 (b) 1/2 (c) ∞ (d) $-\infty$.

2. Write the expression for the n-th term of the following sequences:
   (i) $\{2, 5, 8, 11, \cdots\}$ (ii) $\{2, 3, 2, 3, \cdots\}$
   (iii) $\{1, 3, 6, 10, \cdots\}$ (iv) $\{1, 1, 2, 3, 5, 8, \cdots\}$.

3. Which of the following sequences is convergent?
   (i) $\left\{\frac{(-1)^n}{n}; n \in \mathbb{N}\right\}$ (ii) $\left\{\frac{3n}{2n+1}; n \in \mathbb{N}\right\}$
   (iii) $\{(-1)^n n; n \in \mathbb{N}\}$ (iv) $\{2^{-n}; n \in \mathbb{N}\}$.

4. What is the limit point of the following sequences?
   (i) $\left\{\frac{n+1}{2n+3}; n \in \mathbb{N}\right\}$ (ii) $\left\{\frac{m}{n}; m, n \in \mathbb{N}\right\}$

begininlineMathiii) $\{(-1)^n; n \in \mathbb{N}\}$ (iv) For a non-empty set A, $\{A + \frac{1}{n}; n \in \mathbb{N}\}$.

5. Give an example of a sequence which is bounded but not convergent.
6. A sequence $\{a_n\}$ is defined to be bounded, if there is a number M such that

$$|a_n| \leqslant M, \text{ for every index } n.$$

Show that $\{a_n\}$ is bounded if and only if there are numbers a and b with $a < b$ such that $\{a_n\}$ is a sequence in $[a, b]$.

7. Give an example of a sequence which is monotonic increasing but not convergent.
8. Suppose that the sequence $\{a_n\}$ is monotone. Prove that $\{a_n\}$ converges if and only if $\{a_n^2\}$ converges. Show that this result does not hold without the monotonicity assumption.
9. Show that the sequence $\{s_n\}$ defined by $s_n = \{1 + \frac{1}{n}\}^n$ is convergent and that $\lim_{n\to\infty} \left(1 + \frac{1}{n}\right)^n$ lies between 2 and 3.
10. Show that the sequence of real numbers $\{s_n\}$ defined by $s_1 = 1$, $s_{n+1} = \frac{1}{3+s_n}$ for $n = 1, 2, 3, \cdots$ converges, and determine its limit.
11. Show that the sequence of real numbers $\{s_n\}$ defined by $s_1 = \sqrt{2}$, $s_{n+1} = \sqrt{2 + s_n}$ for $n = 1, 2, 3, \cdots$ converges: $s_n \to 2$ as $n \to \infty$.
12. Show that the monotonic decreasing sequence which is bounded below converges to its greatest lower bound.
13. Does the sequence $\{\sin n\}$ have a convergent subsequence?
14. Prove that a sequence $\{a_n\}$ does not converge to a number a if and only if there are some $\varepsilon > 0$ and a sequence $\{a_{n_k}\}$ such that

$$|a_{n_k} - a| \geqslant \varepsilon, \text{ for every index } k.$$

15. Prove that the sequence $\{\cos n\}$ is divergent.
16. Let $-\infty < a < b < \infty$ and $0 < \lambda < 1$. Define the sequence $\{x_n\}$ by $x_1 = a$, $x_2 = b$ and $x_{n+1} = \lambda x_n + (1 - \lambda)x_{n+1}$ for $n = 1, 2, \cdots$. Show that $\{x_n\}$ converges in $\mathbb{R}$, and then find its limit.
17. Give an example of a sequence $\{x_n\}$ of real numbers for which $\{|x_n|\}$ converges but $\{x_n\}$ does not.
18. Give an example of the bounded sequence which is not a Cauchy sequence.
19. Prove that the sequence $\{\frac{1}{n}\}$ is a Cauchy sequence.
20. Show that the sum of two Cauchy sequences is a Cauchy sequence.
21. Show that the sequence

$$s_n = \frac{1}{n+1} + \frac{1}{n+2} + \frac{1}{n+3} + \cdots + \frac{1}{2n}$$

is convergent.

22. Prove that the sequence

$$s_n = \frac{1 \cdot 3 \cdot 5 \cdot \ldots \cdot (2n-1)}{2 \cdot 4 \cdot 6 \cdot \ldots \cdot (2n)}$$

is convergent and $\lim_{n\to\infty} s_n \leqslant \frac{1}{2}$.

23. Give an example of two divergent sequences whose sum is a convergent sequence.
24. Find $\limsup s_n$ and $\liminf s_n$ for the following sequences:
(i) $\sin \frac{n\pi}{2} + (-1)^n$ (ii) $(\cos \frac{n\pi}{4})^{(-1)^n}$
(iii) $\{s_n\} = \{-2, -1, 1, 2, -2, -1, 1, 2, \cdots\}$.
25. Write the set of all rational numbers in $(0, 1)$ as $\{s_1, s_2, \cdots\}$. Calculate $\limsup_{n\to\infty} s_n$ and $\liminf_{n\to\infty} s_n$.
26. A subsequence $\{s_{n_k}\}$ of a Cauchy sequence $\{s_n\}$ is convergent, then prove that $\{s_n\}$ is also convergent.
27. Show that the sequence $\{x_n\}$ defined as

$$x_n = \int_1^n \frac{\cos t}{t^2} dt$$

is a Cauchy sequence.

28. If $\{s_n\}$ is a convergent sequence in $\mathbb{R}$, then show that

$$\overline{\lim_{n\to\infty}}\, s_n = \limsup_{n\to\infty} s_n = \lim_{n\to\infty} s_n.$$

29. If $\{s_n\}$ is a convergent sequence in $\mathbb{R}$, then show that

$$\underline{\lim_{n\to\infty}}\, s_n = \liminf_{n\to\infty} s_n = \lim_{n\to\infty} s_n.$$

30. Find lim sup, lim inf and lim of the sequence

$$\{s_n\} = \{(-1)^n\}.$$

Chapter 3
Infinite Series of Numbers

In this chapter, we will briefly introduce an infinite series of real numbers and apply our results for sequences to series, or infinite sums. Before discussing infinite series, we would like to explain a few basic definitions. We will also discuss the techniques of the testing nature of infinite series as regards convergence. The convergence and sum of an infinite series are defined in terms of its sequence of finite partial sums.

3.1 Positive Terms Series

Definition 3.1 Given a sequence $\{u_n\}$ of real numbers, then we use expressions such that

$$u_1 + u_2 + u_3 + \cdots + u_n + \cdots \quad \text{and} \quad \sum_{n=1}^{\infty} u_n \quad \text{or sometimes just} \quad \sum u_n.$$

This is called an infinite series or just a series. In other words, we can say that the sum of the terms of a sequence is a series.

The $u_n,\ n = 1, 2, 3, \cdots$ are called the **terms** of the series, and the quantities

$$s_n = u_1 + u_2 + u_3 + \cdots + u_n, \quad n = 1, 2, 3, \cdots$$

are called the **partial sums** of the series.

N. Deo and R. Sakai, *Introduction to Mathematical Analysis*,
https://doi.org/10.1007/978-981-97-6568-3_3

3.1.1 Convergence and Divergence of Series

Assume that $\{u_n\}$ is a sequence of real numbers. If the sequence of partial sums $\{s_n\}$ converges to a real number l (finite and unique), then we say that the infinite series $\sum_{n=1}^{\infty} u_n$ converges to l and write symbolically

$$\sum_{n=1}^{\infty} u_n = l \quad \text{or} \quad \lim_{n\to\infty} s_n = l.$$

If the sequence of partial sums $\{s_n\}$ diverges (i.e., $\{s_n\} \to \infty$ as $n \to \infty$), we say that the series $\sum_{n=1}^{\infty} u_n$ diverges.

Geometrically, as n gets large, the graph of the sequence either levels out at a limit l or the value of s_n gets large (Fig. 3.1).

Example 3.1 (*i*) The series $1 + \frac{1}{2} + \frac{1}{2^2} + \frac{1}{2^3} + \cdots$ is convergent. Since the series is a geometric progressive series

$$\lim_{n\to\infty} s_n = \lim_{n\to\infty} \frac{\left\{1 - \frac{1}{2^n}\right\}}{\left\{1 - \frac{1}{2}\right\}} = \lim_{n\to\infty} \left\{2 - \frac{1}{2^{n-1}}\right\} = 2.$$

(*ii*) The series $1 + 2 + 2^2 + 2^3 + \cdots$ is divergent. Since the series is a geometric progressive series

$$\lim_{n\to\infty} s_n = \lim_{n\to\infty} \frac{(2^n - 1)}{2 - 1} = \lim_{n\to\infty} \left(2^n - 1\right) = \infty.$$

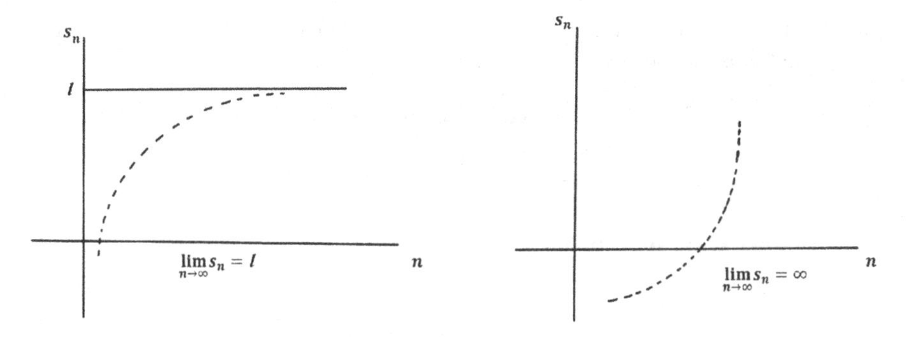

Fig. 3.1 Graphic interpretation for convergence and divergence

3.1.2 Fundamental Properties

(i) The nature of an infinite series $\sum_{n=1}^{\infty} u_n$ remains unaffected if a finite number of terms is added or omitted.

(ii) The nature of an infinite series $\sum_{n=1}^{\infty} u_n$ remains unaffected if each term is multiplied or divided by a fixed non-zero number.

(iii) If two given series $\sum_{n=1}^{\infty} u_n$ and $\sum_{n=1}^{\infty} v_n$ are convergent, then $\sum_{n=1}^{\infty} (u_n \pm v_n)$ are also convergent series.

(iv) The series less than a convergent series is convergent and greater than a divergent series is divergent.

Theorem 3.1 *If the series* $\sum_{n=1}^{\infty} u_n$ *converges, then* $u_n \to 0$ *as* $n \to \infty$*, and the converse is not necessarily true.*

Proof Suppose that $\{s_n\}_{n=1}^{\infty}$ is a sequence of partial sums. Since for all $n > 1$ we have $u_n = s_n - s_{n-1}$, the given series is convergent,

$$\lim_{n\to\infty} s_n = l \text{ and } \lim_{n\to\infty} s_{n-1} = l, \quad l \in \mathbb{R}.$$

Hence

$$\lim_{n\to\infty} u_n = \lim_{n\to\infty} s_n - \lim_{n\to\infty} s_{n-1} = l - l = 0.$$

By example we will show that the converse is not necessarily true.

Let us consider $\sum_{n=1}^{\infty} u_n = \sum_{n=1}^{\infty} \frac{1}{\sqrt{n}}$. Then $u_n \to 0$ as $n \to \infty$.

Now

$$s_n = 1 + \frac{1}{\sqrt{2}} + \frac{1}{\sqrt{3}} + \cdots + \frac{1}{\sqrt{n}} > \frac{n}{\sqrt{n}} = \sqrt{n}.$$

Therefore,

$$s_n \to \infty \text{ as } n \to \infty.$$

Hence, the series $\sum_{n=1}^{\infty} \frac{1}{\sqrt{n}}$ is divergent. ■

A series of the form

$$a + ar + ar^2 + \cdots + ar^n + \cdots \quad (a \neq 0)$$

is called a **geometric series** and the number r is the **common ratio**.

Theorem 3.2 *A geometric series*

$$a + ar + ar^2 + \cdots + ar^n + \cdots \quad a > 0$$

converges if $|r| < 1$ *and diverges if* $|r| \geqslant 1$.

Proof For each positive integer n,

$$s_n = a + ar + ar^2 + \cdots + ar^{n-1}.$$

If $|r| < 1$, i.e., $-1 < r < 1$, then $r^n \to 0$ as $n \to \infty$. Therefore

$$s_n = \frac{a\,(1 - r^n)}{1 - r} \to \frac{a}{1 - r} \quad \text{as } n \to \infty.$$

Hence, the series is convergent.
If $|r| = 1$, then we have $s_n = na \to \infty$ as $n \to \infty$, and if $r = -1$, then we have $s_n = 0$ for the even number n and $s_n = a$ for the odd number n. Hence the series is divergent.
If $r > 1$, then $r^n \to \infty$ as $n \to \infty$. Therefore

$$s_n = \frac{a\,(r^n - 1)}{r - 1} \to \infty \quad \text{as } n \to \infty.$$

If $r < -1$, then we see $r^n \to \pm\infty$ as $n \to \infty$. Hence we see that the series is divergent. ■

In many situations, we cannot find the sum of the first n terms of every series, and therefore we require the methods of determining whether an infinite series is convergent or divergent. Hence some new methods (tests) have been established to overcome this problem of judging convergence and divergence of the series without actually finding out the s_n. Some useful tests of convergence are described here.

3.1.3 Comparison Tests

Suppose that $\sum\limits_{n=1}^{\infty} u_n$ and $\sum\limits_{n=1}^{\infty} v_n$ are two series with non-negative terms such that $0 \leqslant u_n \leqslant v_n$ for all n.

(i) If $\sum\limits_{n=1}^{\infty} v_n$ is convergent, then $\sum\limits_{n=1}^{\infty} u_n$ is convergent.

(ii) If $\sum\limits_{n=1}^{\infty} u_n$ is divergent, then $\sum\limits_{n=1}^{\infty} v_n$ is divergent.

Let

$$s_n = u_1 + u_2 + \cdots + u_n \quad \textit{and} \quad t_n = v_1 + v_2 + \cdots + v_n.$$

Because of $v_n,\ u_n \geqslant 0$ the sequence $\{s_n\}$ and $\{t_n\}$ are increasing sequences.

(i) Since $\sum\limits_{n=1}^{\infty} v_n$ is convergent, therefore $t_n \to T$ (say). Moreover, T is the supremum of $\{t_n\}$

$$t_n \leqslant T, \quad \text{for all } \ n.$$

But, $u_n \leqslant v_n$ for all n, therefore

$$s_n \leqslant t_n, \quad \text{for all } \ n.$$

Hence

$$s_n \leqslant T, \quad \text{for all } \ n.$$

Thus, $\{s_n\}$ is an increasing sequence bounded above. Incidentally, $S \leqslant T$. Finally, the convergence of $\sum\limits_{n=1}^{\infty} v_n$ would imply the convergence of $\sum\limits_{n=1}^{\infty} u_n$.

(ii) Since $\sum\limits_{n=1}^{\infty} u_n$ is divergent, we see $s_n \to \infty$, and $u_n \leqslant v_n$ for all n, therefore

$$s_n \leqslant t_n, \quad \text{for all } \ n.$$

Finally, the divergence of $\sum\limits_{n=1}^{\infty} u_n$ would imply the divergence of $\sum\limits_{n=1}^{\infty} v_n$.

Example 3.2 Since

$$\sqrt{n} \leqslant n \implies \frac{1}{n} \leqslant \frac{1}{\sqrt{n}},$$

we know that the series

$$\sum_{n=1}^{\infty} \frac{1}{n}$$

is divergent. Therefore

$$\sum_{n=1}^{\infty} \frac{1}{\sqrt{n}}$$

is also divergent by test (ii).

Example 3.3 We test $\sum\limits_{n=1}^{\infty} \frac{1}{n^3}$. For all n,

$$n^3 \geqslant n^2 \implies \frac{1}{n^3} \leqslant \frac{1}{n^2}.$$

Thus by the comparison test, we have

$$\sum_{n=1}^{\infty} \frac{1}{n^3},$$

which is convergent.

3.1.4 Comparison Tests (Limit Theorems)

Suppose $\sum\limits_{n=1}^{\infty} u_n$ and $\sum\limits_{n=1}^{\infty} v_n$ are positive term series, such that $\lim\limits_{n\to\infty} \frac{u_n}{v_n} = l$ $(0 \leqslant l \leqslant \infty)$.

(i) If $0 < l < \infty$, then either both series converge or both diverge.

(ii) If $l = 0$, then the convergence of $\sum\limits_{n=1}^{\infty} v_n$ means the convergence of $\sum\limits_{n=1}^{\infty} u_n$. However, when $\sum\limits_{n=1}^{\infty} v_n$ diverges, we do not obtain any information with respect to $\sum\limits_{n=1}^{\infty} u_n$.

(iii) If $l = \infty$, then the divergence of $\sum\limits_{n=1}^{\infty} v_n$ means the divergence of $\sum\limits_{n=1}^{\infty} u_n$. However, when $\sum\limits_{n=1}^{\infty} v_n$ is convergent, we do not obtain any information with respect to $\sum\limits_{n=1}^{\infty} u_n$.

Proof (i) Since $\sum\limits_{n=1}^{\infty} u_n$ and $\sum\limits_{n=1}^{\infty} v_n$ are two given series and all the ratios u_1/v_1, u_2/v_2, $u_3/v_3, \cdots$ are finite, let l_1 and l_2 $(l_1 > l_2 > 0)$ be respectively the greatest and least values of these ratios. Then

$$l_1 \geqslant \frac{u_1 + u_2 + u_3 + \cdots}{v_1 + v_2 + v_3 + \cdots} \geqslant l_2 > 0.$$

Therefore,

$$v_1 + v_2 + v_3 + \cdots < +\infty \Longleftrightarrow u_1 + u_2 + u_3 + \cdots < +\infty,$$

and

$$v_1 + v_2 + v_3 + \cdots = +\infty \Longleftrightarrow u_1 + u_2 + u_3 + \cdots = +\infty.$$

(ii) Let $l = 0$, and let $\sum\limits_{n=1}^{\infty} v_n$ be convergent. Then there exists a number $N > 0$ such that

$$\frac{u_n}{v_n} < 1, \quad \text{for any } n \geqslant N.$$

Hence we see

$$\frac{u_N + u_{N+1} + u_{N+2} + \cdots}{v_N + v_{N+1} + v_{N+2} + \cdots} < 1,$$

so we see that $\sum\limits_{n=1}^{\infty} u_n$ is convergent.

We consider $v_n = \frac{1}{n}$ and $u_n = \frac{1}{n^2}$, or $v_n = \frac{1}{\sqrt{n}}$ and $u_n = \frac{1}{n}$. Then we see that for $l = 0$ the divergence of $\sum\limits_{n=1}^{\infty} v_n$ does not give any information.

(iii) Let $l = +\infty$, and let $\sum\limits_{n=1}^{\infty} v_n$ be divergent. Then there exists a number $N > 0$ such that

$$\frac{u_n}{v_n} > 1, \quad \text{for any } n \geqslant N.$$

Hence we see

$$\frac{u_N + u_{N+1} + u_{N+2+} \cdots}{v_N + v_{N+1} + v_{N+2} + \cdots} > 1,$$

so we see that $\sum\limits_{n=1}^{\infty} u_n$ is divergent.

We consider $v_n = \frac{1}{n^2}$ and $u_n = \frac{1}{n}$, or $v_n = \frac{1}{\sqrt{n^3}}$ and $u_n = \frac{1}{n^2}$. Then we see that for $l = \infty$ the convergence of $\sum\limits_{n=1}^{\infty} v_n$ does not give any information.

■

The infinite series of the form $\sum\limits_{n=1}^{\infty} \frac{1}{n^p}$, p is a real number, are called p-series or p-Harmonic Series.

3.1.5 p-Series Test

A positive term series $\sum\limits_{n=1}^{\infty} \frac{1}{n^p}$ converges if $p > 1$ and diverges if $p \leqslant 1$.

Proof We give a positive term series

$$1+\frac{1}{2^p}+\frac{1}{3^p}+\frac{1}{4^p}+\cdots.$$

(i) When $p>1$,

$$\begin{aligned}\sum_{n=1}^{\infty}\frac{1}{n^p}&=1+\frac{1}{2^p}+\frac{1}{3^p}+\frac{1}{4^p}+\cdots\\&=1+\left(\frac{1}{2^p}+\frac{1}{3^p}\right)+\left(\frac{1}{4^p}+\frac{1}{5^p}+\frac{1}{6^p}+\frac{1}{7^p}\right)+\cdots\\&<1+2\left(\frac{1}{2^p}\right)+4\left(\frac{1}{4^p}\right)+8\left(\frac{1}{8^p}\right)+\cdots\\&=1+\frac{1}{2^{p-1}}+\left(\frac{1}{2^{p-1}}\right)^2+\left(\frac{1}{2^{p-1}}\right)^3+\cdots \text{GP},\end{aligned}$$

where the series is in Geometric Progression (GP) with the common ratio $r=\frac{1}{2^{p-1}}<1$. Hence, the given series is convergent.

(ii) When $p=1$,

$$\begin{aligned}\sum_{n=1}^{\infty}\frac{1}{n}&=1+\frac{1}{2}+\frac{1}{3}+\frac{1}{4}+\cdots\\&=1+\frac{1}{2}+\left(\frac{1}{3}+\frac{1}{4}\right)+\left(\frac{1}{5}+\frac{1}{6}+\frac{1}{7}+\frac{1}{8}\right)+\cdots\\&>1+\frac{1}{2}+2\left(\frac{1}{4}\right)+4\left(\frac{1}{8}\right)+\cdots\\&=1+\frac{1}{2}+\frac{1}{2}+\frac{1}{2}+\cdots \text{GP},\end{aligned}$$

and the common ratio $r=1$. Hence the given series is divergent.

(iii) When $p<1$,

$$n^p<n\Longrightarrow\frac{1}{n^p}>\frac{1}{n}\Longrightarrow\sum_{n=1}^{\infty}\frac{1}{n^p}>\sum_{n=1}^{\infty}\frac{1}{n},\quad\text{for all } n.$$

Since the series $\sum_{n=1}^{\infty}\frac{1}{n}$ is divergent by (Case-(ii)), the given series $\sum_{n=1}^{\infty}\frac{1}{n^p}$ is also divergent.

■

Example 3.4 We test the convergence of the series:

(*i*) We consider $\sum_{n=1}^{\infty} \frac{1}{(3n-1)(3n+1)}$. We put $u_n = \frac{1}{9n^2-1}$. Then

$$u_n = \frac{1}{9n^2 - 1} \leqslant \frac{1}{9n^2 - n^2} < \frac{1}{8n^2},$$

and the series $\sum_{n=1}^{\infty} \frac{1}{8n^2}$ is convergent with $p > 1$ in the p-series test. Therefore, the given series is also convergent.

(*ii*) We consider $\sum_{n=1}^{\infty} \left[\sqrt{n^4+1} - \sqrt{n^4-1}\right]$. We put $u_n = \sqrt{n^4+1} - \sqrt{n^4-1}$. Then

$$\begin{aligned} u_n &= \left[\sqrt{n^4+1} - \sqrt{n^4-1}\right]\left[\frac{\sqrt{n^4+1} + \sqrt{n^4-1}}{\sqrt{n^4+1} + \sqrt{n^4-1}}\right] \\ &= \frac{2}{\sqrt{n^4+1} + \sqrt{n^4-1}}. \end{aligned}$$

Now, taking $v_n = \frac{1}{n^2}$, by the comparison test we have

$$\lim_{n\to\infty} \frac{u_n}{v_n} = \lim_{n\to\infty} \frac{2}{\sqrt{1+\frac{1}{n^4}} + \sqrt{1-\frac{1}{n^4}}} = 1.$$

By the p-series test, the auxiliary series $\sum_{n=1}^{\infty} v_n = \sum_{n=1}^{\infty} \frac{1}{n^2}$ is convergent (since $p = 2 > 1$). Thus, by the comparison test the given series is also convergent.

(*iii*) We consider $\sum_{n=2}^{\infty} \frac{1}{\log n}$ we see $\log n < n$

$$\frac{1}{n} < \frac{1}{\log n} \Longrightarrow \sum_{n=2}^{\infty} \frac{1}{n} < \sum_{n=2}^{\infty} \frac{1}{\log n}.$$

We know that the series $\sum_{n=2}^{\infty} \frac{1}{n}$ is divergent by the p-series test. Thus, the given series is also divergent.

(*iv*) We test the convergence of $\sum_{n=1}^{\infty} n^{\log x}$.

We give

$$\sum_{n=1}^{\infty} n^{\log x} = \sum_{n=1}^{\infty} \frac{1}{n^{-\log x}}.$$

Let $p = -\log x$ and $p \leqslant 1$. Then

$$-\log x \leqslant 1 \Longrightarrow \log x \geqslant -1 \Longrightarrow x \geqslant \frac{1}{e}.$$

Hence, the given series is divergent if $x \geqslant \frac{1}{e}$.
Now we set

$$p = -\log\ x \text{ and } p > 1,$$

then

$$-\log\ x > 1 \Longrightarrow \log\ x < -1 \Longrightarrow x < \frac{1}{e}.$$

Thus, the given series is convergent if $x < \frac{1}{e}$.

(v) We test the convergence of

$$1^{-1} + 2^{-3/2} + 3^{-4/3} + 4^{-5/4} + \cdots .$$

Let us apply the comparison test

$$u_n = n^{-(n+1)/n} = \frac{1}{n^{1+(1/n)}}.$$

Here we see

$$n^{1+(1/n)} = n(n)^{1/n} \ \textit{and}\ (n)^{1/n} \to 1 \ \textit{as}\ n \to \infty.$$

Let us apply the comparison test to u_n and $v_n = 1/n$. Then

$$\lim_{n\to\infty} \frac{u_n}{v_n} = \lim_{n\to\infty} \frac{n}{n(n)^{1/n}} = \lim_{n\to\infty} \frac{1}{(n)^{1/n}} = 1.$$

Since the auxiliary series $\sum_{n=1}^{\infty} v_n$ is divergent, we see that the original series $\sum_{n=1}^{\infty} u_n$ is also divergent.

3.1.6 D'Alembert's Ratio Test

We suppose that $\sum_{n=1}^{\infty} u_n$ is a positive term series and that $\lim_{x\to\infty} \frac{u_n}{u_{n+1}} = l$.

(*i*) If $l > 1$, the series converges absolutely.
(*ii*) If $l < 1$, the series diverges.
(*iii*) If $l = 1$, the test gives no information.

Proof Suppose that the series begins with the specified term

$$u_1 + u_2 + u_3 + \cdots + u_n + \cdots .$$

(*i*) Let $l > 1$. For a fixed $1 < \rho < l$ there is a positive number N such that

$$\frac{u_n}{u_{n+1}} > \rho, \ \ n \geqslant N,$$

then

$$u_N + u_{N+1} + u_{N+2} + \cdots \leqslant u_N + \frac{1}{\rho}u_N + \frac{1}{\rho^2}u_N + \cdots = u_N(1 + \frac{1}{\rho} + \frac{1}{\rho^2} + \cdots).$$

Since $\sum_{k=0}^{\infty} \frac{1}{\rho^k}$ is convergent, $\sum_{k=N}^{\infty} u_k$ is convergent, that is, $\sum_{k=1}^{\infty} u_k$ is convergent.

(*ii*) When $l < 1$, we can find a number ρ as $1 > \rho > l$. Then there is a positive number N such that

$$\frac{u_n}{u_{n+1}} < \rho, \ \ n \geqslant N.$$

Therefore

$$u_N + u_{N+1} + u_{N+2} + \cdots \geqslant u_N + \frac{1}{\rho}u_N + \frac{1}{\rho^2}u_N + \cdots = u_N(1 + \frac{1}{\rho} + \frac{1}{\rho^2} + \cdots).$$

Since $\sum_{k=0}^{\infty} \frac{1}{\rho^k}$ is divergent, $\sum_{k=N}^{\infty} u_k$ is divergent, that is, $\sum_{k=1}^{\infty} u_k$ is also divergent.

(*iii*) When $l = 1$, the test fails. For example, we consider the two series $\sum_{n=1}^{\infty} \frac{1}{n}$ and $\sum_{n=1}^{\infty} \frac{1}{n^2}$.

$\sum_{n=1}^{\infty} \frac{1}{n}$ diverges. Set $u_n = \frac{1}{n}$, then $\lim_{n\to\infty} \frac{u_n}{u_{n+1}} = \lim_{n\to\infty} \frac{n+1}{n} = 1.$

$\sum_{n=1}^{\infty} \frac{1}{n^2}$ converges. Set $u_n = \frac{1}{n^2}$, then $\lim_{n\to\infty} \frac{u_n}{u_{n+1}} = \lim_{n\to\infty} \left(\frac{n+1}{n}\right)^2 = 1.$

■

Example 3.5 We see that for $q > 0$ the series $x^2(\log 2)^q + x^3(\log 3)^q + x^4(\log 4)^q + \cdots$, $(x > 0)$ is convergent if $x < 1$ and divergent if $x \geqslant 1$.

Let $u_n = x^{n+1}(\log(n+1))^q$. Then we have

$$\lim_{n\to\infty} \frac{u_n}{u_{n+1}} = \lim_{n\to\infty} \frac{x^n(\log n)^q}{x^{n+1}\{\log(n+1)\}^q} = \lim_{n\to\infty} \frac{1}{x}\left\{\frac{\log n}{\log(n+1)}\right\}^q = \frac{1}{x}.$$

Hence the given series is convergent if $\frac{1}{x} > 1$, i.e., $x < 1$, and the given series is divergent if $\frac{1}{x} < 1$, i.e., $x > 1$. For $x = 1$ we see that this is a positive term series, and

$$\lim_{n\to\infty} u_n = \lim_{n\to\infty} (\log n)^q = \infty\,(\neq 0)\,.$$

Thus, the given series is divergent when $x = 1$.

Example 3.6 Let us test for the convergence of the series

$$\frac{2^1}{1^2}x + \frac{3^2}{2^3}x^2 + \cdots + \frac{(n+1)^n}{n^{n+1}}x^n + \cdots, \quad x > 0.$$

Here we set

$$u_n = \frac{(n+1)^n}{n^{n+1}}x^n, \quad \text{then} \quad u_{n+1} = \frac{(n+2)^{n+1}}{(n+1)^{n+2}}x^{n+1}.$$

Applying D'Alembert's ratio test,

$$\begin{aligned}\lim_{n\to\infty} \frac{u_n}{u_{n+1}} &= \lim_{n\to\infty} \frac{(n+1)^n}{n^{n+1}} \frac{(n+1)^{n+2}}{(n+2)^{n+1}} \frac{1}{x} \\ &= \lim_{n\to\infty} \frac{\left(1+\frac{1}{n}\right)^{2n}}{\left(1+\frac{2}{n}\right)^n} \frac{\left(1+\frac{1}{n}\right)^2}{\left(1+\frac{2}{n}\right)} \frac{1}{x} = \frac{e^2}{e^2}\frac{1}{x} = \frac{1}{x}.\end{aligned}$$

Hence, if $\frac{1}{x} > 1$, i.e., $x < 1$, the given series is convergent and if $\frac{1}{x} < 1$, i.e., $x > 1$, the given series is divergent.

For $\frac{1}{x} = 1$, i.e., $x = 1$, D'Alembert's ratio test fails. However, we can consider this case as follows.

Now we are applying the comparison test with the p-series test.

Let

$$u_n = \frac{(n+1)^n}{n^{n+1}} \quad \text{and} \quad v_n = \frac{1}{n}.$$

Then

$$\lim_{n\to\infty} \frac{u_n}{v_n} = \lim_{n\to\infty} \frac{n(n+1)^n}{n^{n+1}} = \lim_{n\to\infty} \left(1 + \frac{1}{n}\right)^n = e.$$

The auxiliary series $\sum_{n=1}^{\infty} v_n = \sum_{n=1}^{\infty} \frac{1}{n}$ is divergent (with $p = 1$). Thus, the $\sum_{n=1}^{\infty} u_n$ is also divergent for $x = 1$.

3.1.7 Cauchy's Root Test

If $\sum_{n=1}^{\infty} u_n$ is a positive term series and

$$\lim_{n\to\infty} (u_n)^{1/n} = l,$$

then $\sum_{n=1}^{\infty} u_n$ converges if $l < 1$, diverges if $l > 1$ and fails if $l = 1$.

Proof Let

$$\sum_{n=1}^{\infty} u_n = u_1 + u_2 + u_3 + \cdots + u_n + \cdots$$

be the given infinite series of positive terms. Let us consider an auxiliary series

$$\begin{aligned}\sum_{n=1}^{\infty} v_n &= v_1 + v_2 + v_3 + \cdots + v_n + \cdots \\ &= r + r^2 + r^3 + \cdots + r^n + \cdots .\end{aligned}$$

Then the series $\sum_{n=1}^{\infty} v_n$ will be convergent if the common ratio $r < 1$ and divergent if $r \geqslant 1$.

Now we see

$$\lim_{n\to\infty}\left(\frac{u_n}{v_n}\right)^{1/n} = \lim_{n\to\infty}\frac{u_n^{1/n}}{r} = \frac{l}{r}.$$

If $l < 1$, we take r as $l < r < 1$. Then there exists $N > 0$ such that

$$\left(\frac{u_n}{v_n}\right)^{1/n} < \frac{l}{r} < 1, \quad n \geqslant N.$$

So we have

$$\lim_{n\to\infty}\frac{u_n}{v_n} \leqslant \lim_{n\to\infty}\left(\frac{l}{r}\right)^n = 0.$$

By the comparison test, we see that $\sum_{n=1}^{\infty} u_n$ is convergent.
Let $l > 1$. Then we can take r as $l > r > 1$. Now there exists $N > 0$ such that

$$\left(\frac{u_n}{v_n}\right)^{1/n} > \frac{l}{r} > 1, \quad n \geqslant N.$$

So we have

$$\lim_{n\to\infty}\frac{u_n}{v_n} \geqslant \lim_{n\to\infty}\left(\frac{l}{r}\right)^n = +\infty.$$

By the comparison test, we see that $\sum_{n=1}^{\infty} u_n$ is divergent.
We consider the case of $l = 1$. If we set $u_n = 1/n^p, \ p > 1$, then we see

$$\sum_{n=1}^{\infty} u_n = \sum_{n=1}^{\infty} \frac{1}{n^p} < +\infty,$$

and

$$u_n^{1/n} = \left(\frac{1}{n^{1/n}}\right)^p \to 1 \text{ as } n \to \infty.$$

If we set $u_n = n$, then we see

$$\sum_{n=1}^{\infty} u_n = \sum_{n=1}^{\infty} n = \infty,$$

and then

$$u_n^{1/n} = n^{1/n} \to 1 \text{ as } n \to \infty.$$

■

Example 3.7 We apply Cauchy's root test to the following:

(i) We consider $\sum_{n=1}^{\infty} \frac{n^{n^2}}{(n+\frac{1}{4})^{n^2}}$.

Let

$$u_n = \frac{n^{n^2}}{\left(n+\frac{1}{4}\right)^{n^2}}. \text{ Then } u_n = \left(\frac{1}{1+\frac{1}{4n}}\right)^{n^2}.$$

Therefore

$$(u_n)^{1/n} = \frac{1}{\left(1+\frac{1}{4n}\right)^n} = \frac{1}{\left\{\left(1+\frac{1}{4n}\right)^{4n}\right\}^{1/4}}.$$

Hence

$$\lim_{n\to\infty} (u_n)^{1/n} = \lim_{n\to\infty} \frac{1}{\left\{\left(1+\frac{1}{4n}\right)^{4n}\right\}^{1/4}} = \frac{1}{e^{1/4}} < 1.$$

Thus, the given series is convergent.

(ii) We consider $\sum_{n=1}^{\infty} \left(n^{1/n} - 1\right)^n$.

Let $u_n = (n^{1/n} - 1)^n$. Then $u_n^{1/n} = n^{1/n} - 1 \to 0$ as $n \to \infty$. Thus, the given series is convergent.

When D'Alembert's ratio test fails, in that case we should apply either *Raabe's Test* or the *Integral Test*.

3.1.8 *Raabe's Test*

If $\sum_{n=1}^{\infty} u_n$ is a positive term series and

$$\lim_{n\to\infty} n\left(\frac{u_n}{u_{n+1}} - 1\right) = l,$$

then $\sum_{n=1}^{\infty} u_n$ is a convergent if $l > 1$ and divergent if $l < 1$.

Proof We give a positive term series $\sum_{n=1}^{\infty} u_n = u_1 + u_2 + u_3 + \cdots + u_n + \cdots$. Let us compare the series $\sum_{n=1}^{\infty} u_n$ with the auxiliary series $\sum_{n=1}^{\infty} v_n = \sum_{n=1}^{\infty} \frac{1}{n^p}$. By the $p-$series test, the series $\sum_{n=1}^{\infty} v_n$ is convergent or divergent accordingly as $p > 1$ or $p \leqslant 1$. We have

$$n\left(\frac{v_n}{v_{n+1}} - 1\right) = p + \frac{p(p-1)}{2!}\frac{1}{n} + \frac{p(p-1)(p-2)}{3!}\frac{1}{n^2} + \cdots$$

$$\lim_{n\to\infty} n\left(\frac{v_n}{v_{n+1}} - 1\right) = p. \tag{3.1}$$

Let $l > 1$. Then we take p as $1 < p < l$. Now there is a number $N > 0$ such that by (3.1)

$$n\left(\frac{u_n}{u_{n+1}} - 1\right) > l - \frac{l-p}{3} > p + \frac{l-p}{3} > n\left(\frac{v_n}{v_{n+1}} - 1\right), \quad n \geqslant N.$$

Thus, we have

$$\frac{u_n}{u_{n+1}} > \frac{v_n}{v_{n+1}}, \quad \text{that is,} \quad \frac{u_n}{v_n} > \frac{u_{n+1}}{v_{n+1}}.$$

So, for a finite quantity $l' \geqslant 0$, we have

$$l' = \lim_{n\to\infty} \frac{u_n}{v_n}.$$

Thus, applying the comparison test to $\{u_n\}$ and $\{v_n\}$, we conclude that $\{u_n\}$ is convergent.
When $0 < l < 1$, we can take $0 < l < p < 1$. Then there exists $N > 0$ such that by (3.1)

$$n\left(\frac{u_n}{u_{n+1}} - 1\right) < l + \frac{p-l}{3} < p - \frac{2(p-l)}{3} < n\left(\frac{v_n}{v_{n+1}} - 1\right), \quad n \geqslant N.$$

Then we have

$$\frac{u_n}{u_{n+1}} < \frac{v_n}{v_{n+1}},$$

that is,

$$\frac{u_n}{v_n} < \frac{u_{n+1}}{v_{n+1}}.$$

So we have $l' \leqslant \infty$ where

$$l' = \lim_{n \to \infty} \frac{u_n}{v_n}.$$

Thus applying the comparison test to $\{u_n\}$ and $\{v_n\}$, we see that $\{u_n\}$ is divergent. This test fails for $l = 1$. (In this case we see $\lim_{n \to \infty} \frac{u_n}{u_{n+1}} = 1$), and so by D'Alembert's ratio test (iii) we know this phenomenon.) ■

Example 3.8 We test for the convergence.

(i) We consider the series

$$1 + a + \frac{a(a+1)}{1 \cdot 2} + \frac{a(a+1)(a+2)}{1 \cdot 2 \cdot 3} + \cdots.$$

Ignoring the first term, we have

$$u_n = \frac{a(a+1)(a+2)\cdots(a+n)}{1 \cdot 2 \cdot 3 \cdots (n+1)},$$

and

$$u_{n+1} = \frac{a(a+1)(a+2)\cdots(a+n)(a+n+1)}{1 \cdot 2 \cdot 3 \cdots (n+1)(n+2)}.$$

Therefore

$$\frac{u_n}{u_{n+1}} = \frac{n+2}{a+n+1}.$$

Clearly, $u_n/u_{n+1} \to 1$ *as* $n \to \infty$, so it fails by *D'Alembert's* ratio test. Now applying *Raabe's test*, we obtain

$$\lim_{n \to \infty} n\left(\frac{u_n}{u_{n+1}} - 1\right) = \lim_{n \to \infty} \frac{n(1-a)}{a+n+1} = 1 - a.$$

Hence by *Raabe's test*, if $1 - a > 1$, i.e., $a < 0$, the series is convergent.

When $a = 0$, *Raabe's test* fails, but in this case, except the first term, all the terms of the series become zero. Thus the sum of the series is 1 for $a = 0$, and therefore the series is convergent.

Hence, the series is convergent if $a \leqslant 0$ and divergent if $a > 0$.

(ii) Find the nature of series

$$2 + \frac{3}{2}x + \frac{4}{3}x^3 + \frac{5}{4}x^5 + \cdots, \quad x \geqslant 0.$$

Let

$$u_n = \frac{n+1}{n}x^{2n-1} \quad and \quad u_{n+1} = \frac{n+2}{n+1}x^{2n+1}.$$

Now

$$\frac{u_n}{u_{n+1}} = \frac{(n+1)^2}{n(n+2)}\frac{1}{x^2} \text{ and } \lim_{n\to\infty}\frac{u_n}{u_{n+1}} = \frac{1}{x^2}.$$

By *D'Alembert's* ratio test the series converges when $1/x^2 > 1$ or $1 > x \geqslant 0$. The series diverges when $1/x^2 < 1$ or $1 < x$. However, when $1/x^2 = 1$ or $x = 1$, the test fails. When $x = 1$, we have

$$\frac{u_n}{u_{n+1}} = \frac{(n+1)^2}{n(n+2)} \Longrightarrow \frac{u_n}{u_{n+1}} - 1 = \frac{(n+1)^2}{n(n+2)} - 1 \Longrightarrow n\left(\frac{u_n}{u_{n+1}} - 1\right) = \frac{1}{n+2}.$$

Hence we have the limiting value

$$\lim_{n\to\infty} n\left(\frac{u_n}{u_{n+1}} - 1\right) = 0 < 1.$$

Now applying *Raabe's test*, the series is divergent.

(iii) Find the nature of series

$$x^2 + \frac{2^2}{3\cdot 4}x^4 + \frac{2^2 4^2}{3\cdot 4\cdot 5\cdot 6}x^6 + \frac{2^2 4^2 6^2}{3\cdot 4\cdot 5\cdot 6\cdot 7\cdot 8}x^8 + \cdots.$$

Ignoring the first term, we let

$$u_n = \frac{2^2 4^2 6^2\cdots(2n)^2}{3\cdot 4\cdot 5\cdot 6\cdots(2n+1)(2n+2)}x^{2n+2},$$

and

$$u_{n+1} = \frac{2^2 4^2 6^2\cdots(2n)^2(2n+2)^2}{3\cdot 4\cdot 5\cdot 6\cdots(2n+1)(2n+2)(2n+3)(2n+4)}x^{2n+4}.$$

Then we see

$$\lim_{n\to\infty}\frac{u_n}{u_{n+1}} = \lim_{n\to\infty}\frac{(2n+3)(2n+4)}{(2n+2)^2}\frac{1}{x^2} = \frac{1}{x^2}.$$

By *D'Alembert's* ratio test the series converges when $1/x^2 > 1$ or $1 > |x|$. The series diverges when $1/x^2 < 1$ *or* $1 < |x|$. However, when $1/x^2 = 1$ *or* $|x| = 1$, the test fails. Now applying *Raabe's test*, for $|x| = 1$ we have

$$\frac{u_n}{u_{n+1}} = \frac{(2n+3)(2n+4)}{(2n+2)^2} = \frac{4n^2+14n+12}{4n^2+8n+4}.$$

Therefore,

$$n\left(\frac{u_n}{u_{n+1}} - 1\right) = \frac{n(6n+8)}{4n^2+8n+4}.$$

Then we see

$$\lim_{n\to\infty} n\left(\frac{u_n}{u_{n+1}} - 1\right) = \lim_{n\to\infty} \frac{n(6n+8)}{4n^2+8n+4} = \frac{3}{2} > 1.$$

Thus, the series is convergent. Hence the series is convergent if $|x| \leqslant 1$, and divergent if $|x| > 1$.

3.1.9 *The Integral Test (Cauchy-Maclaurin's Integral Test)*

Consider an integer k and a non-negative monotone decreasing function f defined on the unbounded interval $[k, \infty)$. Then the series

$$\sum_{n=k}^{\infty} f(n)$$

converges if and only if the integral

$$\int_k^{\infty} f(x)dx$$

is finite. In particular, if the integral diverges, then the series diverges as well.

In other ways, we can write it as follows.

Suppose that $f(x)$ is a continuous, positive monotonic decreasing function on the interval $[k, \infty)$ and let $u_n = f(n)$. Then the series $\sum\limits_{n=k}^{\infty} u_n$, i.e., $\sum\limits_{n=k}^{\infty} f(n)$ is convergent if and only if the improper integral $\int_k^{\infty} f(x)dx$ is convergent.

And if the improper integral $\int_k^{\infty} f(x)dx$ is divergent, then the series $\sum\limits_{n=k}^{\infty} f(n)$ is divergent.

Example 3.9 (i) By the Integral test, we examine the convergence of the series

$$\sum_{n=2}^{\infty} \frac{1}{n(\log n)^p}.$$

For all positive values of p, $u_n = \frac{1}{n(\log n)^p}$ is a monotonic decreasing function. Let

$$f(x) = \frac{1}{x(\log x)^p}.$$

When $p > 1$,

$$\begin{aligned}
\int_2^{\infty} \frac{dx}{x(\log x)^p} &= \int_{\log 2}^{\infty} \frac{dt}{t^p} \quad (\because \log x = t) \\
&= \lim_{n\to\infty} \int_{\log 2}^{n} \frac{dt}{t^p} \\
&= \lim_{n\to\infty} \left[\frac{1}{(1-p)t^{p-1}}\right]_{\log 2}^{n} \\
&= \lim_{n\to\infty} \frac{1}{(1-p)} \left[\frac{1}{n^{p-1}} - \frac{1}{(\log 2)^{p-1}}\right] \\
&= \frac{1}{(p-1)} \frac{1}{(\log 2)^{p-1}}.
\end{aligned}$$

Thus, if $p > 1$, then the integral converges.
And, if $p < 1$, then

$$\int_2^{\infty} \frac{dx}{x(\log x)^p} = \lim_{n\to\infty} \frac{1}{1-p}\left[n^{1-p} - (\log 2)^{1-p}\right] = \infty.$$

When $p = 1$, we have

$$\int_2^{\infty} \frac{1}{x \log x} dx = \left[\log(\log x)\right]_2^{\infty} = \infty.$$

Consequently,

$$\sum_{n=2}^{\infty} \frac{1}{n(\log n)^p}$$

is convergent if $p > 1$ and divergent if $p \leqslant 1$.

(ii) We show that the series

$$\sum_{n=1}^{\infty} \frac{1}{n^p}$$

is convergent if $p > 1$ and divergent if $p \leqslant 1$.
For all positive values of p, we have $u_n = \frac{1}{n^p}$, which is a monotonic decreasing function on the interval $[1, \infty)$.
Now let $f(x) = \frac{1}{x^p}$. For $p > 1$ we have

$$\int_1^\infty \frac{1}{x^p}dx = \lim_{n\to\infty}\int_1^n \frac{1}{x^p}dx$$
$$= \lim_{n\to\infty}\left[\frac{1}{(1-p)x^{p-1}}\right]_1^n$$
$$= \lim_{n\to\infty}\frac{1}{(1-p)}\left(\frac{1}{n^{p-1}}-1\right)$$
$$= \frac{1}{p-1}.$$

Thus, this integral converges for $p > 1$.
And if $p < 1$, then

$$\int_1^\infty \frac{dx}{x^p} = \lim_{n\to\infty}\frac{1}{1-p}\left[n^{1-p}-1\right] = \infty.$$

Consequently, $\sum_{n=1}^{\infty}\frac{1}{n^p}$ is convergent if $p > 1$, and divergent if $p < 1$.

When $p = 1$, we have

$$\int_1^\infty \frac{1}{x^p}dx = \lim_{n\to\infty}\int_1^n \frac{1}{x}dx = \lim_{n\to\infty}\left[\log x\right]_1^n = \lim_{n\to\infty}\left[\log n - \log 1\right]$$
$$= \lim_{n\to\infty}\log n = \infty.$$

Thus, $\sum_{n=1}^{\infty}\frac{1}{n^p}$ is divergent if $p = 1$.

(iii) We test the convergence of the series

$$\sum_{n=1}^{\infty} ne^{-n^2}.$$

Let $f(x) = xe^{-x^2}$ and

$$\int_1^\infty xe^{-x^2}dx = \lim_{n\to\infty}\int_1^n xe^{-x^2}dx$$
$$= \lim_{n\to\infty}\frac{1}{2}\int_1^{n^2} e^{-t}dt \quad (\text{put } t = x^2 \Rightarrow dt = 2xdx)$$
$$= \lim_{n\to\infty}\frac{1}{2}\left\{-e^{-t}\right\}_1^{n^2} = \frac{1}{2e}.$$

Hence, the series is convergent.

3.1.10 Logarithmic Test

The positive term series $\sum_{n=1}^{\infty} u_n$ is convergent or divergent according as

$$\lim_{n\to\infty} n \log \frac{u_n}{u_{n+1}} = l > 1 \ \text{ or } \ < 1.$$

Proof Let the given series of positive terms be

$$\sum_{n=1}^{\infty} u_n = u_n + u_n + u_n + \cdots + u_n + \cdots .$$

Compare this series with the auxiliary series $\sum_{n=1}^{\infty} v_n = \sum_{n=1}^{\infty} \frac{1}{n^p}$ which, by the p-series test, is convergent or divergent according as $p > 1$ or $p \leqslant 1$. We put

$$v_n = \frac{1}{n^p}, \quad \text{and} \quad v_{n+1} = \frac{1}{(n+1)^p},$$

then we see

$$\begin{aligned} \frac{v_n}{v_{n+1}} = \left(1 + \frac{1}{n}\right)^p \Rightarrow \log \frac{v_n}{v_{n+1}} &= p \log\left(1 + \frac{1}{n}\right) \\ &= p\left[\frac{1}{n} - \frac{1}{2n^2} + \frac{1}{3n^3} - \cdots\right]. \end{aligned}$$

Hence

$$\lim_{n\to\infty} n \log \frac{v_n}{v_{n+1}} = p. \tag{3.2}$$

Let $l > 1$. Then we take p as $1 < p < l$. By (3.2) there exists a number $N > 0$ such that for $n \geqslant N$

$$1 < n \log \frac{v_n}{v_{n+1}} < n \log \frac{u_n}{u_{n+1}},$$

so we have

$$\frac{v_n}{v_{n+1}} < \frac{u_n}{u_{n+1}}, \text{ i.e., } \frac{u_{n+1}}{v_{n+1}} < \frac{u_n}{v_n}.$$

Hence, there exists $\alpha \geqslant 0$ such that

$$\lim_{n\to\infty} \frac{u_n}{v_n} = \alpha.$$

Thus, applying the comparison test to $\{u_n\}$ and $\{v_n\}$, we see that $\{u_n\}$ is convergent.

Let $l < 1$. Then we take p as $0 < l < p < 1$. By (3.2) there exists a number $N > 0$ such that for $n \geqslant N$

$$n \log \frac{u_n}{u_{n+1}} < n \log \frac{v_n}{v_{n+1}} < 1,$$

so we have

$$\frac{u_n}{u_{n+1}} < \frac{v_n}{v_{n+1}}, \text{ i.e., } \frac{u_n}{v_n} < \frac{u_{n+1}}{v_{n+1}}.$$

Hence, there exists $\beta \leqslant \infty$ such that

$$\lim_{n\to\infty} \frac{u_n}{v_n} = \beta.$$

Thus, applying the comparison test to $\{u_n\}$ and $\{v_n\}$, we see that $\{u_n\}$ is divergent. If $l = 1$, then we see $u_n/u_{n+1} \to 1$ as $n \to \infty$. Thus, we have no information from this test. ■

Remark 3.1 (i) Logarithmic test fails when

$$\lim_{n\to\infty} n \log \frac{u_n}{u_{n+1}} = 1.$$

(ii) Logarithmic test is applied when D'Alembert's ratio test fails, and it is preferred when e or its power is involved.

Example 3.10 We test the convergence or divergence of the series

$$x + \frac{2^2x^2}{2!} + \frac{3^3x^3}{3!} + \frac{4^4x^4}{4!} + \cdots, (x > 0).$$

We apply the D'Alembert's ratio test to the series. Let

$$u_n = \frac{n^n x^n}{n!} \text{ and } u_{n+1} = \frac{(n+1)^{n+1}x^{n+1}}{(n+1)!}.$$

Then we have

$$\frac{u_n}{u_{n+1}} = \frac{n^n}{(n+1)^n x} = \frac{1}{\left(1+\frac{1}{n}\right)^n x}.$$

Therefore

$$\lim_{n\to\infty} \frac{u_n}{u_{n+1}} = \frac{1}{ex}.$$

If $1/(ex) > 1 \Longrightarrow x < 1/e$, then the series is convergent, and if $1/ex < 1 \Longrightarrow x > 1/e$, the series is divergent.

If $1/(ex) = 1 \Longrightarrow x = 1/e$, then D'Alembert's ratio test fails. Now, we apply the logarithmic test. From

$$\frac{u_n}{u_{n+1}} = \frac{e}{\left(1+\frac{1}{n}\right)^n},$$

we see

$$\begin{aligned}\log \frac{u_n}{u_{n+1}} &= \log e - n\log\left(1+\frac{1}{n}\right)\\ &= 1 - n\left(\frac{1}{n} - \frac{1}{2n^2} + \frac{1}{3n^3} - \frac{1}{4n^4} + \frac{1}{5n^5} + \cdots\right)\\ &= \frac{1}{2n} - \frac{1}{3n^2} + \frac{1}{4n^3} - \frac{1}{5n^4}\cdots.\end{aligned}$$

Therefore

$$n\log\frac{u_n}{u_{n+1}} = \frac{1}{2} - \frac{1}{3n} + \frac{1}{4n^2} - \frac{1}{5n^3}\cdots.$$

Here we see

$$\left|-\frac{1}{3n} + \frac{1}{4n^2} - \frac{1}{5n^3}\cdots\right| \leqslant \frac{1}{n}\left(\frac{1}{3} + \frac{1}{4} + \frac{1}{5}\cdots\right) \leqslant \frac{\log n}{n} \to 0. \tag{3.3}$$

Hence

$$\lim_{n\to\infty}\left(n\log\frac{u_n}{u_{n+1}}\right) = \frac{1}{2} < 1,$$

and the series is divergent for $x = 1/e$.

Example 3.11 We test the convergence of the series

$$\frac{a+x}{1!} + \frac{(a+2x)^2}{2!} + \frac{(a+3x)^3}{3!} + \cdots.$$

Now let $f(x) = 1/x^p$. For $p \neq 1$ we have

$$u_n = \frac{(a+nx)^n}{n!};\ u_{n+1} = \frac{\{a+(n+1)x\}^{n+1}}{(n+1)!},$$

so that

$$\frac{u_n}{u_{n+1}} = \frac{(a+nx)^n}{n!} \frac{(n+1)!}{\{a+(n+1)x\}^{n+1}} = \frac{(n+1)(a+nx)^n}{\{a+(n+1)x\}^{n+1}}$$
$$= \frac{(n+1)(nx)^n\left(1+\frac{a}{nx}\right)^n}{\{(n+1)x\}^{n+1}\left(1+\frac{a}{(n+1)x}\right)^{n+1}} = \frac{\left(1+\frac{a/x}{n}\right)^n}{x\left(1+\frac{1}{n}\right)^n\left(1+\frac{a/x}{n+1}\right)^{n+1}}.$$

$$\text{Therefore,} \quad \lim_{n\to\infty} \frac{u_n}{u_{n+1}} = \frac{1}{ex}\frac{e^{a/x}}{e^{a/x}} = \frac{1}{ex}.$$

Hence, if $\frac{1}{ex} > 1$, i.e., $x < \frac{1}{e}$, then the series is convergent, and if $\frac{1}{ex} < 1$, i.e., $x > \frac{1}{e}$, then the series is divergent.

Now we apply *Logarithmic test* to the case of $x = 1/e$, and then

$$\frac{u_n}{u_{n+1}} = \frac{e\left(1+\frac{ae}{n}\right)^n}{\left(1+\frac{1}{n}\right)^n\left(1+\frac{ae}{n+1}\right)^{n+1}}.$$

Therefore,

$$\begin{aligned}
\log \frac{u_n}{u_{n+1}} &= \log e - n\log\left(1+\frac{1}{n}\right) + n\log\left(1+\frac{ae}{n}\right) \\
&\quad - (n+1)\log\left(1+\frac{ae}{n+1}\right) \\
&= 1 - n\left(\frac{1}{n} - \frac{1}{2n^2} + \frac{1}{3n^3} - \frac{1}{4n^4} + \frac{1}{5n^5} - \cdots\right) \\
&\quad + n\left(\frac{ae}{n} - \frac{a^2e^2}{2n^2} + \frac{a^3e^3}{3n^3} - \frac{a^4e^4}{4n^4} + \frac{a^5e^5}{5n^5} - \cdots\right) \\
&\quad - (n+1)\left(\frac{ae}{n+1} - \frac{a^2e^2}{2(n+1)^2} + \frac{a^3e^3}{3(n+1)^3}\right. \\
&\quad \left. - \frac{a^4e^4}{4(n+1)^4} + \frac{a^5e^5}{5(n+1)^5} - \cdots\right) \\
&= \left(\frac{1}{2n} - \frac{1}{3n^2} + \frac{1}{4n^3} - \frac{1}{5n^4} + \cdots\right) \\
&\quad + \left(-\frac{a^2e^2}{2n} + \frac{a^3e^3}{3n^2} - \frac{a^4e^4}{4n^3} + \frac{a^5e^5}{5n^4} - \cdots\right) \\
&\quad + \left(\frac{a^2e^2}{2(n+1)} - \frac{a^3e^3}{3(n+1)^2} + \frac{a^4e^4}{4(n+1)^3} - \frac{a^5e^5}{5(n+1)^4} + \cdots\right).
\end{aligned}$$

Hence, noting (3.3), we conclude

$$\lim_{n\to\infty}\left(n\log\frac{u_n}{u_{n+1}}\right)=\frac{1}{2}.$$

Thus, for $x = 1/e$ the given series is divergent.

Theorem 3.3 (The Cauchy Criterion for Series)
The series $\sum\limits_{n=1}^{\infty} u_n$ *converges if and only if for a given* $\varepsilon > 0$, *there exists* $N \in \mathbb{N}$ *such that, whenever* $n \geqslant m > N$, *it follows that*

$$|u_{m+1}+u_{m+2}+\cdots+u_n|<\varepsilon. \tag{3.4}$$

Proof The partial sum is

$$s_n = u_1 + u_2 + \cdots + u_n,$$

and we observe that

$$|s_n - s_m| = |u_{m+1}+u_{m+2}+\cdots+u_n| < \varepsilon, \quad n \geqslant m > N,$$

and apply the Cauchy Criterion condition for sequences.

The Cauchy Criterion condition leads to economical proofs of several basic facts about the series. ∎

3.2 Alternating Series

If $\{u_n\}$ is a sequence of positive numbers, then the series

$$\sum_{n=1}^{\infty}(-1)^n u_n$$

is called an alternating series because the signs of the terms alternate between + and −. If the terms in an alternating series have decreasing absolute values and converge to zero, then the series converges however slowly its terms approach zero.

There is a simple test to determine that the certain alternating series converge.

Theorem 3.4 (Leibniz's alternating series test) *Let* $\sum\limits_{n=1}^{\infty}(-1)^n u_n$ *be the alternating series which converges if*

$$u_1 \geqslant u_2 \geqslant \cdots \geqslant u_n \geqslant \cdots \geqslant 0 \quad \textit{and} \quad \lim_{n\to\infty} u_n = 0.$$

Proof It is sufficient to show that the series satisfies the Cauchy Criterion condition (3.4). For each $\varepsilon > 0$ there exists $N \in \mathbb{N}$ such that

$$\left|\sum_{k=m}^{n} (-1)^k u_k\right| \leqslant u_N < \varepsilon, \quad n \geqslant m > N. \tag{3.5}$$

To prove the condition (3.5), we consider $n \geq m$ and define as follows:

$$A = u_m - u_{m+1} + u_{m+2} - u_{m+3} + \cdots \pm u_n,$$

so that

$$\sum_{k=m}^{n} (-1)^k u_k = (-1)^m A. \tag{3.6}$$

If $n - m$ is odd, the last term of A is $-u_n$. So

$$A = \left[u_m - u_{m+1}\right] + \left[u_{m+2} - u_{m+3}\right] + \cdots + \left[u_{n-1} - u_n\right] \geqslant 0,$$

and

$$A = u_m - \left[u_{m+1} - u_{m+2}\right] - \left[u_{m+3} - u_{m+4}\right] - \cdots - \left[u_{n-2} - u_{n-1}\right] - u_n \leqslant u_m.$$

If $n - m$ is even, the last term of A is $+u_n$. So

$$A = \left[u_m - u_{m+1}\right] + \left[u_{m+2} - u_{m+3}\right] + \cdots + \left[u_{n-2} - u_{n-1}\right] + u_n \geqslant 0,$$

and

$$A = u_m - \left[u_{m+1} - u_{m+2}\right] - \left[u_{m+3} - u_{m+4}\right] - \cdots - \left[u_{n-1} - u_n\right] \leqslant u_m.$$

For both of the cases we have $0 \leqslant A \leqslant u_m$. Thus, from (3.6) we obtain

$$\left|\sum_{k=m}^{n} (-1)^k u_k\right| = A \leqslant u_m.$$

From (3.5) we have

$$\left|\sum_{k=m}^{n} (-1)^k u_k\right| \leqslant u_m \leqslant u_N, \quad \text{whenever } n \geqslant m > N,$$

that is, $\lim_{m\to\infty}\sum_{k=m}^{\infty}(-1)^k u_k = 0$. Thus we see that $\sum_{k=1}^{\infty}(-1)^k u_k$ is convergent. ■

Example 3.12 We test the convergence of the series

$$x - \frac{x^2}{2^2} + \frac{x^3}{3^2} - \cdots .$$

We consider each case of

$$(i)\ x = 0,\quad (ii)\ x > 0,\quad (iii)\ x < 0.$$

Case of (i): The series converges to 0 trivially.
Case of (ii): We put

$$u_n = \frac{x^n}{n^2},\quad n \in \mathbb{N},$$

then for $0 < x \leqslant 1$ we see

$$u_1 > u_2 > u_3 > \cdots,\quad \lim_{n\to\infty} u_n = 0.$$

Therefore, by Leibniz alternating series test the series converges.
Let $1 < x$. We put

$$s_n = \sum_{k=1}^{n}(-1)^{k-1}\frac{x^k}{k^2},$$

then

$$s_{2n} = \sum_{k=1}^{n}\left\{\frac{x^{2k-1}}{(2k-1)^2} - \frac{x^{2k}}{(2k)^2}\right\} = \sum_{k=1}^{n}\frac{x^{2k-1}}{(2k)^2}\left\{\frac{1}{\left(1-\frac{1}{2k}\right)} - x\right\}.$$

And

$$s_{2n+1} = \sum_{k=1}^{n}\left\{-\frac{x^{2k}}{(2k)^2} + \frac{x^{2k+1}}{(2k+1)^2}\right\} = \sum_{k=1}^{n}\frac{x^{2k}}{(2k)^2}\left\{-1 + \frac{x}{\left(1+\frac{1}{2k}\right)^2}\right\}.$$

Since $1 < x$ for $N,\ n$ large enough, we see

$$\sum_{k=N}^{n}\left\{\frac{x^{2k-1}}{(2k-1)^2} - \frac{x^{2k}}{(2k)^2}\right\} = \sum_{k=N}^{n}\frac{x^{2k-1}}{(2k)^2}\left\{\frac{1}{\left(1-\frac{1}{2k}\right)} - x\right\} < 0,$$

and

$$\sum_{k=N}^{n}\left\{-\frac{x^{2k}}{(2k)^2}+\frac{x^{2k+1}}{(2k+1)^2}\right\}=\sum_{k=N}^{n}\frac{x^{2k}}{(2k)^2}\left\{-1+\frac{x}{\left(1+\frac{1}{2k}\right)^2}\right\}>0.$$

On the other hand, we see

$$\lim_{k\to\infty}\frac{x^{2k-1}}{(2k)^2}=+\infty.$$

Hence, we have

$$\lim_{n\to\infty}s_{2n}=-\infty,\quad\text{and}\quad\lim_{n\to\infty}s_{2n+1}=+\infty,$$

that is, the series diverges.

Case of (iii): For $x<0$ we see

$$x-\frac{x^2}{2^2}+\frac{x^3}{3^2}-\cdots=-\left\{|x|+\frac{|x|^2}{2^2}+\frac{|x|^3}{3^2}-\cdots\right\}.$$

We put

$$v_n=\frac{|x|^n}{n^2},$$

then we see

$$\lim_{n\to\infty}\frac{v_n}{v_{n+1}}=\lim_{n\to\infty}\left(1+\frac{1}{n}\right)^2\frac{1}{|x|}=\frac{1}{|x|}.$$

We use D'Alembert's ratio test. If $-1<x<0$, then the series converges, and if $x<-1$, then the series diverges. When $x=-1$, D'Alembert's ratio test fails. However, using Leibniz's test, for $x=-1$, we know that the series converges.

Example 3.13 The alternating series

$$1-3+\frac{1}{2}-\frac{3}{2}+\frac{1}{3}-1+\frac{1}{4}-\frac{3}{4}+\frac{1}{5}-\frac{3}{5}+\cdots+\frac{1}{n}-\frac{3}{n}+\cdots$$

diverges. Even though $u_n\to 0$, but $\{u_n\}$ is not monotone decreasing.

Since we see that even though its term u_n satisfies $u_n\to 0$ as $n\to\infty$, but $\{u_n\}$ is not monotone decreasing, it does not satisfy the condition in Leibniz alternating series test. We consider the partial sum s_{2n}, s_{2n+1}. We have

$$s_{2n} = -2\sum_{k=1}^{2n}\frac{1}{k}, \text{ and } s_{2n+1} = -2\sum_{k=1}^{2n}\frac{1}{k} + \frac{1}{2n+1}.$$

Therefore we see

$$\lim_{n\to\infty} s_n = -\infty.$$

3.3 Absolute and Conditional Convergence

In this section, we investigate what happens to a convergent series when its terms are replaced by their absolute values.

Definition 3.2 An infinite series $\sum_{n=1}^{\infty} u_n$ converges *absolutely* if and only if $\sum_{n=1}^{\infty} |u_n|$ converges. If $\sum_{n=1}^{\infty} u_n$ converges but $\sum_{n=1}^{\infty} |u_n|$ diverges, then the series $\sum_{n=1}^{\infty} u_n$ is said to *conditionally* converge.

Remark 3.2 If $\sum_{n=1}^{\infty} |u_n|$ converges, then $\sum_{n=1}^{\infty} u_n$ is also convergent.

The series

$$\sum_{n=1}^{\infty} u_n = \sum_{n=1}^{\infty}\frac{(-1)^{n-1}}{n} = 1 - \frac{1}{2} + \frac{1}{3} - \frac{1}{4} + \frac{1}{5} - \frac{1}{6} + \cdots$$

is convergent, but does not converge absolutely. In fact, we see

$$\sum_{k=1}^{2n} u_k = \sum_{k=1}^{n}\left(\frac{1}{2k-1} - \frac{1}{2k}\right) = \sum_{k=1}^{n}\frac{1}{2k(2k-1)}$$

and

$$\sum_{k=1}^{2n+1} u_k = \sum_{k=1}^{n}\frac{1}{2k(2k-1)} + \frac{1}{2n+1}.$$

Thus, the series $\left\{\sum_{n=1}^{\infty} u_n\right\}$ is convergent. But it does not converge absolutely. Therefore the given series $\sum_{n=1}^{\infty}\frac{(-1)^{n-1}}{n}$ is conditionally convergent.

The series

$$\sum_{n=1}^{\infty} u_n = \sum_{n=1}^{\infty} (-1)^n \frac{1}{n\,(n+1)}$$

converges *absolutely*, since the series

$$\sum_{n=1}^{\infty} |u_n|$$

converges.

Example 3.14 We test the convergence or divergence of the series

$$\sum_{n=1}^{\infty} (-1)^{n+1} \frac{x^n}{n}.$$

We consider

$$\sum_{n=1}^{\infty} \left| (-1)^{n+1} \frac{x^n}{n} \right| \tag{3.7}$$

and use the ratio test. We see

$$\left| \frac{u_n}{u_{n+1}} \right| = \left| \frac{(-1)^{n+1} x^n}{n} \cdot \frac{n+1}{(-1)^{n+2} x^{n+1}} \right| = \frac{n+1}{n\,|x|} \to \frac{1}{|x|} \quad as\ n \to \infty.$$

If $|x| < 1$, then the series (3.7) is convergent and the original series is *absolutely* convergent. If $|x| > 1$, the series (3.7) is divergent. Then for $|x| = 1$ we test the convergence or divergence with respect to the original series. When $x = 1$, we have

$$1 - \frac{1}{2} + \frac{1}{3} - \frac{1}{4} + \cdots,$$

which is convergent (by Leibniz's alternating series test), that is, it is conditionally convergent. When $x = -1$, we have

$$-1 - \frac{1}{2} - \frac{1}{3} - \frac{1}{4} - \cdots,$$

which is divergent (by the p-series test).

3.4 Rearrangement of Terms

We know very well that the addition of a finite number of terms in any order or any arrangement does not affect the sum. But in the case of an infinite series, its sum (it exists) is affected by the rearrangement of terms and a divergent series may become convergent. For example, the series

$$\sum_{n=0}^{\infty}(-1)^n = 1 - 1 + 1 - 1 + 1 - 1 + \cdots$$

is a divergent series. However, if we arrange the series as

$$(1-1)+(1-1)+(1-1)+\cdots,$$

then we have the sum 0. Moreover, if we arrange the series as

$$1-(1-1)-(1-1)-(1-1)-\cdots,$$

then we have the sum 1. Now we consider the series

$$\sum_{n=1}^{\infty}\frac{(-1)^{n-1}}{n} = 1 - \frac{1}{2} + \frac{1}{3} - \frac{1}{4} + \frac{1}{5} - \frac{1}{6} + \frac{1}{7} - \frac{1}{8} + \frac{1}{9} - \frac{1}{10} + \frac{1}{11} - \frac{1}{12} + \cdots, \tag{3.8}$$

where we put two negative even terms between each of the positive odd terms. Note also that both the positive and negative parts of the alternating harmonic series diverge to infinity, since

$$1 + \frac{1}{3} + \frac{1}{5} + \frac{1}{7} + \frac{1}{9} + \frac{1}{11} \cdots > \frac{1}{2} + \frac{1}{4} + \frac{1}{6} + \frac{1}{8} + \frac{1}{10} + \frac{1}{12} + \cdots \tag{3.9}$$

$$> \frac{1}{2}\left(1 + \frac{1}{2} + \frac{1}{3} + \frac{1}{4} + \frac{1}{5} + \frac{1}{6} + \cdots\right), \tag{3.10}$$

and the harmonic series diverges. This is what allows us to change the sum by rearranging the series.

Again we consider an arrangement of the series (3.8) in the following way

$$1 - \frac{1}{2} - \frac{1}{4} + \frac{1}{3} - \frac{1}{6} - \frac{1}{8} + \cdots + \frac{1}{2n-1} - \frac{1}{4n-2} - \frac{1}{4n} + \cdots. \tag{3.11}$$

Let s_n be the n-th partial sum of the original series (3.8) which is convergent to the value s (by Leibniz's alternating series test). Let t_n be the n-th partial sum of the rearranged series (3.11). Then we have

$$
\begin{aligned}
t_{n,2} &= \left(1 - \frac{1}{2} - \frac{1}{4}\right) + \left(\frac{1}{3} - \frac{1}{6} - \frac{1}{8}\right) + \cdots + \left(\frac{1}{2n-1} - \frac{1}{4n-2} - \frac{1}{4n}\right) \\
&= \left(\frac{1}{2} - \frac{1}{4}\right) + \left(\frac{1}{6} - \frac{1}{8}\right) + \left(\frac{1}{10} - \frac{1}{12}\right) + \cdots + \left(\frac{1}{4n-2} - \frac{1}{4n}\right) \\
&= \frac{1}{2}\left(1 - \frac{1}{2} + \frac{1}{3} - \cdots + \frac{1}{2n-1} - \frac{1}{2n}\right) = \frac{1}{2}s_{2n}. \qquad (3.12)
\end{aligned}
$$

Since $s_{2n} \to s$, it follows that $t_{n,2} \to s/2$.
Hence the sum of the series has been affected by the above rearrangement which gives the sum $s/2$.

Theorem 3.5 *If a series is absolutely convergent, then every rearrangement of the series converges to the same sum.*

Proof Let the sequence $\{a_n\}_{n=1}^{\infty}$ be absolutely convergent, and let $\{b_m\}_{m=1}^{\infty}$ be a rearrangement of $\{a_n\}_{n=1}^{\infty}$. For a given n there exists $N_b \in \mathbb{N}$ such that

$$
\{a_k\}_{k=1}^{n} \subset \{b_j\}_{j=1}^{N_b}.
$$

Then we set

$$
s_n = \sum_{k=1}^{n} a_k, \quad t_m = \sum_{j=1}^{m} b_j, \quad \{b_j\}_{j=1}^{N_b} - \{a_k\}_{k=1}^{n} = \{a_{l_i}\}_{i=1}^{m-n}, \text{ where } l_1 > n.
$$

We see that for $m \geqslant N_b$

$$
|t_m - s_n| = |\sum_{i=1}^{m-n} a_{l_i}| \leqslant \sum_{i=1}^{m-n} |a_{l_i}| \leqslant \sum_{k=n+1}^{\infty} |a_k|.
$$

Since $\sum\limits_{k=n+1}^{\infty} |a_k| \to 0$ as $n \to \infty$, we have

$$
\lim_{m\to\infty} t_m = \lim_{n\to\infty} s_n.
$$

■

Theorem 3.6 *Let*

$$
s = \sum_{n=1}^{\infty} \frac{(-1)^{n-1}}{n} \text{ and } s_n = \sum_{k=1}^{n} \frac{(-1)^{k-1}}{k}.
$$

(a) Then the series converges to $s = \lim\limits_{n\to\infty} s_n = \log 2$..

(b) Let $p = 2, 3, \cdots$ *be fixed, and let us consider the rearrangement of* s. *Put*

$$t_{n,p} = \sum_{k=1}^{n} \left(\frac{1}{2k-1} - \frac{1}{2\{(k-1)p+1\}} - \frac{1}{2\{(k-1)p+2\}} - \cdots - \frac{1}{2pk} \right).$$

Then we have

$$\lim_{n \to \infty} t_{n,p} = \log 2 - \frac{1}{2} \log p.$$

Proof We show (a). We consider the geometric series

$$\frac{1}{1-x} = 1 + x + x^2 + x^3 + \cdots, \quad |x| < 1.$$

Then we see that for $|x| < 1$

$$\int_0^x \frac{1}{1-x} dx = \int_0^x (1 + x + x^2 + x^3 + \cdots) dx,$$

that is,

$$-\log(1-x) = x + \frac{1}{2}x^2 + \frac{1}{3}x^3 + \frac{1}{4}x^4 + \cdots.$$

Here we change x to $-x$, then we have

$$-\log(1+x) = -x + \frac{1}{2}x^2 - \frac{1}{3}x^3 + \cdots,$$

that is,

$$\log(1+x) = x - \frac{1}{2}x^2 + \frac{1}{3}x^3 - \cdots. \tag{3.13}$$

This equation holds for $-1 < x \leqslant 1$.
Now we consider the series $s = 1 - \frac{1}{2} + \frac{1}{3} - \frac{1}{4} + \cdots$. This series is convergent. In fact, since

$$s_{2n} = \left(1 - \frac{1}{2}\right) + \left(\frac{1}{3} - \frac{1}{4}\right) + \cdots + \left(\frac{1}{2n-1} - \frac{1}{2n}\right), \quad s_{2n+1} = s_{2n} + \frac{1}{2n+1},$$

we have

$$s_{2n} = \sum_{k=1}^{n} \frac{1}{2k(2k-1)}, \quad s_{2n+1} = \sum_{k=1}^{n} \frac{1}{2k(2k-1)} + \frac{1}{2n+1}.$$

Thus, the series s_n converges. Hence, from (3.11) and (3.13) we have

$$\log 2 = 1 - \frac{1}{2} + \frac{1}{3} - \frac{1}{4} + \cdots .$$

(*b*) To prove (*b*) we consider the series

$$t_{n,p} = \sum_{k=1}^{n} \left\{ \frac{1}{2k-1} - \frac{1}{2\{(k-1)p+1\}} - \frac{1}{2\{(k-1)p+2\}} - \cdots - \frac{1}{2pk} \right\}$$

and

$$s_{n,p} = \frac{1}{2n+1} + \frac{1}{2n+3} + \frac{1}{2n+5} + \cdots + \frac{1}{2pn-1}.$$

We see

$$\begin{aligned} &\frac{1}{2n+3} + \frac{1}{2n+5} + \cdots + \frac{1}{2pn-1} \\ &< \int_{n}^{pn-1} \frac{1}{2x+1} dx \\ &< \frac{1}{2n+1} + \frac{1}{2n+3} + \cdots + \frac{1}{2pn-1}, \end{aligned}$$

so we have

$$\begin{aligned} &\frac{1}{2n+3} + \frac{1}{2n+5} + \cdots + \frac{1}{2pn-1} \\ &< \frac{1}{2} \log \frac{p-(1/2n)}{1+(1/2n)} \\ &< \frac{1}{2n+1} + \frac{1}{2n+3} + \cdots + \frac{1}{2pn-1}, \end{aligned}$$

that is,

$$s_{n,p} - \frac{1}{2n+1} < \frac{1}{2} \log \frac{p-(1/2n)}{1+(1/2n)} < s_{n,p}.$$

Thus, we have

$$\lim_{n\to\infty} s_{n,p} = \frac{1}{2} \log p.$$

Since

$$s_{2pn} := \sum_{k=1}^{2pn} \frac{(-1)^{n-1}}{k} = t_{n,p} + s_{n,p}, \quad \lim_{n\to\infty} s_{2pn} = \log 2 \ \ (\text{by}(a)),$$

we have

$$\log 2 = \lim_{n\to\infty} t_{n,p} + \lim_{n\to\infty} s_{n,p} = \lim_{n\to\infty} t_{n,p} + \frac{1}{2}\log p.$$

Therefore, we conclude

$$\lim_{n\to\infty} t_{n,p} = \log 2 - \frac{1}{2}\log p.$$

We remark that (3.11) with $p = 2$ means

$$\lim_{n\to\infty} t_{n,p} = \frac{1}{2}\log 2 = \frac{1}{2}\lim_{n\to\infty} s_{2n},$$

(see (3.12)). Furthermore, we see with $p = 4$

$$\lim_{n\to\infty} t_{n,p} = \log 2 - \frac{1}{2}\log 4 = 0.$$

■

3.5 Exercises

1. Check the convergence of the following series.
   (a) $\sum_{n=1}^{\infty} \frac{n^2}{n!}$ (b) $\sum_{n=1}^{\infty} \frac{2n+1}{1+n^2}$ (c) $\sum_{n=1}^{\infty} \frac{2}{1+3^n}$ (d) $\sum_{n=1}^{\infty} (\frac{10}{11})^n$
   (e) $\sum_{n=1}^{\infty} \frac{n^p}{e^n}$, $p > 0$ (f) $\sum_{n=1}^{\infty} \frac{1}{n^2+n^3}$ (g) $\sum_{n=1}^{\infty} \frac{1}{n} + \frac{1}{2^n}$ (h) $\sum_{n=1}^{\infty} \frac{x^n}{n^n}$
   (i) $\sum_{n=1}^{\infty} \sin(\frac{1}{n})$ (j) $\sum_{n=1}^{\infty} \log(\frac{n}{n+1})$ (k) $\sum_{n=1}^{\infty} \frac{n}{n+1}x^n$.
2. Check the convergence of the following series.
   (a) $\sum_{n=1}^{\infty} nx^n$, $x > 0$ (b) $\sum_{n=1}^{\infty} \frac{n^{1/n}}{(n+1)^2 x^n}$ (c) $\sum_{n=1}^{\infty} \frac{(2n)!}{(n!)^2} x^{2n}$
   (d) $\sum_{n=1}^{\infty} \frac{1}{n\log n \log\log n}$ (e) $\sum_{n=1}^{\infty} \frac{(1+1/n)^{2n}}{e^n}$.
3. Show that the series $\sum_{n=1}^{\infty} \cos(\frac{1}{n})$ is divergent.
4. Show that if $\sum_{n=1}^{\infty} a_n$ converges, then $\sum_{n=1}^{\infty} \left(\frac{1+\sin(a_n)}{2}\right)^n$ converges.
5. Show that the series $\sum_{n=1}^{\infty} \left(\tan^{-1}(n)\tan^{-1}(n+1)\right)$ converges.
6. Let $s_1 = \sqrt{2}$ and $s_{n+1} = \sqrt{2+\sqrt{s_n}}$. Then prove that the sequence $< s_n >$ converges and that $s_n < 2$ for all $n \leqslant 1$.
7. Show that the series $\sum_{n=1}^{\infty} \frac{1}{(\log n)^n}$, $n \geqslant 2$ is convergent.

8. Consider that A contains 100 cc of milk and B contains 100 cc of water. 5 cc of the liquid in A is transferred to B, the mixture is thoroughly stirred and 5 cc of the mixture in B is transferred back into A. Each such two-way transfer is called a dilution. Let a_n be the percentage of water dilution with the understanding that $a_0 = 0$.
(a) Prove that $a_1 = \frac{100}{21}$ and that, in general, $a_n = \frac{100}{21} + \frac{19}{21}a_{n-1}$ for $n = 1, 2, 3, \cdots$.
(b) Using (a), prove that $a_n = 50\left[1 - (\frac{19}{21})^n\right]$ for $n = 1, 2, 3, \cdots$. Find the $\lim_{n\to\infty} a_n$.
9. Test the convergence of the series $1 + \frac{2}{5}x + \frac{6}{9}x^2 + \frac{30}{33}x^4 + \cdots$.
10. Test the convergence of the series

$$\frac{1}{1\cdot 2\cdot 3} + \frac{x}{4\cdot 5\cdot 6} + \frac{x^2}{7\cdot 8\cdot 9} + \cdots .$$

Chapter 4
Limits, Continuity, and Differentiability

Calculus is a branch of mathematics which mainly deals with the study of change in the value of a function as the points in the domain change. This chapter should study the limits, continuity, and differentiability of real-valued functions defined in a certain domain.

The idea of limits of functions underlies the entire subject of calculus. Without an understanding of limits, the concepts of derivative and integral cannot be made rigorous. Derivatives and Integrals are the core practical aspects of calculus. They were the first things investigated by Archimedes and developed by G.W. Leibniz as well as Isaac Newton. The process involved examining smaller and smaller pieces to get a sense of a progression toward a goal. This process was not formalized algebraically at the time, however. Bolzano, Weierstrass, and other mathematicians later developed and formalized these operations' theoretical underpinnings. These core concepts in this area are Limits, Continuity, and Differentiability. Derivatives and Integrals are defined in terms of limits. Continuity and Differentiability are important because almost every theorem in Calculus begins with the condition that the function is continuous and differentiable.

4.1 The Limit of a Function

First, we define the statement $\lim\limits_{x \to x_0} f(x) = L$ (to be read as the limit of $f(x)$, as x tends to x_0, equals L). To indicate that f is a real-valued function with the domain $\mathfrak{D}(f)$, we write $f : \mathfrak{D}(f) \to \mathbb{R}$. A notable point is that in taking the limit "as x approaches x_0." Thus, we do not require the value of $f(x_0)$, nor even whether $f(x_0)$ exists. We only care about $f(x)$ for values of x "close to" (but different from) x_0. Thus, we do not require that x_0 be in the domain of f. But we do require that x_0 be a limit point of the domain– otherwise values of x in the domain of f could not get "close to" x_0. Now, saying that $f(x)$ gets close to L is saying that $|f(x) - L|$ gets

N. Deo and R. Sakai, *Introduction to Mathematical Analysis*,
https://doi.org/10.1007/978-981-97-6568-3_4

small. Similarly, saying that x gets close to (but not equal to) x_0 is saying that $|x - x_0|$ gets small without equaling 0. This is made precise by the following definitions.)

Definition 4.1 Let $f : \mathfrak{D}(f) \to \mathbb{R}$, and x_0 be a limit point of $\mathfrak{D}(f)$. Then we say that $f(x)$ approaches the limit L as x approaches x_0, and write (Fig. 4.1)

$$\lim_{x \to x_0} f(x) = L.$$

If f is defined on some deleted neighborhood of x_0, for every $\varepsilon > 0$ there is a $\delta > 0$ such that

$$|f(x) - L| < \varepsilon, \text{ if } 0 < |x - x_0| < \delta.$$

Definition 4.2 Let $f : \mathfrak{D}(f) \to \mathbb{R}$. Then we say that $f(x)$ approaches the limit L as x approaches x_0 from the right if for a given $\varepsilon > 0$, and there is $\delta > 0$ such that

$$|f(x) - L| < \varepsilon, \text{ if } x_0 < x < x_0 + \delta.$$

In this case, we write

$$\lim_{x \to x_0+} f(x) = L.$$

Definition 4.3 Let $f : \mathfrak{D}(f) \to \mathbb{R}$. Then we say that $f(x)$ approaches the limit L as x approaches x_0 from left if for a given $\varepsilon > 0$, and there is $\delta > 0$ such that

$$|f(x) - L| < \varepsilon, \quad \text{if } x_0 - \delta < x < x_0.$$

In this case, we write

$$\lim_{x \to x_0-} f(x) = L.$$

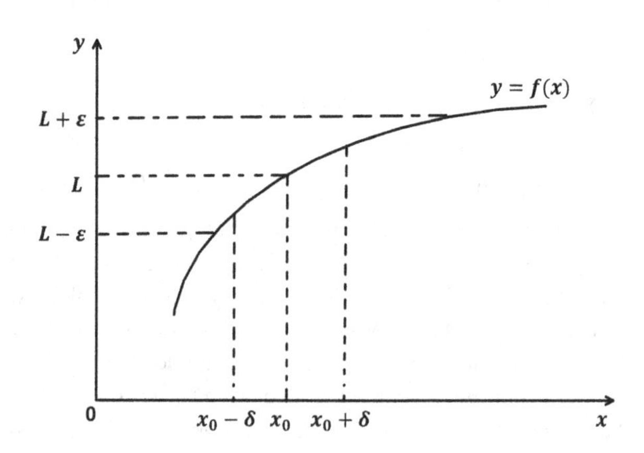

Fig. 4.1 Limit of a function at x_0

Example 4.1 We show that

$$\lim_{x\to 4}\frac{2x^2-32}{x-4}=16.$$

Consider

$$f(x)=\frac{2x^2-32}{x-4}.$$

Let $\varepsilon>0$. Note that $4\notin\mathfrak{D}(f)$ and when $x\neq 4$,

$$f(x)=\frac{2(x-4)(x+4)}{x-4}=2x+8.$$

Choose $\delta=\varepsilon/2$. Then

$$\begin{aligned}0<|x-4|<\delta &\Longrightarrow x\neq 4 \textit{ and } |x-4|<\delta\\ &\Longrightarrow f(x)=2x+8 \textit{ and } |x-4|<\frac{\varepsilon}{2}\\ &\Longrightarrow |f(x)-16|=|(2x+8)-16|=|2x-8|\\ &\Longrightarrow |f(x)-16|=2|x-4|<2\cdot\frac{\varepsilon}{2}=\varepsilon.\end{aligned}$$

Hence,

$$\lim_{x\to 4}\frac{2x^2-32}{x-4}=16.$$

Example 4.2 We show that

$$\lim_{x\to 3}x^2=9.$$

Let $\varepsilon>0$. Choose $\delta=\min\{4,\varepsilon/10\}$. Then

$$\begin{aligned}|x-3|<\delta &\Longrightarrow |x-3|<4 \text{ and } |x-3|<\frac{\varepsilon}{10}\\ &\Longrightarrow -4<x-3<4 \text{ and } |x-3|<\frac{\varepsilon}{10}\\ &\Longrightarrow 2<x+3<10 \text{ and } |x-3|<\frac{\varepsilon}{10}\\ &\Longrightarrow |x+3|<10 \text{ and } |x-3|<\frac{\varepsilon}{10}\\ &\Longrightarrow |x+3|\,|x-3|<10\cdot\frac{\varepsilon}{10}\\ &\Longrightarrow \left|x^2-9\right|<\varepsilon.\end{aligned}$$

Theorem 4.1 *If* $\lim_{x \to x_0} f(x)$ *exists, then it is unique, that is, if*

$$\lim_{x \to x_0+} f(x) = L \quad and \quad \lim_{x \to x_0-} f(x) = L', \tag{4.1}$$

then $L = L'$.

Proof See Theorem 2.2 of Chap. 2. ∎

On the other hand, both $\lim_{x \to x_0+} f(x)$ and $\lim_{x \to x_0-} f(x)$ may exist without being equal to each other. For example, if

$$f(x) = \begin{cases} x^2, & \text{if } 0 \leqslant x < 1, \\ 6 - x^2, & \text{if } 1 \leqslant x \leqslant 2, \end{cases}$$

then $\lim_{x \to 1+} f(x) = 5$ and $\lim_{x \to 1-} f(x) = 1$.

4.1.1 Some Theorems on Limits

Theorem 4.2 *If* $\lim_{x \to a} f(x) = L \neq 0$, *then there exist* $k \in (0, \infty)$ *and* $\delta > 0$ *such that*

$$0 < |x - a| < \delta \implies |f(x)| > k.$$

Proof Since $\lim_{x \to a} f(x) = L \neq 0$, for $\varepsilon = \frac{1}{2}|L|$ there exists $\delta > 0$ such that

$$0 < |x - a| < \delta \implies |f(x) - L| < \frac{1}{2}|L|$$

$$\implies |L| - |f(x)| < \frac{1}{2}|L|, \quad i.e., \quad |f(x)| > \frac{1}{2}|L|.$$

Hence, we can take a number $k = \frac{1}{2}|L| \in (0, \infty)$ as required. ∎

Theorem 4.3 *If* $\lim_{x \to a} f(x) = L$ *and* $\lim_{x \to a} g(x) = M$, *then*

(i) $\lim_{x \to a} [f(x) \pm g(x)] = L \pm M$.
(ii) $\lim_{x \to a} [f(x)g(x)] = LM$.
(iii) $\lim_{x \to a} \dfrac{f(x)}{g(x)} = \dfrac{L}{M}$, *provided* $M \neq 0$.

Proof See Theorem 2.6 of Chap. 2. ∎

4.2 Continuity

If we consider the function is continuous, in simple language, it means that we can draw the graph without lifting our pen and we get the result without "breaking" or "jumping" curves. That is thought to be the reason we use the word "continuous," to describe such functions, even though that is not the mathematical meaning of the word. In beginning calculus, an attempt is made at being more rigorous, that is, a function is defined to be continuous at a point x_0 if three conditions hold:

(i) $f(x_0)$ exists;
(ii) $\lim_{x \to x_0} f(x)$ exists;
(iii) $\lim_{x \to x_0} f(x) = f(x_0)$.

The definition we are going to use is equivalent to these three conditions only when x_0 is a limit point of the domain of f.

Definition 4.4 Let $f : \mathfrak{D}(f) \to \mathbb{R}$ and $x_0 \in \mathfrak{D}(f)$. Then f is continuous at x_0 if for each $\varepsilon > 0$ there is a $\delta > 0$ such that

$$|f(x) - f(x_0)| < \varepsilon, \quad \text{whenever } |x - x_0| < \delta.$$

In this case, we write

$$\lim_{x \to x_0} f(x) = f(x_0).$$

If f is not continuous at x_0, then we say that f is **discontinuous** at x_0.

Remark 4.1 If x_0 is a limit point of $\mathfrak{D}(f)$, then f is continuous at x_0 if and only if $\lim_{x \to x_0} f(x) = f(x_0)$.

Graphically continuity of a function at a point x_0 means that there is no break in the graph of the curve $y = f(x)$ at $x = x_0$ and given, however small, $\varepsilon > 0 \; \exists \; \delta > 0$ such that $f(x) > f(x_0) - \varepsilon$ at $x = x_0 - \delta$ and $f(x) < f(x_0) + \varepsilon$ at $x = x_0 + \delta$, where $x \in (x_0 - \delta, x_0 + \delta)$ (Fig. 4.2).

We may also visualize this definition as follows (Fig. 4.3):

Definition 4.5 Let $f : \mathfrak{D}(f) \to \mathbb{R}$. A function f defined on a semi-open interval $[x_0, b)$ is continuous from the right at x_0 if for each $\varepsilon > 0$ there is some $\delta > 0$ such that

$$|f(x) - f(x_0)| < \varepsilon, \quad \text{whenever } x_0 \leqslant x < x_0 + \delta.$$

In this case, we write

$$\lim_{x \to x_0+} f(x) = f(x_0).$$

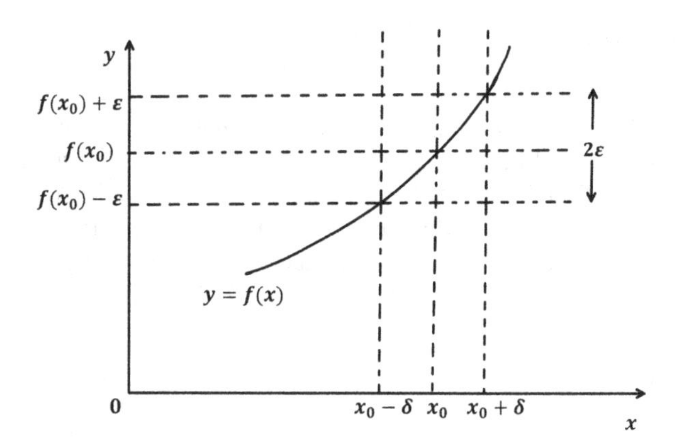

Fig. 4.2 Continuity of f at x_0

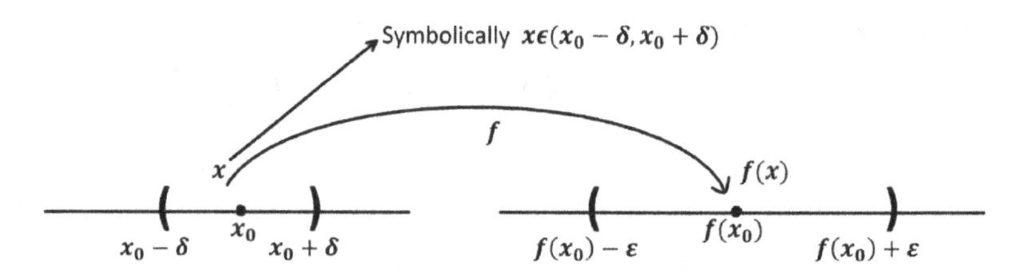

Fig. 4.3 Continuity of f at x_0

Definition 4.6 Let $f : \mathfrak{D}(f) \to \mathbb{R}$. A function f defined on a semi-open interval $(a, x_0]$ is continuous from the left at x_0 if for each $\varepsilon > 0$ there is some $\delta > 0$ such that

$$|f(x) - f(x_0)| < \varepsilon, \text{ whenever } x_0 - \delta < x \leqslant x_0.$$

In this case, we write

$$\lim_{x \to x_0-} f(x) = f(x_0).$$

Example 4.3 We prove that the function $f(x) = 4x^2 - 9x - 7$ is continuous at $x_0 = 3$. We have $f(3) = 2$.

(i) Let $\varepsilon > 0$. We have to obtain $\delta > 0$ such that

$$\begin{aligned} |x-3| < \delta &\Longrightarrow \left|4x^2 - 9x - 7 - 2\right| < \varepsilon \\ &\Longrightarrow \left|4x^2 - 9x - 9\right| < \varepsilon \\ &\Longrightarrow |(4x+3)(x-3)| < \varepsilon \\ &\Longrightarrow |4x+3|\,|x-3| < \varepsilon. \end{aligned}$$

Now, if we take $\delta = 1$, then

$$
\begin{aligned}
|x-3|<1 \Longrightarrow -1<x-3<1 \Longrightarrow 2<x<4 &\\
\Longrightarrow 8<4x<16 &\\
\Longrightarrow 11<4x+3<19. &\quad (4.2)
\end{aligned}
$$

Thus, we want to make sure that $\delta \leqslant 1$ and $\delta < \varepsilon/19$.

(ii) Let $\varepsilon > 0$. Choose $\delta = \min\{1, \varepsilon/19\}$, then

$$
\begin{aligned}
|x-3|<\delta &\Longrightarrow |x-3|<1 \text{ and } |x-3|<\frac{\varepsilon}{19}\\
&\Longrightarrow 11<4x+3<19 \text{ and } |x-3|<\frac{\varepsilon}{19}\\
&\Longrightarrow |4x+3|<19 \text{ and } |x-3|<\frac{\varepsilon}{19}\\
&\Longrightarrow |4x+3|\,|x-3|<19.\frac{\varepsilon}{19}\\
&\Longrightarrow \left|4x^2-9x-9\right|<\varepsilon\\
&\Longrightarrow \left|(4x^2-9x-7)-2\right|<\varepsilon.
\end{aligned}
$$

Hence, the function $f(x) = 4x^2 - 9x - 7$ is continuous at $x_0 = 3$.

Example 4.4 We define the function $f : \mathbb{R} \to \mathbb{R}$ by

$$
f(x) = \begin{cases} 1, & \text{if } x \geqslant 0, \\ 3, & \text{if } x < 0. \end{cases}
$$

Look at the following Fig. 4.4 of the graph of this function.

The function $f : \mathbb{R} \to \mathbb{R}$ is continuous at each point x_0 except for $x_0 = 0$.

Example 4.5 We define the function $f : \mathbb{R} \to \mathbb{R}$ by

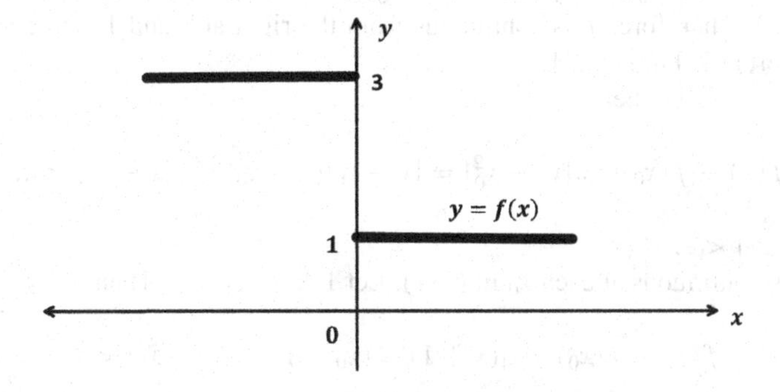

Fig. 4.4 The function $f : \mathbb{R} \to \mathbb{R}$ is not continuous at $x = 0$

$$f(x) = \begin{cases} 1, & \text{if } x \text{ is rational,} \\ 0, & \text{if } x \text{ is irrational.} \end{cases}$$

This function is called *Dirichlet's function*. There is no point x_0 in R at which Dirichlet's function is continuous. Indeed, given a point x_0 in R, by the sequential density of the rationals and irrationals, there is a sequence $\{u_n\}$ of rational numbers that converge to x_0 and also a sequence $\{v_n\}$ of irrational numbers that converge to x_0. But $\{f(u_n)\}$ is a constant sequence all of whose terms equal 1, while $\{f(v_n)\}$ is a constant sequence all of whose terms equal 0. Thus,

$$\lim_{n\to\infty} f(u_n) = 1 \neq 0 = \lim_{n\to\infty} f(v_n).$$

Since both of the sequences $\{u_n\}$ and $\{v_n\}$ converge to x_0, it is not possible for f to be continuous at x_0. Observe that one expression of the discontinuous nature of Dirichlet's function is that there is no way to graph it.

Remark 4.2 Dirichlet's function (cf. Example 4.5) is given as follows:

$$f(x) = \lim_{m\to\infty} (\lim_{n\to\infty} \{\cos(m!\pi x)\}^{2n}).$$

In fact, for $x \in \mathbb{Q}$ and m large enough, there exists $k \in \mathbb{N}$ such that

$$m!\pi x = 2k\pi.$$

Consequently, we have $f(x) = 1$. If $x \in \mathbb{R}\backslash\mathbb{Q}$, then we see $-1 < m!\pi x < 1$ for any $m \in \mathbb{N}$. Thus, we have $f(x) = 0$.

Example 4.6 Let f be defined on $[0, 2]$ by

$$f(x) = \begin{cases} x^2, & \text{if } 0 \leqslant x < 1, \\ x+1, & \text{if } 1 \leqslant x \leqslant 2. \end{cases}$$

Then $f(0+) = 0 = f(0)$, $f(1-) = 0 \neq f(1) = 2$, $f(1+) = 2 = f(1)$, $f(2-) = 3 = f(2)$. Therefore, f is continuous from the right at 0 and 1, and continuous from the left at 2, but not at 1.
Let $0 < x,\ x_0 < 1$. Then

$$|f(x) - f(x_0)| = |x^2 - x_0^2| = |x - x_0||x + x_0| \leqslant 2|x - x_0| < \varepsilon,$$

from $|x - x_0| < \varepsilon$.
Hence, f is continuous at each x_0 in $(0, 1)$. Let $1 < x, x_0 < 2$. Then

$$|f(x) - f(x_0)| = |(x+1) - (x_0 - 1)| = |x - x_0| < \varepsilon,$$

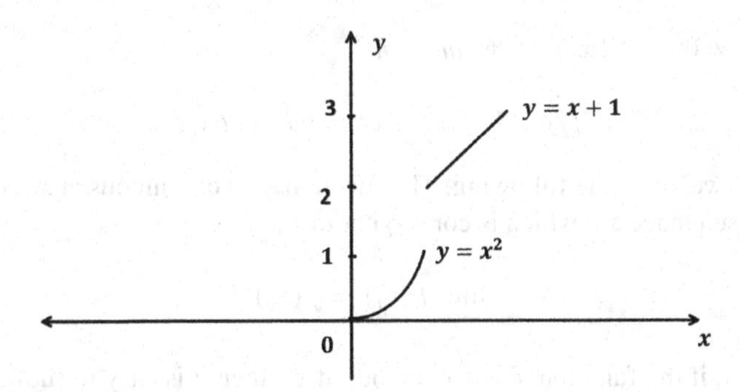

Fig. 4.5 Figure for Example 4.6

from $|x - x_0| < \varepsilon$. Hence, f is continuous at each x_0 in $(1, 2)$ (Fig. 4.5).

4.2.1 Sums, Products, and Quotients of Continuous Functions

Given two functions $f : S \to \mathbb{R}$ and $g : S \to \mathbb{R}$, we define the sum $f + g : S \to \mathbb{R}$ and the product $fg : S \to \mathbb{R}$ by

$$(f + g)(x) = f(x) + g(x) \ \ and \ \ (fg)(x) = f(x)g(x), \ \forall x \in S.$$

Moreover, if $g(x) \neq 0$ for all x in S, the quotient $f/g : S \to \mathbb{R}$ is defined by

$$(f/g)(x) = \frac{f(x)}{g(x)}, \ \forall x \in S.$$

The following theorem is an analog, and also a consequence, of the sum, product, and quotient properties of convergent sequences.

4.2.2 Some Theorems on Continuous Functions

Theorem 4.4 *The functions $f : S \to \mathbb{R}$ and $g : S \to \mathbb{R}$ are continuous at the point x_0 in S. Then the sum*

$$f + g : S \to \mathbb{R} \ \text{ is continuous at } x_0, \tag{4.3}$$

the product

$$fg : S \to \mathbb{R} \ \text{ is continuous at } x_0, \tag{4.4}$$

and, if $g(x) \neq 0$ for all x in S, the quotient

$$f/g : S \to \mathbb{R} \text{ is continuous at } x_0. \tag{4.5}$$

Proof First, we show the following. The function f is continuous at x_0 if and only if for every sequence x_n, which is convergent to x_0,

$$\lim_{n\to\infty} f(x_n) = f(x_0) \tag{4.6}$$

holds. In fact, if the function f is continuous at x_0, then it is easy to show (4.6).

Conversely, we suppose (4.6). Then if $f(x)$ is not continuous at x_0, we see that for some $\varepsilon_0 > 0$ and any $n \in \mathbb{N}$ there exists x_n such that

$$|x_n - x_0| < \frac{1}{n} \text{ and } |f(x_n) - f(x_0)| \geqslant \varepsilon_0 \ (> 0).$$

This means that even though $x_n \to x_0$, we do not obtain (4.6). But this fact contradicts our assumption.

Now, we can prove the theorem using the sequence $\{x_n\}$ which converges to x_0. Let

$$\lim_{n\to\infty} f(x_n) = f(x_0) \text{ and } \lim_{n\to\infty} g(x_n) = g(x_0).$$

The sum property of convergent sequences implies that

$$\lim_{n\to\infty} \left[f(x_n) + g(x_n)\right] = f(x_0) + g(x_0),$$

and the product property of convergent sequences implies that

$$\lim_{n\to\infty} \left[f(x_n)\, g(x_n)\right] = f(x_0)\, g(x_0).$$

If $g(x) \neq 0$ for all x in S, the quotient property of convergent sequences implies that

$$\lim_{n\to\infty} \frac{f(x_n)}{g(x_n)} = \frac{f(x_0)}{g(x_0)}.$$

■

The Polynomial Property of convergent sequences is precisely the assertion that a polynomial is continuous. Thus, by the quotient property of continuous functions, we have the following corollary describing a general class of continuous functions.

Corollary 4.1 *Let $p : \mathbb{R} \to \mathbb{R}$ and $q : \mathbb{R} \to \mathbb{R}$ be polynomials. Then the quotient $p/q : S \to \mathbb{R}$ is continuous, where $S = \{x \in \mathbb{R} : q(x) \neq 0\}$.*

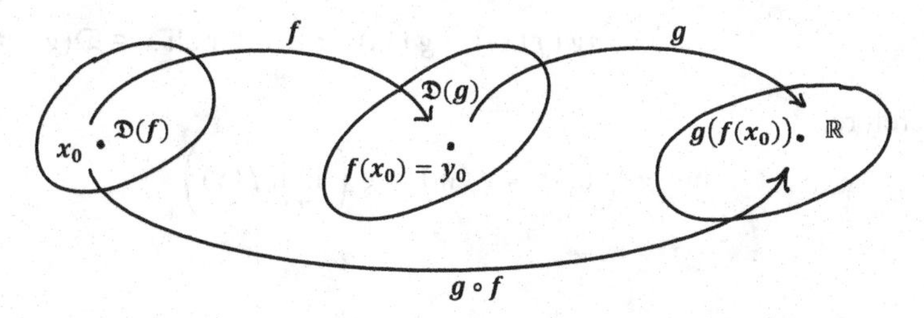

Fig. 4.6 The composition of f and g

4.2.3 Compositions of Continuous Functions

Theorem 4.5 *(i) Let $f : \mathfrak{D}(f) \to \mathbb{R}$ be continuous at x_0 and $g : \mathfrak{D}(g) \to \mathbb{R}$ be continuous at $f(x_0)$. Then the composite function $g \circ f$ is continuous at x_0.*
(ii) Let $\lim_{x \to x_0} f(x) = y_0 \in \mathfrak{D}(g)$ and g be continuous at y_0. Then (Fig. 4.6)

$$\lim_{x \to x_0} g(f(x)) = g\left(\lim_{x \to x_0} f(x)\right) \in g(y_0).$$

Proof (i) Let f be a continuous function at x_0 and g be continuous at $f(x_0)$. Let $\varepsilon > 0$. Since g is continuous at $f(x_0)$, $\exists\, \delta > 0$ such that

$$|y - f(x_0)| < \delta \implies |g(y) - g(f(x_0))| < \varepsilon, \quad \text{for all } y \in \mathfrak{D}(f). \quad (4.7)$$

Since f is continuous at $f(x_0)$, there exists $\delta' > 0$ such that

$$\begin{aligned} |x - x_0| < \delta' &\implies |f(x) - f(x_0)| < \delta \\ &\implies |g(f(x)) - g(f(x_0))| < \varepsilon \\ &\implies |(g \circ f)(x) - (g \circ f)(x_0)| < \varepsilon, \quad \forall x \in \mathfrak{D}(g \circ f), \end{aligned}$$

the composition $(g \circ f)$ is continuous at x_o.
(ii) Let

$$\lim_{x \to x_0} f(x) = y_0 \in \mathfrak{D}(g),$$

and g be continuous at y_0. For $\varepsilon > 0$ there exists $\delta > 0$ such that

$$|y - y_0| < \delta \implies |g(y) - g(y_0)| < \varepsilon, \quad \text{for all } y \in \mathfrak{D}(g).$$

Since $\lim_{x \to x_0} f(x) = y_0$, there exists $\delta' > 0$ such that

$$0 < |x - x_0| < \delta' \implies |f(x) - y_0| < \delta$$

$$\Longrightarrow |g(f(x)) - g(y_0)| < \varepsilon, \quad \text{for all } x \in \mathfrak{D}(g \circ f).$$

Therefore,

$$\lim_{x \to x_0} g(f(x)) = g(y_0) = g\left(\lim_{x \to x_0} f(x)\right).$$

■

4.3 The Intermediate Value Theorem

Theorem 4.6 [Intermediate Value Theorem]
Suppose that f is continuous on an interval $I = [a, b]$, $f(a) \neq f(b)$, and λ is between $f(a)$ and $f(b)$, then we can find a ξ between a and b such that $\lambda = f(\xi)$.

Proof The graphical interpretation of this theorem is that when the horizontal line $y = \lambda$ separates the points $(a, f(a))$ and $(b, f(b))$, the graph of f must cross the line at some point whose x-coordinate is between a and b. Obviously the graph must also cross any vertical line that separates the points, but we do not need a theorem to tell us that. Since we know the x-coordinate, we can just evaluate f to locate the point on a given vertical line. The intermediate value theorem deals with something that can be a real problem: finding a value of x that satisfies the equation $f(x) = \lambda$ (Fig. 4.7).

Suppose that $f(a) < \lambda < f(b)$. The set

$$S = \{x \mid a \leqslant x \leqslant b \text{ and } f(x) \leqslant \lambda\}$$

is non-empty and bounded. Let $\xi = \sup S$. We will show that $f(\xi) = \lambda$. If $f(\xi) > \lambda$, then $\xi > a$, and since f is continuous at ξ, for $\varepsilon > 0$ there exists $\delta > 0$ such that

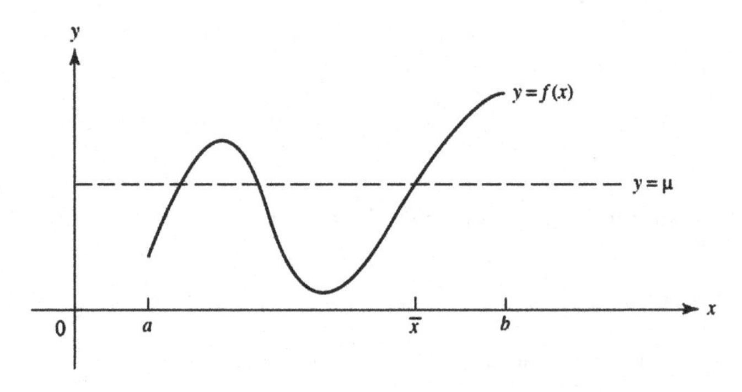

Fig. 4.7 Intermediate value theorem

$$|f(x) - f(\xi)| < \varepsilon \text{ if } \xi - \delta < x < \xi + \delta.$$

Hence, if $f(\xi) > \lambda$, then we see that for $\varepsilon > 0$ small enough,

$$f(x) > \lambda \quad \text{if} \quad \xi - \delta < x \leqslant \xi. \tag{4.8}$$

But, by the definition of ξ we can find $x \in S$ such that $\xi - \delta < x \leqslant \xi$, and so

$$f(x) \leqslant \lambda. \tag{4.9}$$

Therefore, (4.8) contradicts (4.9). If $f(\xi) < \lambda$, then we have

$$f(x) < \lambda \quad \text{if} \quad \xi \leqslant x \leqslant \xi + \delta. \tag{4.10}$$

Especially, we see $f(\xi) < \lambda$. From (4.10) we can find $x \in S$ such as $\xi < x$, but this contradicts the definition of ξ. Consequently, we have $f(\xi) = \lambda$.

For the case of $f(b) < \lambda < f(a)$ the proof can be obtained by applying this result to $-f$. ■

Theorem 4.7 *If f is continuous on a finite closed interval $[a, b]$, then f is bounded on $[a, b]$.*

Theorem 4.8 *Suppose that f is continuous on a finite closed interval $[a, b]$. Let*

$$\alpha = \inf_{a \leqslant x \leqslant b} f(x) \text{ and } \beta = \sup_{a \leqslant x \leqslant b} f(x).$$

Then α and β are respectively the minimum and maximum of f on $[a, b]$; that is, there are points x_1 and x_2 in $[a, b]$ such that $f(x_1) = \alpha$ and $f(x_2) = \beta$.

4.4 Uniform Continuity

Continuity is not quite strong enough for integration, therefore one needs a kind of uniform continuity, independent of x.

Definition 4.7 Let $f : \mathfrak{D}(f) \to \mathbb{R}$. A function f is uniformly continuous if for given $\varepsilon > 0$, there exists $\delta > 0$ such that

$$|f(y) - f(x)| < \varepsilon, \quad \text{whenever} \quad |x - y| < \delta.$$

The only difference from mere continuity is that the δ appears before the x and consequently cannot depend on x, and δ must work for all x. For example, $f(x) = 1/x$ is continuous on $(0, 1)$ but not uniformly continuous.

4.5 Types of Discontinuity

If a function is not continuous at a point, then it is said to be discontinuous at that point and the point is called a point of discontinuity of this function.

(i) A function f is said to have a **removable discontinuity** at a point a if and only if $\lim_{x\to a} f(x)$ exists but is not equal to $f(a)$, i.e., if

$$f(a+0) = f(a-0) \neq f(a).$$

Example 4.7 The function f defined as

$$f(x) = \begin{cases} \frac{x^2-9}{x-3}, & \text{if } x \neq 3, \\ 3, & \text{if } x = 3, \end{cases}$$

is continuous for all $x \in \mathbb{R} - \{3\}$. At $x = 3$ it has a removable discontinuity.

(ii) If both $f(a+0)$ and $f(a-0)$ exist but are not equal, then we say that f has a discontinuity of the **first kind or ordinary discontinuity** at a. The point a is said to be a point of discontinuity from the left or right according as

$$f(a-0) \neq f(a) = f(a+0) \text{ or } f(a-0) = f(a) \neq f(a+0).$$

Example 4.8 The function f defined as

$$f(x) := \begin{cases} 1, \text{ if } x < 1, \\ 2, \text{ if } x \geqslant 1, \end{cases}$$

has a *discontinuity of the first kind* from the left at $x = 1$.
The function f defined as

$$f(x) = \begin{cases} \frac{1}{1-e^{-1/x}}, & \text{if } x \neq 0, \\ 0, & \text{if } x = 0. \end{cases}$$

Here both $f(0+0) = 1$ and $f(0-0) = 0$ exist, but not equal so that $f(x)$ is discontinuous at $x = 0$ and the function has a *discontinuity of the first kind* at $x = 0$.

(iii) A function f is said to have a discontinuity of *the second kind* at a, if both the limits $f(a+0)$ and $f(a-0)$ do not exist. The point a is point of a **discontinuity of the second kind** from the left or right according as $f(a-0)$ or $f(a+0)$ does not exist.

Example 4.9 The function f defined as

$$f(x) = \sin\frac{1}{x};\ x \neq 0$$

has a discontinuity of the second kind at $x = 0$.

(iv) A function is said to have a **mixed discontinuity** at a if f has a discontinuity of the second kind on one side of a, and on the other side a discontinuity of the first kind, or may be continuous.

Example 4.10 The function

$$f(x) = e^{1/x}\sin\frac{1}{x};\ x \neq 0$$

has a mixed discontinuity at $x = 0$.

(v) If one or more of the functional limits $\overline{f(a+0)}$, $\underline{f(a+0)}$, $\overline{f(a-0)}$ and $\overline{f(a-0)}$ is $+\infty$ or $-\infty$ and f is discontinuous at a, then we say f has an **infinite discontinuity** at a. Evidently, if f has a discontinuity at a and is unbounded in every neighborhood of a, then f will have an infinite discontinuity at a.

Example 4.11 The function f defined as

$$f(x) = \frac{1}{x-a};\ x \neq a$$

is continuous for all $x \neq a$. It has an infinite discontinuity at $x = a$.

4.6 Differentiability

The idea of a derivative may be defined as follows. Let us suppose that a graph of a function looks like a straight line. Now we can talk about the slope of this line. The slope tells us how fast the value of the function is changing at the particular point, i.e., the rate of change of the function at that particular point. If the slope of the line is positive, then the function is increasing; if the slope is negative, then the function is decreasing. Moreover, the magnitude of the slope is an indication of how fast the function is increasing or decreasing. Of course, we are considering any function that has corners or discontinuities.

A function f is differentiable at an interior point x_0 of its domain if the difference quotient

$$\frac{f(x) - f(x_0)}{x - x_0},\quad x \neq x_0$$

approaches a limit as x approaches x_0, in which the limit is called the derivative of f at x_0, and is denoted by $f'(x_0)$. Thus,

$$f'(x_0) = \lim_{x \to x_0} \frac{f(x) - f(x_0)}{x - x_0}, \quad x \neq x_0. \tag{4.11}$$

It is sometimes convenient to let $x = x_0 + h$ and write (4.11) as

$$f'(x_0) = \lim_{h \to 0} \frac{f(x_0 + h) - f(x_0)}{h}.$$

Let f be defined on an open set S. If f is differentiable at every point of S, we say that f is differentiable on S, then f' is a function on S. We say that f is continuously differentiable on S if f' is continuous on S. If f is differentiable on a neighborhood of x_0, it is reasonable to ask if f' is differentiable at x_0. If so, we denote the derivative of f' at x_0 by f''. This is the second derivative of f at x_0, and it is also denoted by $f^{(2)}$. Continuing inductively, if $f^{(n-1)}$ is defined on a neighborhood of x_0, then the n-th derivative of f at x_0, and is denoted by $f^{(n)}$, is the derivative of $f^{(n-1)}$ at x_0. For convenience we define the zeroth derivative of f to be f itself. Thus

$$f^{(0)} = f.$$

We assume that you are familiar with the other standard notations for derivatives. For example,

$$f^{(2)} = f'', \ f^{(3)} = f''',$$

and so on, and

$$\frac{d^n f}{dx^n} = f^{(n)}.$$

Example 4.12 A function is continuous at a point without being differentiable. For example, if

$$f(x) = |x|, \quad x \in \mathbb{R},$$

then f is a continuous function. For $x > 0$, f is differentiable, and $f'(x) = 1$ for all $x > 0$.
For $x < 0$, f is differentiable, and $f'(x) = -1$ for all $x < 0$. Thus, f *is continuous but not differentiable at* 0 (Fig. 4.8).

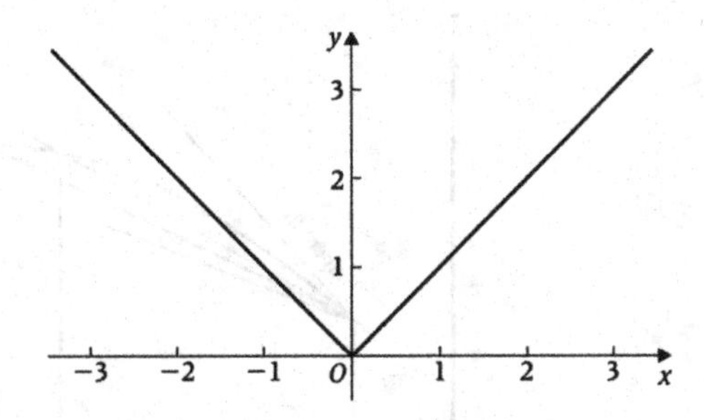

Fig. 4.8 A continuous function is not differentiable at 0

4.7 Interpretation of the Derivative

If $f(x)$ is the position of a particle at time $x \neq x_0$, the difference quotient

$$\frac{f(x) - f(x_0)}{x - x_0}$$

is the average velocity of the particle between times x_0 and x. As x approaches x_0, the average applies to shorter and shorter intervals. Therefore, it makes sense to regard the limit (4.11), if it exists, as the particles instantaneous velocity at time x_0. This interpretation may be useful even if x is not time, so we often regard $f'(x_0)$ as the instantaneous rate of change of $f(x)$ at x_0, regardless of the specific nature of the variable x. The derivative also has a geometric interpretation. The equation of the line through two points $(x_0, f(x_0))$ and $(x_1, f(x_1))$ on the curve $y = f(x)$ is

$$y = f(x_0) + \frac{f(x_1) - f(x_0)}{x_1 - x_0}(x - x_0).$$

Now, this line tends to

$$y = f(x_0) + f'(x_0)(x - x_0) \tag{4.12}$$

as x_1 approaches x_0. This is the tangent to the curve $y = f(x)$ at the point $(x_0, f(x_0))$. The figure below depicts the situation for various values of x_1 (Fig. 4.9).

Here is a less intuitive definition of the tangent line. If the function

$$T(x) = f(x_0) + m(x - x_0)$$

approximates f so well near x_0, that is,

$$\lim_{x \to x_0} \frac{f(x) - T(x)}{x - x_0} = 0,$$

then we say that the line $y = T(x)$ is tangent to the curve $y = f(x)$ at $(x_0, f(x_0))$. This tangent line exists if and only if $f'(x_0)$ exists, and if m is uniquely determined

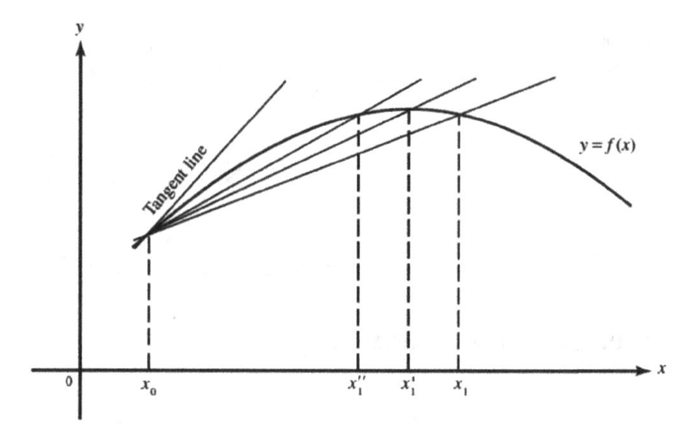

Fig. 4.9 The tangent line is the limiting position of the cords

by $m = f'(x_0)$. Thus, (4.12) is the equation of the tangent line. We will use the following lemmas to study differentiable functions.

Theorem 4.9 *If f is differentiable at x_0, then f is continuous at x_0.*

The converse of this theorem is false, since a function may be continuous at a point without being differentiable at the point.

Example 4.13 $f(x) = |x|$ is continuous at 0 but not differentiable at 0.

Theorem 4.10 *If f and g are differentiable at x_0, then so are $f + g$, $f - g$, and fg, with*

(a) $(f + g)'(x_0) = f'(x_0) + g'(x_0)$.
(b) $(f - g)'(x_0) = f'(x_0) - g'(x_0)$.
(c) $(fg)'(x_0) = f'(x_0)g(x_0) - f(x_0)g'(x_0)$.
(d) $\left(\frac{f}{g}\right)'(x_0) = \frac{f'(x_0)g(x_0) - f(x_0)g'(x_0)}{[g(x_0)]^2}$; *if* $g(x_0) \neq 0$.

Theorem 4.11 *Suppose that g is differentiable at x_0 and f is differentiable at $g(x_0)$. Then the composite function $h = f \circ g$ defined by*

$$h(x) = f\big(g(x)\big)$$

is differentiable at x_0 with

$$h'(x_0) = f'\big(g(x_0)\big)g'(x_0).$$

4.8 One-Sided Derivatives

One-sided limits of difference quotients such as

$$\lim_{x \to x_0+} \frac{f(x) - f(x_0)}{x - x_0}$$

and

$$\lim_{x \to x_0-} \frac{f(x) - f(x_0)}{x - x_0}$$

are called one-sided or right and left-hand derivatives. That is, if f is defined on $[a, b]$, the right-hand derivative of f at x_0 is defined to be

$$f'_+(x_0) = \lim_{x \to x_0+} \frac{f(x) - f(x_0)}{x - x_0},$$

and the left-hand derivative of f at x_0 is defined to be

$$f'_-(x_0) = \lim_{x \to x_0-} \frac{f(x) - f(x_0)}{x - x_0}.$$

Then f is differentiable at x_0, if and only if $f'_+(x_0)$ and $f'_-(x_0)$ exist and they are equal, that is,

$$f'(x_0) = f'_+(x_0) = f'_-(x_0).$$

Theorem 4.12 (Rolle's Theorem)
Suppose that f is continuous on the closed interval $[a, b]$ and differentiable on the open interval (a, b), and $f(a) = f(b)$. Then $f'(c) = 0$ for some c in the open interval (a, b).

Proof Since f is continuous on $[a, b]$, f attains minimum α and maximum β at two numbers x_1 and x_2 lying between a and b such that (Fig. 4.10)

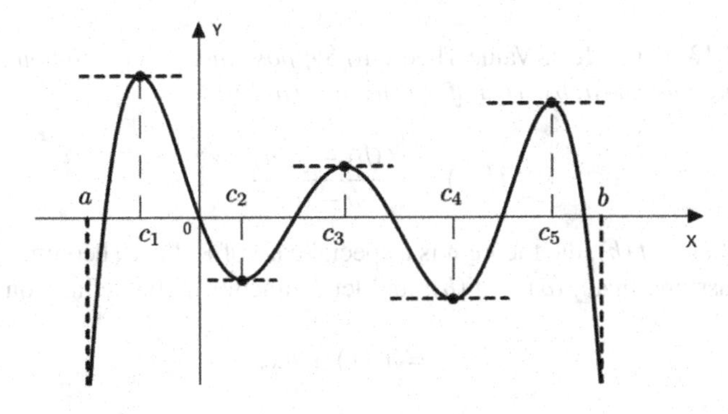

Fig. 4.10 Rolle's Theorem

$$f(x_1) = \alpha \text{ and } f(x_2) = \beta.$$

Case 1: If $\alpha = \beta$, then f is a constant function so that $f' = 0$, for all $x \in (a, b)$.
Case 2: If $\alpha \neq \beta$, then $\beta = f(x_2) \geqslant f(x_2 + h)$ for values of h, both positive and negative. Then

$$\frac{f(x_2 + h) - f(x_2)}{h} \leqslant 0, \text{ for } h > 0,$$

and

$$\frac{f(x_2 + h) - f(x_2)}{h} \geqslant 0, \text{ for } h < 0.$$

Since f is differentiable in (a, b) as $h \to 0$, we have

$$f'(x_2) \leqslant 0 \text{ and } f'(x_2) \geqslant 0.$$

Hence $f'(x_2) = 0$ for $c = x_2$. Similarly, if $\alpha = f(x_1) \leqslant f(x_1 + h)$ for values of h both positive and negative, it follows that

$$\frac{f(x_1 + h) - f(x_1)}{h} \geqslant 0, \text{ for } h > 0,$$

and

$$\frac{f(x_1 + h) - f(x_1)}{h} \leqslant 0, \text{ for } h < 0.$$

As $h \to 0$, we have

$$f'(x_1) \leqslant 0 \text{ and } f'(x_1) \geqslant 0.$$

Hence, $f'(x_1) = 0$. ■

4.9 The Mean-Value Theorem

Theorem 4.13 (The Mean-Value Theorem) *Suppose that f is continuous on $[a, b]$ and differentiable on (a, b). Then, for some $c \in (a, b)$,*

$$f'(c) = \frac{f(b) - f(a)}{b - a}.$$

Proof If $f(a) = f(b)$, the theorem is a special case of Rolle's Theorem. Hence we assume that $f(a) \neq f(b)$, and let a function F be defined on $[a, b]$ as follows:

$$F(x) = f(x) + hx,$$

where h is a constant. Then F is continuous on $[a, b]$ and differentiable on (a, b). We choose the constant h so that $F(a) = F(b)$. In fact, we take h as follows:

$$f(a) + ha = f(b) + hb$$

and so

$$h = -\frac{f(b) - f(a)}{b - a}.$$

Since F satisfies the conditions of Rolle's Theorem, $F'(c) = 0$ for some $c \in (a, b)$. Therefore, we see

$$F'(c) = f'(c) + h = 0,$$

and the theorem follows. ∎

4.9.1 The Geometric Interpretation of the Mean-Value Theorem

Let P and Q be two points on the curve $y = f(x)$. The expression

$$\frac{f(b) - f(a)}{b - a}$$

is the slope of the chord PQ.

Thus, for some $c \in (a, b)$, the tangent to f at c is parallel to PQ (Fig. 4.11).

Theorem 4.14 (The Generalized Mean-Value Theorem) *If f and g are continuous on the closed interval $[a, b]$ and differentiable on the open interval (a, b), then*

$$[g(b) - g(a)]f'(c) = [f(b) - f(a)]g'(c)$$

for some c in (a, b).

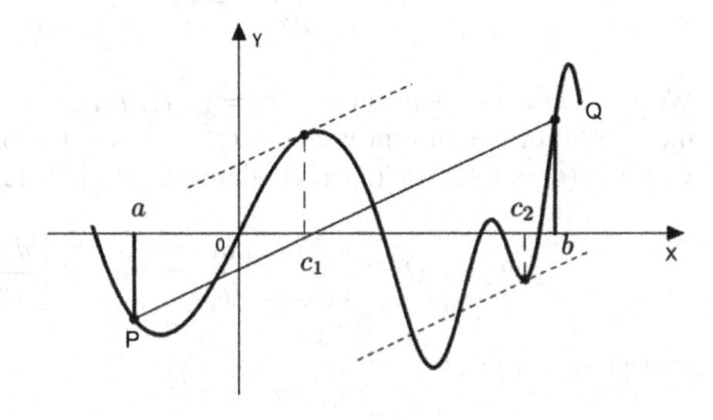

Fig. 4.11 The Mean value Theorem

Proof The function

$$h(x) = [g(b) - g(a)]f(x) - [f(b) - f(a)]g(x)$$

is continuous on $[a, b]$ and differentiable on (a, b), and

$$h(a) = h(b) = g(b)f(a) - f(b)g(a).$$

Therefore, Rolle's Theorem implies that $h'(c) = 0$ for some c in (a, b). Since

$$h'(c) = [g(b) - g(a)]f'(c) - [f(b) - f(a)]g'(c),$$

this implies

$$[g(b) - g(a)]f'(c) = [f(b) - f(a)]g'(c).$$

■

An elegant application of the Mean-Value Theorem is its use in proving L'Hopital's rule for evaluating indeterminate forms of the type $0/0$ or ∞/∞. For this, we need the theorems in a slightly strengthened form which uses two functions simultaneously. In the statement, we call the independent variable t, as the independent variable, as it makes the proof a bit more natural.

Corollary 4.2 (Cauchy's Mean-Value Theorem) *If $f(t)$ and $g(t)$ are continuous on the closed interval $[a, b]$ and differentiable on the open interval (a, b), let $g'(t) \neq 0$ for all $t \in (a, b)$. Then there exists $c \in (a, b)$ such that*

$$\frac{f(b) - f(a)}{g(b) - g(a)} = \frac{f'(c)}{g'(c)}.$$

Proof Now we set

$$x = g(t), \quad y = f(t),$$

then we have

$$\frac{dy}{dx} = \frac{dy/dt}{dx/dt} = \frac{f'(t)}{g'(t)}.$$

We consider the smooth curve $F(t) = (g(t), f(t)),\ a \leqslant t \leqslant b$, then $\left[\frac{dy}{dx}\right]_{t=c}$ means the tangent of the tangent vector $F'(t) = (g'(t), f'(t))$ at $t = c$. Then the tangent vector $F'(t) = (g'(t), f'(t))$ is continuous. Hence there exists $c \in (a, b)$ such that

$$\text{slope } PQ = \frac{f(b) - f(a)}{g(b) - g(a)} = \text{tangent}\left[\frac{dy}{dx}\right]_{t=c} = \frac{f'(c)}{g'(c)},$$

so we have the result. ■

Corollary 4.3 (Lagrange's Mean-Value Theorem) *If f is differentiable on (a, b) and $x_1, x_2 \in (a, b)$, then f is continuous on the closed interval with endpoints x_1 and x_2 and differentiable on its interior. Hence, the Mean-Value Theorem implies that*

$$f(x_2) - f(x_1) = f'(c)(x_2 - x_1)$$

for some c between x_1 and x_2.

Proof Put $g(x) = x$ in Theorem 4.14, and we get the desired result. ■

4.10 The Consequences of the Mean-Value Problem

Theorem 4.15 *If $f'(x) = 0$ for all x in (a, b), then f is constant on (a, b).*

Proof Let $a < x_1 < x_2 < b$. Then by Corollary 4.3 we see

$$f(x_2) - f(x_1) = (x_2 - x_1) f'(x) = 0,$$

for some x such that $x_1 < x < x_2$. ■

Theorem 4.16 *If f exists as well as does not change the sign on (a, b), and if f is differentiable on the open interval (a, b). Then we have the following:*

(i) If $f'(x) > 0$ for all $x \in (a, b)$, then f is a strictly increasing function on $[a, b]$.
(ii) If $f'(x) < 0$ for all $x \in (a, b)$, then f is a strictly decreasing function on $[a, b]$.

The converses of these two cases are false.

Proof (i) Suppose f is not an increasing function on (a, b), and $a < b$ but $f(a) \geqslant f(b)$. Since f is differentiable on the open interval (a, b), by Theorem 4.13 there exists $c \in (a, b)$ such that

$$f'(c) = \frac{f(b) - f(a)}{b - a}.$$

Since $f(a) \geqslant f(b)$, we have $f(b) - f(a) \leqslant 0$ but $b - a > 0$. We see

$$f'(c) = \frac{f(b) - f(a)}{b - a} \leqslant 0.$$

However, $f'(x) > 0$ for all $x \in (a, b)$. This is a contradiction, and therefore f is an increasing function on (a, b).

(ii) Similarly, we can prove this case.

■

Theorem 4.17 *Let f be differentiable on the open interval (a, b). Then we have the following:*

(i) If $f'(x) \geqslant 0$ for all $x \in (a, b)$, then f is a monotonically increasing function on (a, b).
(ii) If $f'(x) \leqslant 0$ for all $x \in (a, b)$, then f is a monotonically decreasing function on (a, b).

Proof Since f is differentiable over the open interval (a, b), we have the following.

(i) We see $f'(x) \geqslant 0$ for all $x \in (a, b)$. Let $a \leqslant x_1 < x_2 \leqslant b$. By Theorem 4.13, f is differentiable over the open interval (x_1, x_2), then there exists $c \in (x_1, x_2)$ such that

$$f'(c) = \frac{f(x_2) - f(x_1)}{x_2 - x_1} \geqslant 0, \quad (\text{since } f'(c) \geqslant 0).$$

But $x_2 - x_1 > 0$ since $x_2 > x_1$. Thus, $f(x_2) - f(x_1) \geqslant 0$, that is,

$$f(x_2) \geqslant f(x_1).$$

Thus, f is a monotonically decreasing function over (a, b).
(ii) Similarly, we can prove this case.

■

4.11 Taylor's Theorem

We will now discuss Taylor's Theorem which is an extension of the Mean-Value Theorem and it relates to a function and its derivatives. Taylor's Theorem has one of the most important applications in the theory of approximation. Taylor's Theorem provides a way to approximate a function f by a simpler function that is $(n + 1)$ times differentiable in a neighborhood of a point c by a polynomial of degree $< n$ in the powers of $(x - c)$, whose coefficients can be determined by the derivatives $f', f'', \cdots, f^{(n)}$ at c. This polynomial will be called the n-th Taylor polynomial for f at c and will be denoted by $P_n(x)$.

4.11.1 Taylor Polynomials

Suppose that f is differentiable at the point c. Then the equation of the tangent to $y = f(x)$ at the point c is

$$y = f(c) + (x - c) f'(c).$$

Suppose that f is n-times differentiable at the point c. Then the n derivatives at the point c of the polynomial

$$P_n(x) = f(c) + \frac{1}{1!}(x-c) f'(c) + \frac{1}{2!}(x-c)^2 f''(c) + \cdots + \frac{1}{n!}(x-c)^n f^{(n)}(c)$$

are the *same* as those of f, that is,

$$P_n^{(j)}(c) = f^{(j)}(c), \quad j = 0, 1, \cdots, n.$$

The following theorem provides an estimate for the error function

$$E_n(x) = f(x) - P_n(x).$$

Theorem 4.18 (The Taylor's Theorem) *Let $f : [a, b] \to \mathbb{R}$ be such that $f^{(n-1)}$ is continuous on $[a, b]$ and $f^{(n)}$ exists on (a, b). Then there exists $c \in (a, b)$ such that*

$$f(b) = f(a) + \frac{f'(a)}{1!}(b-a) + \frac{f''(a)}{2!}(b-a)^2 + \cdots + \frac{f^{(n-1)}(a)}{(n-1)!}(b-a)^{n-1} + \frac{f^{(n)}(c)}{n!}(b-a)^n.$$

Proof Let the function F be defined by

$$F(x) = f(b) - f(x) - \frac{f'(x)}{1!}(b-x) - \frac{f''(x)}{2!}(b-x)^2 - \cdots - \frac{f^{(n-1)}(x)}{(n-1)!}(b-x)^{n-1}. \tag{4.13}$$

Then

$$F'(x) = -\frac{f^{(n)}(x)}{(n-1)!}(b-x)^{n-1}. \tag{4.14}$$

Define

$$g(x) = F(x) - \left(\frac{b-x}{b-a}\right)^n F(a).$$

It is easy to check that $g(a) = g(b) = 0$ and apply Rolle's Theorem 4.12 to g. There exists some $c \in (a, b)$ such that

$$g'(c) = F'(c) + n\frac{(b-c)^{n-1}}{(b-a)^n} F(a) = 0. \tag{4.15}$$

From (4.14) and (4.15) we obtain that

$$\frac{(b-c)^{n-1}}{(n-1)!}f^{(n)}(c) = n\frac{(b-c)^{n-1}}{(b-a)^n}F(a),$$

that is,

$$F(a) = \frac{(b-a)^n}{n!}f^{(n)}(c). \tag{4.16}$$

Putting $x = a$ in (4.13) and substituting the value of $F(a)$ as given in (4.16), we get the desired result. ■

We notice that Taylor's Theorem is a generalization of Lagrange's Mean-Value Theorem 4.13. If $f''(x)$ exists on (a, b), then for $a < x < b$, we have

$$f(x) = f(a) + \frac{f'(a)}{1!}(x-a) + \frac{f''(c)}{2!}(x-a)^2, \quad a < c < x. \tag{4.17}$$

The right-hand side of (4.17) is a polynomial of degree two.

4.12 The L'Hopital's Rule

Now we discuss a very useful device for finding the limit of a ratio of functions, known as L'Hopital's Rule. L'Hopital's Rule is a general method of evaluating indeterminate forms such as $\frac{0}{0}$ or $\frac{\infty}{\infty}$. We need to differentiate the numerator as well as the denominator and then take the limit.

Theorem 4.19 (The L'Hopital's Rule: 0/0 Form) *Suppose that f and g are differentiable in a deleted neighborhood I of a. If*

(i) $\lim_{x\to a} f(x) = 0$,
(ii) $\lim_{x\to a} g(x) = 0$,
(iii) for every $x \in I,\ g'(x) \neq 0$*, and*
(iv) $\lim_{x\to a} \frac{f'(x)}{g'(x)}$ *exists,*

then

$$\lim_{x\to a}\frac{f(x)}{g(x)} = \lim_{x\to a}\frac{f'(x)}{g'(x)}.$$

Proof We extend the definitions of f and g, i.e., f and g are defined at $x = a$ by $f(a) = g(a) = 0$. From assumptions (i) and (ii), these results in continuous functions are defined on the neighborhood $I \cup \{a\}$ of the point $x = a$ from Cauchy's Mean-Value Theorem 4.2.

Suppose $x \in I$ and $a < x$. By Theorem 4.14 there exists $a < c < x$ such that

$$[f(x) - f(a)]\, g'(c) = [g(x) - g(a)]\, f'(c). \tag{4.18}$$

Since $f(a) = g(a) = 0$,

$$f(x)g'(c) = g(x)f'(c), \quad \text{for } x > a \text{ in } I. \tag{4.19}$$

We can express (4.19) in the form

$$\frac{f(x)}{g(x)} = \frac{f'(c)}{g'(c)}. \tag{4.20}$$

Now $x \to a,\ c \to a$, since $a < c < x$. Therefore

$$\lim_{x\to a} \frac{f(x)}{g(x)} = \lim_{c\to a} \frac{f'(c)}{g'(c)} = \lim_{x\to a} \frac{f'(x)}{g'(x)}.$$

Thus,

$$\lim_{x\to a} \frac{f(x)}{g(x)} = \lim_{x\to a} \frac{f'(x)}{g'(x)}.$$

Similarly, we have the result for $a > x$.

Alternate Proof. If $f,\ g,\ f'$ and g' are all continuous over an open interval containing a. Since $\lim\limits_{x\to a} f(x) = 0 = \lim\limits_{x\to a} g(x)$ and both f and g are continuous at a, we have $f(a) = 0 = g(a)$. Therefore

$$\begin{aligned}
\lim_{x\to a} \frac{f(x)}{g(x)} &= \lim_{x\to a} \frac{f(x) - f(a)}{g(x) - g(a)} \\
&= \lim_{x\to a} \frac{\frac{f(x)-f(a)}{x-a}}{\frac{g(x)-g(a)}{x-a}} \\
&= \frac{\lim\limits_{x\to a} \frac{f(x)-f(a)}{x-a}}{\lim\limits_{x\to a} \frac{g(x)-g(a)}{x-a}} && \text{(by limit of a quotient)} \\
&= \frac{f'(a)}{g'(a)} && \text{(by definition of the derivative)} \\
&= \frac{\lim\limits_{x\to a} f'(x)}{\lim\limits_{x\to a} g'(x)} && \text{(by continuity of } f' \text{ and } g'\text{)} \\
&= \lim_{x\to a} \frac{f'(x)}{g'(x)}. && \text{(by limit of a quotient)}
\end{aligned}$$

■

Example 4.14 Evaluate

$$\lim_{x \to 0} \frac{1 + x - e^x}{x^2}.$$

Let $f(x) = 1 + x - e^x$ and $g(x) = x^2$. Then $\lim\limits_{x \to a} \frac{f(x)}{g(x)}$ and $\lim\limits_{x \to a} \frac{f'(x)}{g'(x)}$ give 0/0. A second attempt at application of the theorem shows that

$$\lim_{x \to 0} \frac{f''(x)}{g''(x)} = \lim_{x \to 0} = \frac{-e^x}{2} = -\frac{1}{2}.$$

Hence $\lim\limits_{x \to 0} \frac{f''(x)}{g''(x)} = -\frac{1}{2}$, so that $\lim\limits_{x \to 0} \dfrac{f(x)}{g(x)} = -\dfrac{1}{2}$.

Example 4.15 Evaluate

$$\lim_{x \to 0} \frac{\log(1 + 4x)}{x}.$$

Let $f(x) = \log(1 + 4x)$ and $g(x) = x$. Then

$$\lim_{x \to 0} f(x) = \lim_{x \to 0} g(x) = 0, \ f'(x) = \frac{4}{1 + 4x}, \text{ and } g'(x) = 1.$$

Hence

$$\lim_{x \to 0} \frac{\log(1 + 4x)}{x} = \lim_{x \to 0} \frac{4}{1 + 4x} = 4.$$

Corollary 4.4 *The L'Hopital's Rule holds for two-sided (limits from the left as well as limits from the right). If the two-sided limits are assumed to exist in the hypotheses of Theorem 4.19, then we may conclude the existence of the two-sided limit of f/g.*

Corollary 4.5 *For limits as $x \to -\infty$ or $+\infty$, a theorem which is similar to Theorem 4.19 holds.*

Proof Suppose that

$$\lim_{x \to +\infty} f(x) = \lim_{x \to +\infty} g(x) = 0,$$

and

$$\lim_{x \to +\infty} \frac{f'(x)}{g'(x)} = l, \ g'(x) \neq 0 \ \text{ on } I.$$

Let $t = 1/x$. Define the function F and G by

$$F(t) = f\left(\frac{1}{t}\right), \ G(t) = g\left(\frac{1}{t}\right).$$

Then

$$F'(t) = -\frac{1}{t^2} f'\left(\frac{1}{t}\right), \ G'(t) = -\frac{1}{t^2} g'\left(\frac{1}{t}\right).$$

Hence

$$\frac{F'(t)}{G'(t)} \to l, \ F(t) \to 0 \text{ and } G(t) \to 0, \text{ as } t \to 0^+.$$

Thus we apply Theorem 4.19 to $\frac{F(t)}{G(t)}$, and the result is established. The argument, when $x \to -\infty$ or $+\infty$, is similar. ■

We can also use the L'Hopital's Rule to evaluate the limits of quotients $\frac{f(x)}{g(x)}$ in which $f(x) \to \infty$ and $g(x) \to \infty$. Limits of this form are classified as indeterminate forms of type $\frac{\infty}{\infty}$. Actually, we are not dividing ∞ by ∞ that is impossible, since ∞ is not a real number, rather, $\frac{\infty}{\infty}$ is used to represent a quotient of limits, each of which is $-\infty$ or $+\infty$.

Theorem 4.20 (The L'Hopital's Rule: ∞/∞Form) *Suppose that f and g are differentiable in a deleted neighborhood I of c. If*

(i) $\lim_{x\to a} f(x) = \infty$,
(ii) $\lim_{x\to a} g(x) = \infty$,
(iii) for every $x \in I$, $g'(x) \neq 0$, *and*
(iv) $\lim_{x\to a} \frac{f'(x)}{g'(x)}$ *exists, then*

$$\lim_{x\to a} \frac{f(x)}{g(x)} = \lim_{x\to a} \frac{f'(x)}{g'(x)}.$$

For $a = \pm\infty$ *or* $\lim_{x\to a} g(x) = -\infty$, *the analogous statements are valid.*

Theorem 4.20 holds for limits from the left as well as limits from the right. The proof is more complicated than the proof of Theorem 4.19, although it is the same in principle. We omit the details.

4.13 Exercises

1. Find the limit of the following
   (i) $\lim_{x\to 0} x^2 \cos(\frac{1}{x})$ (ii) $\lim_{x\to 0} x \sin(\frac{1}{x})$ (iii) $\lim_{x\to\infty} x \sin(\frac{1}{x})$.
2. Find the limit of $f(x) = \frac{x^2-4}{x-2}$ by using the $\epsilon - \delta$ definition of the limit.

3. $f(x) = x/(2x + |x|),\ x \neq 0$. Find the left and right limits of f at $x = 0$.
4. Let k be a natural number. Prove that

$$\lim_{x \to 1} \frac{x^k - 1}{x - 1} = k.$$

5. For each number x, define $f(x)$ to be the largest integer that is less than or equal to x. Graph the function $f : \mathbb{R} \to \mathbb{R}$. Given a number x_0, examine $\lim_{x \to x_0} f(x)$.
6. Let $f(x) = (1 - \cos \alpha x)/(x \sin x)$ when $x \neq 0$, and let $f(x) = 1/2$ when $x = 0$. Find the value of α as $f(x)$ is continuous at $x = 0$.
7. Prove that $f(x) = x - [x],\ x \in \mathbb{R}$ is discontinuous at all integer points, where $[x]$ is the greatest integer $\leqslant x$.
8. A function $f(x) = |x - 1| + |x| + |x + 1|$ is defined in $(0, 2)$, then which of the following is true? The function $f(x)$

   (i) is continuous at $x = 1$.
   (ii) has a removable discontinuity at $x = 1$.
   (iii) has a jump discontinuity at $x = 1$.
   (iv) may not be continuous at $x = 1$.

9. Examine the validity of hypothesis and the conclusion of Rolle's Theorem and Lagrange's Mean-Value Theorem.

   (i) $|x|$ on $[-1, 1]$.
   (ii) $\log x$ on $[1/2, 2]$.
   (iii) $x^{\frac{1}{3}},\ x \in [-1, 1]$.

10. Let a function $f(x) = x^2 \sin(1/x)$, for $x \neq 0$, and let $f(x) = 0$ for $x = 0$. Then, show that $f(x)$ is differentiable at $x = 0$ but $\lim_{x \to 0} f'(x) = f'(0)$.
11. Show that the function $f(x) = |x| + |x - 1|$ is not differentiable at exactly two points.
12. Let $f(x) = 3x$ for $x \in \mathbb{Q}$ and $f(x) = 0$. If $x \in \mathbb{Q}^c$, then show that f is continuous only at $x = 0$.
13. Suppose that the function $g : \mathbb{R} \to \mathbb{R}$ is differentiable at $x = 0$. Also suppose that for each natural number n, $g(\frac{1}{n}) = 0$. Prove that $g(0) = 0$ and $g'(0) = 0$.
14. Let $f : \mathbb{R} \to \mathbb{R}$ be continuous. Assume that $f(r) = 0$ for $r \in \mathbb{Q}$. Show that $f = 0$.
15. Let $f : \mathbb{R} \to \mathbb{R}$ be differentiable. Let $n \in \mathbb{N}$. Fix $a \in \mathbb{R}$. Find

$$\lim_{x \to a} \frac{a^n f(x) - x^n f(a)}{x - a}.$$

16. If f is continuous at x_0 and $f(x_0) > \mu$, then $f(x) > \mu$ for all x in some neighborhood of x_0.

    (a) State a result analogous to the above for the case where $f(x_0) < \mu$.
    (b) Prove: If $f(x) \leqslant \mu$ for all x in S and x_0 is a limit point of S at which f is continuous, then $f(x_0) \leqslant \mu$.
    (c) State results analogous to (a), (b) for the case where f is continuous from the right or left at x_0.

Chapter 5
Metric Spaces

Roughly speaking, a metric space is any set provided with a sensible notion of the "distance" between points.

In mathematics, a metric space is a set where a notion of distance (called a metric) between elements of the set is defined. A set with a metric is called a metric space.

The ways in which distance is measured and the sets involved may be very diverse. For instance, in real numbers, the distance is measured by the absolute difference of any two real values x and y. Whereas in normed linear spaces, the distance is measured in terms of "norm," represented by $\|x\|$. This simply means "the length of x." Now, we can again write that $\|x - y\|$ is the distance from x to y. There are many other kinds of "distance" measurements in real life. For example, the set could be the sphere, and we could measure distance either along great circles or along straight lines through the globe. If you want to travel from New Delhi to Tokyo, the distance is not $\|$New Delhi $-$ Tokyo$\|$, because you cannot fly through the mantle of the earth. Instead, you must go around, and the "true" distance is the length of the great circle segment joining New Delhi and Tokyo. In this example, we have defined a "distance function" which is different from New Delhi to Tokyo. We take any two points x, y on earth, and then we give $dist(x, y) = \|x - y\|$ as the length of the shortest great circle segment from x to y.

5.1 Definitions of Metric Spaces and Examples

"The metric space is the generalization of absolute-value function or the concept of distance."

Now, the geometric interpretation of $|x - y|$ is the distance from x to y, for $x, y \in \mathbb{R}$. We define the "distance function" ρ by

$$\rho\,(x, y) = |x - y| \quad (x, y \in \mathbb{R}).$$

N. Deo and R. Sakai, *Introduction to Mathematical Analysis*,
https://doi.org/10.1007/978-981-97-6568-3_5

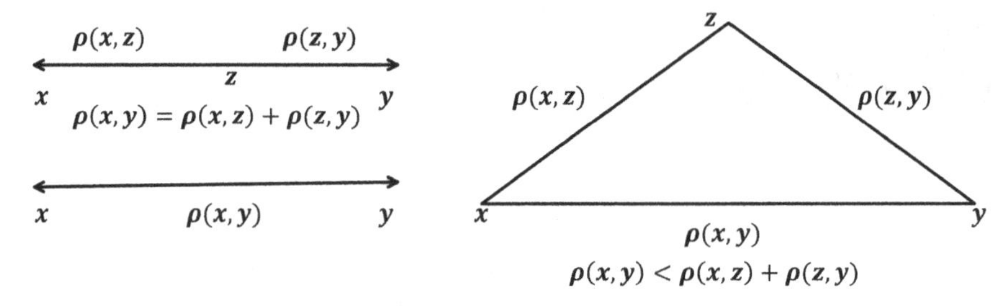

Fig. 5.1 Geometric interpretation of metric on a plane

The properties of the absolute-value function can be expressed in the "distance function" form as follows:

(i) $|0| = 0 \Longrightarrow \rho(x, x) = 0$, i.e., the distance from a point to itself is 0.
(ii) $|a| > 0 \, (a \in \mathbb{R}, a \neq 0) \Longrightarrow \rho(x, y) > 0 \, (x \neq y)$, i.e., the distance between two distinct points is strictly positive.
(iii) $|a| = |-a| \Longrightarrow \rho(x, y) = \rho(y, x)$, i.e., the distance from x to y is equal to the distance from y to x.
(iv) $|a + b| \leqslant |a| + |b| \Longrightarrow \rho(x, y) \leqslant \rho(x, z) + \rho(z, y)$, i.e., the triangle inequality.

The following definition is a consequence of properties (i)-(iv).

Definition 5.1 Let X be a non-empty set. A function ρ defined on $X \times X$ into $\mathbb{R}$ $(\rho : X \times X \to \mathbb{R})$ is said to be a metric on X if it satisfies the following conditions:

$(M1)$ $\rho(x, X) = 0, \;$ if $\forall x \in X$; *(by definiteness)*
$(M2)$ $\rho(x, y) > 0, \;$ if $x \neq y, \; \forall x, y \in X$; *(by positivity)*
$(M3)$ $\rho(x, y) = \rho(y, x), \; \forall x, y \in X$; *(by symmetry)*
$(M4)$ $\rho(x, y) \leqslant \rho(x, z) + \rho(z, y), \; \forall x, y, z \in X$; *(by triangle inequality).*

If ρ is a metric for X, then the ordered pair (X, ρ) is called a *metric space*. A metric ρ is also called a distance function.

The meaning of the first axiom $(M1)$; the distance from a point to itself is zero.

The meaning of the second axiom $(M2)$; the distance between the two distinct points is strictly positive.

The meaning of the third axiom $(M3)$; the distance does not depend on the order of the points x and y.

The meaning of the fourth axiom $(M4)$; the sums of the lengths of the two sides of the triangle are greater than or equal to the length of the third side (Fig. 5.1).

To illustrate the concept of a metric space and the process of verifying the axioms of a metric, in particular, the triangle inequality $(M4)$, we give another example (space l^p), is the most important one of them in applications.

5.1.1 The l^p-Spaces

Let $p \geqslant 1$ be a fixed real number. By definition, each element in the space l^p is a sequence $x = \{\xi_i\} = \{\xi_1, \xi_2, \cdots\}$ of numbers such that $|\xi_1|^p + |\xi_2|^p + \cdots$ converges; thus

$$\sum_{i=1}^{\infty} |\xi_i|^p < \infty, \tag{5.1}$$

and the metric is defined by

$$\rho(x, y) = \left(\sum_{i=1}^{\infty} |\xi_i - \eta_i|^p \right)^{1/p}, \tag{5.2}$$

where $y = \{\eta_i\}$ and $\sum_{i=1}^{\infty} |\eta_i|^p < \infty$. If we take only real sequences satisfying (5.1), we get the real space l^p. In the case $p = 2$ we have the famous *Hilbert sequence space* l^2 with the metric defined as

$$\rho(x, y) = \sqrt{\sum_{i=1}^{\infty} |\xi_i - \eta_i|^2}. \tag{5.3}$$

Now we prove that l^p-space is a metric space. Clearly, (5.2) satisfies the first three properties of Metric Spaces, provided that the series on the right converges. We will prove that it converges and that $(M4)$ is satisfied. Proceeding step-wise, we will derive the following $(i) - (iv)$.

(i) **The Auxiliary Inequality**: Let $p > 1$ and q be defined as

$$\frac{1}{p} + \frac{1}{q} = 1, \tag{5.4}$$

where p and q are called *conjugate exponents*. From (5.4) we have

$$\frac{p+q}{pq} = 1 \implies pq = p + q \implies (p-1)(q-1) = 1. \tag{5.5}$$

Thus

$$q - 1 = \frac{1}{p-1} \implies q = \frac{p}{p-1}, \tag{5.6}$$

so that

$$v = u^{p-1} \implies u = v^{q-1}.$$

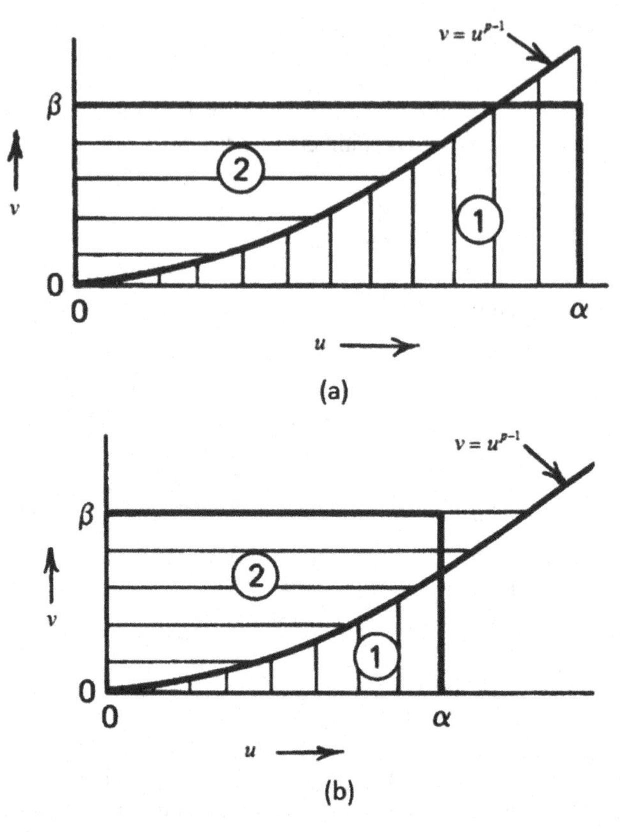

Fig. 5.2 The comparison of Area ① corresponds to the first integral and Area ② to the second integral of inequality (5.7)

Let α and β be any positive numbers. Since $\alpha\beta$ is the area of the rectangle (see the above (Fig. 5.2), we obtain the following inequality.

$$\alpha\beta \leqslant \int_0^{\alpha} u^{p-1}du + \int_0^{\beta} v^{q-1}dv = \frac{\alpha^p}{p} + \frac{\beta^q}{q}. \tag{5.7}$$

The graphical interpretation of inequality (5.7), the area of the entire shaded region is ① + ② $= \frac{\alpha^p}{p} + \frac{\beta^q}{q}$, which is always larger than the area of the rectangle $\alpha\beta$.

(ii) **The Hölder's Inequality:** Let

$$\{\xi_i\} = \{\xi_1, \xi_2, \cdots\} \quad and \quad \{\eta_i\} = \{\eta_1, \eta_2, \cdots\}$$

be two sequences.

$$S = \left(|\xi_1|^p + |\xi_2|^p + \cdots\right)^{1/p} = \left(\sum |\xi_i|^p\right)^{1/p}$$

$$\textit{and} \quad T = \left(|\eta_1|^q + |\eta_2|^q + \cdots\right)^{1/q} = \left(\sum |\eta_i|^q\right)^{1/q}. \tag{5.8}$$

Putting

$$\alpha = \frac{|\xi_i|}{S} \textit{ and } \beta = \frac{|\eta_i|}{T}$$

in (5.7), we obtain

$$\frac{|\xi_i|}{S}\frac{|\eta_i|}{T} \leqslant \frac{1}{p}\left(\frac{|\xi_i|}{S}\right)^p + \frac{1}{q}\left(\frac{|\eta_i|}{T}\right)^q. \tag{5.9}$$

Adding inequality (5.9), and using notation $\sum$ and from (5.8) and (5.4), we have

$$\begin{aligned}\sum \frac{|\xi_i \eta_i|}{ST} &\leqslant \frac{1}{p}\sum\left(\frac{|\xi_i|}{S}\right)^p + \frac{1}{q}\sum\left(\frac{|\eta_i|}{T}\right)^q \\ &= \frac{1}{p} + \frac{1}{q}.\end{aligned} \tag{5.10}$$

Finally, (5.10) means

$$\sum |\xi_i \eta_i| \leqslant ST, \tag{5.11}$$

which is Hölder's inequality. We write Hölder's inequality in the usual form

$$\sum |\xi_i \eta_i| \leqslant \left(\sum |\xi_i|^p\right)^{1/p}\left(\sum |\eta_i|^q\right)^{1/q}. \tag{5.12}$$

Let

$$S = \left(\int |f(t)|^p dt\right)^{1/p} \textit{ and } \quad T = \left(\int |g(t)|^q dt\right)^{1/q}, \tag{5.13}$$

and

$$\alpha = \frac{|f(t)|}{S} \textit{ and } \beta = \frac{|g(t)|}{T}$$

and proceed as (5.12). Then we get

$$\int |f(t)g(t)|dt \leqslant ST. \tag{5.14}$$

Thus

$$\int |f(t)g(t)|dt \leqslant \left(\int |f(t)|^p dt\right)^{1/p}\left(\int |g(t)|^q dt\right)^{1/q}. \tag{5.15}$$

If $p = 2$, then $q = 2$, and (5.12) yields the **Cauchy-Schwarz inequality** for sums

$$\sum |\xi_i \eta_i| \leqslant \left(\sum |\xi_i|^2\right)^{1/2} \left(\sum |\eta_i|^2\right)^{1/2}. \tag{5.16}$$

Again if $p = 2$, then $q = 2$, and Hölder's Inequality (5.15) becomes the **Cauchy-Schwarz inequality** for integral

$$\int |f(t)g(t)|dt \leqslant \left(\int |f(t)|^2 dt\right)^{1/2} \left(\int |g(t)|^2 dt\right)^{1/2}. \tag{5.17}$$

(iii) **The Minkowski's Inequality:** In the year 1896, for finite sums the following inequality was given by H. Minkowski:

$$\left(\sum |\xi_i + \eta_i|^p\right)^{1/p} \leqslant \left(\sum |\xi_i|^p\right)^{1/p} + \left(\sum |\eta_i|^p\right)^{1/p}. \tag{5.18}$$

For $p = 1$ the inequality follows readily from the triangle inequality for numbers. For any real number $p > 1$ we may write

$$\begin{aligned}\sum |\xi_i + \eta_i|^p &= \sum |\xi_i + \eta_i||\xi_i + \eta_i|^{p-1} \\ &\leqslant \sum |\xi_i||\xi_i + \eta_i|^{p-1} + \sum |\eta_i||\xi_i + \eta_i|^{p-1}.\end{aligned} \tag{5.19}$$

Applying to the Hölder's Inequality (5.12) and from (5.6), we have

$$\begin{aligned}\sum |\xi_i + \eta_i|^p &\leqslant \left(\sum |\xi_i|^p\right)^{1/p} \left(\sum |\xi_i + \eta_i|^{(p-1)q}\right)^{1/q} \\ &\quad + \left(\sum |\eta_i|^p\right)^{1/p} \left(\sum |\xi_i + \eta_i|^{(p-1)q}\right)^{1/q} \\ &\leqslant \left[\left(\sum |\xi_i|^p\right)^{1/p} + \left(\sum |\eta_i|^p\right)^{1/p}\right] \left(\sum |\xi_i + \eta_i|^p\right)^{1/q}.\end{aligned}$$

Therefore

$$\left(\sum |\xi_i + \eta_i|^p\right)^{1-1/q} \leqslant \left(\sum |\xi_i|^p\right)^{1/p} + \left(\sum |\eta_i|^p\right)^{1/p}.$$

Thus

$$\left(\sum |\xi_i + \eta_i|^p\right)^{1/p} \leqslant \left(\sum |\xi_i|^p\right)^{1/p} + \left(\sum |\eta_i|^p\right)^{1/p}.$$

(iv) **The Triangle Inequality:** From (5.18) it follows that for $x, y \in l^p$ the series in (5.2) converges. Minkowski's Inequality (5.18) also yields the triangle inequality. In fact, taking any $x, y, z \in l^p$, writing $z = \{\zeta_i\}$ and using the triangle inequality for numbers, from Minkowski's Inequality (5.18), we have

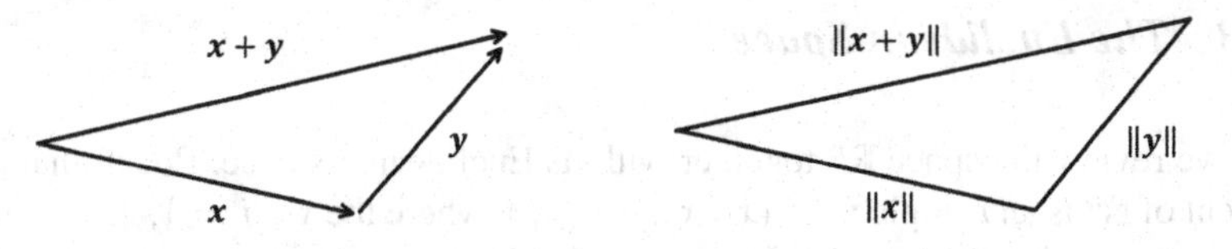

Fig. 5.3 Illustration of triangle inequality (N4)

$$
\begin{aligned}
\rho(x, y) &= \left(\sum |\xi_i - \eta_i|^p\right)^{1/p} \\
&\leqslant \left(\sum \{|\xi_i - \zeta_i| + |\zeta_i - \eta_i|\}^p\right)^{1/p} \\
&\leqslant \left(\sum |\xi_i - \zeta_i|^p\right)^{1/p} + \left(\sum |\zeta_i - \eta_i|^p\right)^{1/p} \\
&= \rho(x, z) + \rho(z, y).
\end{aligned}
$$

Therefore, l^p is a metric space.

5.1.2 Normed Linear Spaces

A norm on a linear space X is a real-valued function, whose value at an $x \in X$ is denoted by $\|x\|$, with the properties:

$(N1)$ $\|x\| = 0$, if $x = 0$.
$(N2)$ $\|x\| > 0$, if $x \neq 0$.
$(N3)$ $\|\alpha x\| = |\alpha| \, \|x\|$, $\alpha \in \mathbb{R}$.
$(N4)$ $\|x + y\| \leqslant \|x\| + \|y\|$, $\forall x, y \in X$ (triangle inequality),

where x and y are arbitrary real valued in X. A normed linear space is also called a normed vector space.

A norm on X defines a metric ρ on X as follows:

$$
\rho(x, y) = \|x - y\|, \quad x, y \in X \tag{5.20}
$$

and is called the metric induced by the norm. The normed space just defined is denoted by $(X, \|\cdot\|)$ or simply by X.

The meanings of properties $(N1)$ and $(N2)$ state that all vectors have positive lengths except the zero vector which has length zero. $(N3)$ means that when a vector is multiplied by a scalar, its length is multiplied by the absolute value of the scalar. $(N4)$ means that the length of one side of a triangle cannot exceed the sum of the lengths of the two other sides (Fig. 5.3).

Thus, the normed linear space is a metric space.

5.1.3 The Euclidean Space

Now we review the space $\mathbb{R}^n$ together with its Euclidean distance. Recall that each element of $\mathbb{R}^n$ is an n-tuple $\mathbf{x} = (x_1, x_2, \cdots, x_n)$, where the x_i, $i = 1, 2, \cdots, n$ are real numbers. The elements of $\mathbb{R}^n$ are called points or vectors, and we are familiar with the operations like addition of vectors and multiplication by scalars.

The most familiar metric space is the Euclidean space $\mathbb{R}^n$, which is defined as: The Euclidean metric is the function $\rho : \mathbb{R}^n \times \mathbb{R}^n \to \mathbb{R}$ that assigns to any two vectors in Euclidean n-space $\mathbf{x} = (x_1, x_2, \cdots, x_n)$ and $\mathbf{y} = (y_1, y_2, \cdots, x_n)$. The distance between $\mathbf{x}$ and $\mathbf{y}$ can be calculated using the standard formula.

$$\begin{aligned}\rho(\mathbf{x}, \mathbf{y}) &= \rho((x_1, x_2, \cdots, x_n), (y_1, y_2, \cdots, y_n)) \\ &= \sqrt{(x_1 - y_1)^2 + \cdots + (x_n - y_n)^2} \\ &= \sqrt{\sum_{i=1}^{n} (x_i - y_i)^2}.\end{aligned}$$

Example 5.1 The Euclidean space $\mathbb{R}^n$, especially $\mathbb{R}^1 = \mathbb{R}$ (the real line). The function $\rho : \mathbb{R} \times \mathbb{R} \to \mathbb{R}$ defined by

$$\rho(\mathbf{x}, \mathbf{y}) = |x - y|, \quad \forall x, y \in \mathbb{R},$$

is a metric on the set $\mathbb{R}$ of all real numbers. This metric ρ is called the usual metric .

Example 5.2 In the coordinate plane, let $X = \mathbb{R}^2$(Euclidean plane) be the set of all points. For $\mathbf{x} = (x_1, x_2)$ and $\mathbf{y} = (y_1, y_2)$ in $\mathbb{R}^2$, ρ defined as follows:

(i) $\rho(\mathbf{x}, \mathbf{y}) = \sqrt{(x_1 - y_1)^2 + (x_2 - y_2)^2}$,
(ii) $\rho^*(\mathbf{x}, \mathbf{y}) = \max\{|x_1 - y_1|, |x_2 - y_2|\}$,
(iii) $\rho^{**}(\mathbf{x}, \mathbf{y}) = |x_1 - y_1| + |x_2 - y_2|$,

then all the spaces $(\mathbb{R}^2, \rho)$, $(\mathbb{R}^2, \rho^*)$, $(\mathbb{R}^2, \rho^{**})$ are metric spaces (Fig. 5.4).

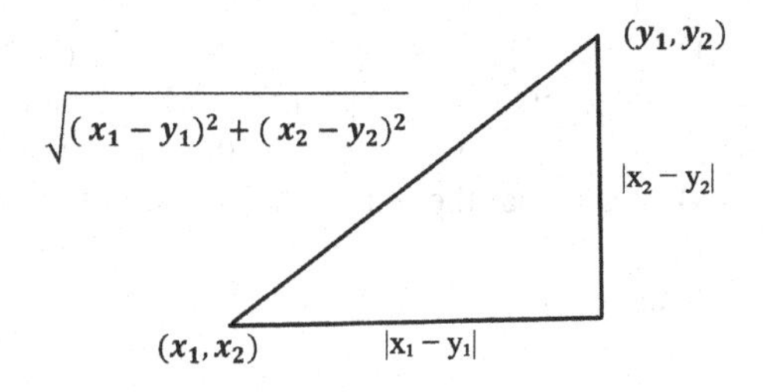

Fig. 5.4 Illustration of Euclidean metrics on the plane

Example 5.3 Show that the function ρ defined by

$$\rho(\mathbf{x}, \mathbf{y}) = \sqrt{\sum_{i=1}^{n} (x_i - y_i)^2}, \quad \forall \mathbf{x} = (x_1, x_2, \cdots, x_n), \mathbf{y} = (y_1, y_2, \cdots, y_n) \in \mathbb{R}^n.$$

in the set $\mathbb{R}^n$ of all ordered n-tuples is a metric space $(\mathbb{R}^n, \rho)$.

$(M1)$ $\rho(\mathbf{x}, \mathbf{y}) \geqslant 0, \quad \forall \mathbf{x}, \mathbf{y} \in \mathbb{R}^n.$
$(M2)$ $\rho(\mathbf{x}, \mathbf{y}) = 0 \Longleftrightarrow \mathbf{x} = \mathbf{y} \quad \forall \mathbf{x}, \mathbf{y} \in \mathbb{R}^n.$

$$\rho(\mathbf{x}, \mathbf{y}) = 0 \Longleftrightarrow \sqrt{\sum_{i=1}^{n} (x_i - y_i)^2} = 0 \Longleftrightarrow \sum_{i=1}^{n} (x_i - y_i)^2 = 0$$
$$\Longleftrightarrow x_i = y_i, \ i = 1, 2, \cdots, n \Longleftrightarrow \mathbf{x} = \mathbf{y}.$$

$(M3)$ $\rho(\mathbf{x}, \mathbf{y}) = \rho(\mathbf{y}, \mathbf{x}), \quad \forall \mathbf{x}, \mathbf{y} \in \mathbb{R}^n.$

$$\rho(\mathbf{x}, \mathbf{y}) = \sqrt{\sum_{i=1}^{n} (x_i - y_i)^2} = \sqrt{\sum_{i=1}^{n} (y_i - x_i)^2} = \rho(\mathbf{y}, \mathbf{x}).$$

$(M4)$ $\rho(\mathbf{x}, \mathbf{y}) \leqslant \rho(\mathbf{x}, \mathbf{z}) + \rho(\mathbf{z}, \mathbf{y}), \quad \forall \mathbf{x}, \mathbf{y}, \mathbf{z} \in \mathbb{R}^n.$
By using the *Cauchy-Schwarz inequality* (5.16), we prove this condition

$$\left[\rho(\mathbf{x}, \mathbf{z}) + \rho(\mathbf{z}, \mathbf{y})\right]^2 = \left[\sqrt{\sum_{i=1}^{n} (x_i - z_i)^2} + \sqrt{\sum_{i=1}^{n} (z_i - y_i)^2}\right]^2$$
$$= \sum_{i=1}^{n} (x_i - z_i)^2 + \sum_{i=1}^{n} (z_i - y_i)^2 + 2\sqrt{\sum_{i=1}^{n} (x_i - z_i)^2}\sqrt{\sum_{i=1}^{n} (z_i - y_i)^2}$$
$$\geqslant \sum_{i=1}^{n} (x_i - z_i)^2 + \sum_{i=1}^{n} (z_i - y_i)^2 + 2\sum_{i=1}^{n} (x_i - z_i)(z_i - y_i)$$
$$= \sum_{i=1}^{n} [(x_i - z_i) + (z_i - y_i)]^2$$
$$= \sum_{i=1}^{n} (x_i - y_i)^2 = \left[\rho(\mathbf{x}, \mathbf{y})\right]^2.$$

We can prove $(M4)$ in other ways with the help of *Minkowski's Inequality* (5.18) for $p = 2$, define

$$\begin{aligned}\rho(\mathbf{x}, \mathbf{y}) &= \left\{\sum_{i=1}^{n}(x_i - y_i)^2\right\}^{1/2} \\ &= \left\{\sum_{i=1}^{n}(x_i - z_i + z_i - y_i)^2\right\}^{1/2} \\ &\leqslant \left\{\sum_{i=1}^{n}(x_i - z_i)^2\right\}^{1/2} + \left\{\sum_{i=1}^{n}(z_i - y_i)^2\right\}^{1/2} \\ &= \rho(\mathbf{x}, \mathbf{z}) + \rho(\mathbf{z}, \mathbf{y}).\end{aligned}$$

Example 5.4 Let X be an arbitrary non-empty set, and let the function ρ be defined by

$$\rho(x, y) = \begin{cases} 0, & \text{if } x = y, \\ 1, & \text{if } x \neq y. \end{cases} \tag{5.21}$$

The metric ρ is called the discrete (trivial) metric on X. The space (X, ρ) is called the discrete metric space or the trivial metric space.

The geometric interpretation of the *discrete metric* is that all points are equally discrete each other. When X is a finite set, we can draw a diagram, see the following figure (Fig. 5.5).

Sampling a discrete metric space {a,b,c,d,e}, the distance between any two points is 1.

Let $\rho : X \times X \to \mathbb{R}$ be defined by (5.21), we have

$(M1)$ $\rho(x, y) \geqslant 0,\ \forall x, y \in X.$

If $x = y$, then $\rho(x, y) = 0$ and if $x \neq y$, then $\rho(x, y) = 1$

$$\therefore \rho(x, y) \geqslant 0,\ \forall x, y \in X.$$

$(M2)$ $\rho(x, y) = 0 \Longleftrightarrow x = y,\ \forall x, y \in X.$

If $x = y$, then by definition of ρ, $\rho(x, y) = 0$ and if $\rho(x, y) = 0$, then $x = y$

$$\therefore \rho(x, y) = 0 \Longleftrightarrow x = y,\ \forall x, y \in X.$$

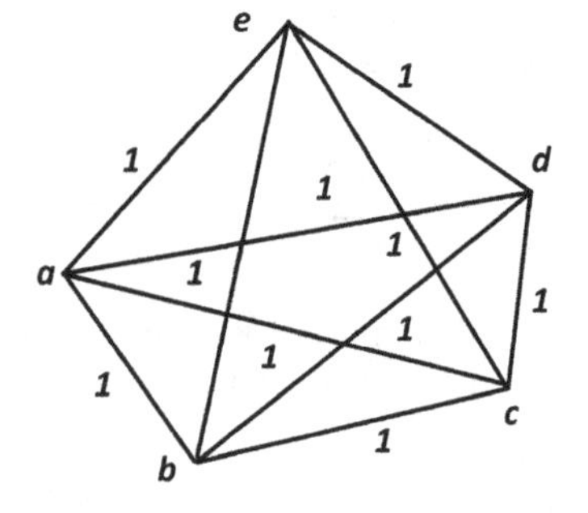

Fig. 5.5 Geometric interpretation of discrete metric

$(M3)$ $\rho(x, y) = \rho(y, x)$, $\forall x, y \in X$.
If $x = y$, then $\rho(x, y) = 0$. Again $x = y \Longrightarrow y = x \Longrightarrow \rho(y, x) = 0$

$$\therefore \rho(x, y) = 0 = \rho(y, x).$$

If $x \neq y$, then $\rho(x, y) = 1$. Again $x \neq y \Longrightarrow y \neq x \Longrightarrow \rho(y, x) = 1$

$$\therefore \rho(x, y) = 1 = \rho(y, x).$$

$(M4)$ $\rho(x, y) \leqslant \rho(x, z) + \rho(z, y)$, $\forall x, y, z \in X$.

(i) If all three x, y, z are equal, that is, $x = y = z$, then $\rho(x, y) = 0$, $\rho(x, z) = 0$, $\rho(z, y) = 0$.

$$\therefore \rho(x, y) = \rho(x, z) + \rho(z, y).$$

(ii) If $x \neq y \neq z$, then $\rho(x, y) = 1$, $\rho(x, z) = 1$, $\rho(z, y) = 1$.

$$\therefore \rho(x, y) < \rho(x, z) + \rho(z, y) \Longrightarrow 1 < 1 + 1.$$

(iii) If $x \neq y, x = z, z \neq y$, then $\rho(x, y) = 1$, $\rho(x, z) = 0$, $\rho(z, y) = 1$.

$$\therefore \rho(x, y) = \rho(x, z) + \rho(z, y) \Longrightarrow 1 = 0 + 1.$$

(iv) If $x \neq y, x \neq z, z = y$, then $\rho(x, y) = 1$, $\rho(x, z) = 1$, $\rho(z, y) = 0$.

$$\therefore \rho(x, y) = \rho(x, z) + \rho(z, y) \Longrightarrow 1 = 1 + 0.$$

Hence, for every condition we have

$$\rho(x, y) \leqslant \rho(x, z) + \rho(z, y).$$

Example 5.5 French train track system is one of the best examples of a metric space. France is a centralized country, therefore every train that goes from one French city to another has to pass through Paris. This is slightly exaggerated, but not too much, as the map shows (Fig. 5.6).

This French railroad network motivates us to construct a French railroad metric. Let (X, ρ) be a metric space (France), and fix $p \in X$ (Paris). Define a new metric ρ_p on X as

$$\rho_p(x, y) = \begin{cases} 0, & x = y, \\ \rho(x, p) + \rho(p, y), & otherwise, \end{cases}$$

for $x, y \in X$. Then $\left(X, \rho_p\right)$ is again a metric space.

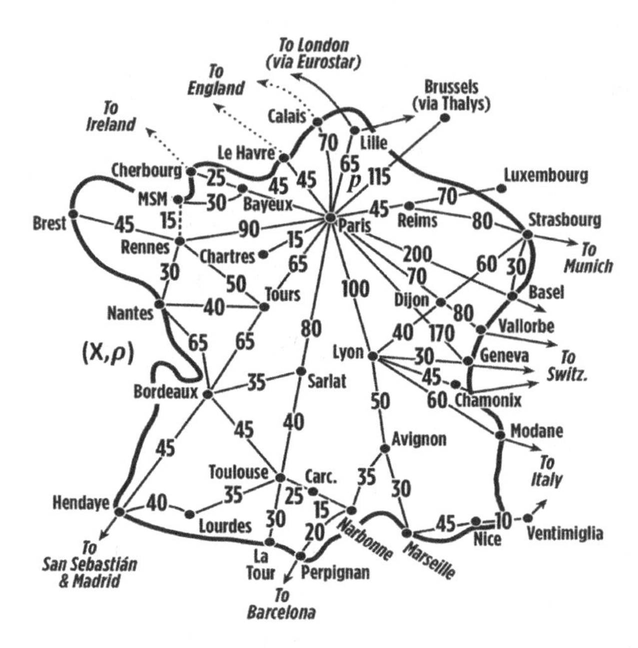

Fig. 5.6 Map of the French railroad network

Example 5.6 In a metric space (X, ρ), let $X = \mathbb{R}^2$ be the set of all points in the coordinate plane. For $x = (x_1, x_2)$ and $y = (y_1, y_2)$ in X, ρ defined as follows:

$$\rho(x, y) = \begin{cases} |x_1 - y_1|, & x_2 = y_2, \\ |x_1| + |y_1| + |x_2 - y_2|, & x_2 \neq y_2. \end{cases}$$

Example 5.7 Let (X, ρ) be any metric space. Then ρ^* defined as:

$$\rho^*(x, y) = \frac{\rho(x, y)}{1 + \rho(x, y)}, \quad \forall x, y \in X.$$

Show that ρ^* is a metric on X.

$(M1)$ $\rho^*(x, y) \geqslant 0, \ \forall x, y \in X.$

$$\text{In fact,} \quad \rho(x, y) \geqslant 0 \text{ and } 1 + \rho(x, y) \geqslant 1,$$

$$\text{therefore} \quad \frac{\rho(x, y)}{1 + \rho(x, y)} \geqslant 0 \implies \rho^*(x, y) \geqslant 0.$$

$(M2)$ $\rho^*(x, y) = 0 \iff x = y, \ \forall x, y \in X.$

$$\text{In fact,} \quad \rho^*(x, y) = 0 \Longleftrightarrow \frac{\rho(x, y)}{1 + \rho(x, y)} = 0 \quad (\text{by Definition of } \rho^*),$$
$$\Longleftrightarrow \rho(x, y) = 0$$
$$\Longleftrightarrow x = y \quad (\text{by Definition 5.1}).$$

$(M3)$ $\rho^*(x, y) = \rho^*(y, x)$, $\forall x, y \in X$.

$$\begin{aligned} \text{In fact,} \quad \rho^*(x, y) &= \frac{\rho(x, y)}{1 + \rho(x, y)} \\ &= \frac{\rho(y, x)}{1 + \rho(y, x)} \quad (\text{since } \rho(x, y) = \rho(y, x)) \\ &= \rho^*(y, x). \end{aligned}$$

$(M4)$ $\rho^*(x, y) \leqslant \rho^*(x, z) + \rho^*(z, y)$, $\forall x, y, z \in X$.

$$\begin{aligned} \text{In fact,} \quad \rho^*(x, y) &= \frac{\rho(x, y)}{1 + \rho(x, y)} = 1 - \frac{1}{1 + \rho(x, y)} \\ &\leqslant 1 - \frac{1}{1 + \rho(x, z) + \rho(z, y)} = \frac{\rho(x, z) + \rho(z, y)}{1 + \rho(x, z) + \rho(z, y)} \\ &= \frac{\rho(x, z)}{1 + \rho(x, z) + \rho(z, y)} + \frac{\rho(z, y)}{1 + \rho(x, z) + \rho(z, y)} \\ &\leqslant \frac{\rho(x, z)}{1 + \rho(x, z)} + \frac{\rho(z, y)}{1 + \rho(z, y)} = \rho^*(x, z) + \rho^*(z, y). \end{aligned}$$

Thus, ρ^* is a metric on X.

Example 5.8 For any metric space (X, ρ), prove that

(i) $\rho(x, y) \geqslant |\rho(x, z) - \rho(z, y)|$.
(ii) $|\rho(x, y) - \rho(x', y')| \leqslant \rho(x, x') + \rho(y, y')$,

where x, x', y and $y' \in X$.

(i) Let $\rho : X \times X \to \mathbb{R}$. Then we have

$$\begin{aligned} \rho(x, z) &\leqslant \rho(x, y) + \rho(y, z) \quad (\text{by Definition 5.1}) \\ &= \rho(x, y) + \rho(z, y) \quad (\text{by Definition 5.1}) \\ \Longrightarrow \rho(x, z) - \rho(z, y) &\leqslant \rho(x, y). \end{aligned} \tag{5.22}$$

Again

$$\begin{aligned} \rho(z, y) &\leqslant \rho(z, x) + \rho(x, y) \quad (\text{by Definition 5.1}) \\ &= \rho(x, z) + \rho(x, y) \quad (\text{by Definition 5.1}) \\ \Longrightarrow \rho(z, y) - \rho(x, z) &\leqslant \rho(x, y). \end{aligned} \tag{5.23}$$

From (5.22) and (5.23), we get

$$\rho(x, y) \geqslant |\rho(x, z) - \rho(z, y)|.$$

(ii) Now

$$\begin{aligned}\left|\rho(x, y) - \rho\left(x', y'\right)\right| &= \left|\rho(x, y) - \rho\left(y, x'\right) + \rho\left(y, x'\right) - \rho\left(x', y'\right)\right| \\ &\leqslant \left|\rho(x, y) - \rho\left(y, x'\right)\right| + \left|\rho\left(y, x'\right) - \rho\left(x', y'\right)\right| \\ &\leqslant \rho\left(x, x'\right) + \rho\left(y, y'\right) \quad \text{(from the above } (i)).\end{aligned}$$

5.1.4 The Pseudo-Metric (The Semi-Metric)

A pseudo-metric space is a generalized metric space in which the distance between two distinct points can be zero. In the same way as every normed space is a metric space, every seminormed space is a pseudo-metric space.

Definition 5.2 A function $\rho : X \times X \to \mathbb{R}$ is called a *pseudo-metric* for X if and only if ρ satisfies the axioms $(M1)$, $(M3)$, $(M4)$ in Definition 5.1 and the following axiom:
$(M'2)$ $\rho(x, x) = 0,\ \forall x \in X$; in other words, if $x = y$, then $\rho(x, y) = 0$ but not conversely, that is, $\rho(x, y) = 0$ need not imply $x = y$.

For example, let us define a mapping $\rho : \mathbb{R} \times \mathbb{R} \to \mathbb{R}$ such that

$$\rho(x, y) = \left|x^2 - y^2\right|,\ \forall x, y \in \mathbb{R},$$

where ρ is a pseudo-metric on $\mathbb{R}$.

$$\begin{aligned}\rho(x, y) = 0 &\Longrightarrow \left|x^2 - y^2\right| = 0 \\ &\Longrightarrow x^2 - y^2 = 0 \\ &\Longrightarrow x = \pm y.\end{aligned}$$

Thus $\rho(x, y) = 0$ does not necessarily imply $x = y$. Hence, $(\mathbb{R}, \rho)$ is a pseudo-metric space, but not a metric space.

Remark 5.1 (i) Every metric is a pseudo-metric but a pseudo-metric is not necessarily a metric.
(ii) $(M3)$ and $(M4)$ imply that both a metric and a pseudo-metric are real positive functions, i.e.,

$$\rho(x, x) \leqslant \rho(x, z) + \rho(z, x) = 2\rho(x, z),$$

since $\rho(x,x) = 0 \Longrightarrow \rho(x,z) \geqslant 0, \ \forall x, z \in X$.

(iii) Postulates $(M1)$, $(M2)$ and $(M3)$ are called the *Hausdorff postulate.*

5.1.5 Quasi-metric Spaces

Definition 5.3 A function $\rho : X \times X \to \mathbb{R}$ is called a quasi-metric for X if and only if ρ satisfies the following axioms:

$(M1)$ $\rho(x, y) \geqslant 0$, for every $x, y \in X$.
$(M2)$ $\rho(x, y) = 0$, if and only if $x = y$.
$(M3)$ $\rho(x, y) = \rho(y, x)$, for every $x, y \in X$.
(M^*4) $\rho(x, z) \leqslant K \max\{\rho(x, y), \rho(y, z)\}$, for every $x, y, z \in X$ and some $K \geqslant 1$.

The function ρ is also known as a $K-$quasi-metric. The axiom (M^*4) is a generalized version of the ultra-metric triangle inequality (for $K = 1$).

5.2 Limits of Sequences in (X, ρ)

A sequence $\{a_n\}$ of points in a metric space (X, ρ) is a function from $\mathbb{N}$ to X, i.e., $f : \mathbb{N} \to X$, where $a_n = f(n)$. Before discussing the limit of a sequence in a metric space, we must recall the definition of the limit of a sequence of real numbers from the second chapter.

Definition 5.4 Suppose $\{a_n\}$ is a sequence of real numbers. We say that $\{a_n\}$ approaches the limit L as n approaches ∞, if for every $\varepsilon > 0$, there is a positive integer n_0 such that

$$|a_n - L| < \varepsilon, \quad \text{when } n \geqslant n_0(\varepsilon).$$

Now we formulate the corresponding definition for a metric space as follows.

Definition 5.5 In a metric space (X, ρ), the sequence $\{a_1, a_2, a_3, \cdots, a_n, \cdots\}$ of points of X *converges* to the *limit* $L \in X$. If $\rho(a_n, L) \to 0$ (i.e., if the sequence $\rho(a_1, L), \rho(a_2, L), \rho(a_3, L), \cdots$ of *real numbers* converges to 0) as $n \to \infty$, and if given $\varepsilon > 0$, there exists a positive integer n_0 such that

$$\rho(a_n, L) < \varepsilon, \quad \text{when } n \geqslant n_0(\varepsilon).$$

We may write $a_1, a_2, a_3, \cdots \to L$ or simply $a_n \to L$ as $n \to \infty$ (Fig. 5.7).

Remark 5.2 By comparing the above with the definition of convergence in $\mathbb{R}$ (*or* $\mathbb{C}$), we find that $a_n \to L$ if and only if $\lim\limits_{n\to\infty} \rho(a_n, L) = 0$, where ρ denotes the usual metric in $\mathbb{R}$ (or $\mathbb{C}$).

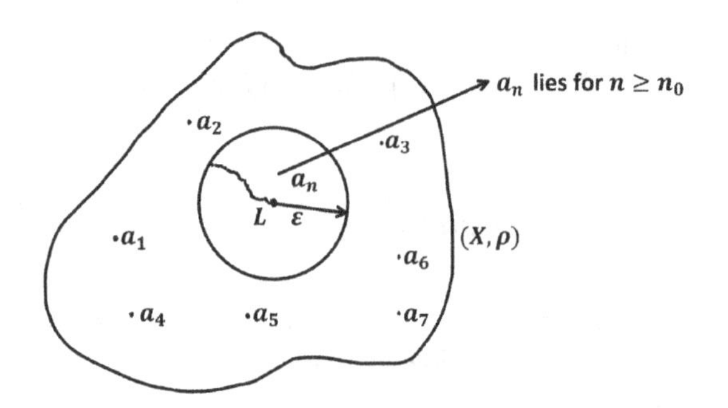

Fig. 5.7 Convergence of sequence in (X, ρ)

Example 5.9 By the definition the sequence

$$(0, 1), \left(\frac{2}{3}, \frac{1}{2}\right), \left(\frac{8}{9}, \frac{11}{27}\right), \cdots, \left(1 - \frac{1}{3^{n-1}}, \frac{n^2+2}{3n^2}\right), \cdots$$

converges to $\left(1, \frac{1}{3}\right)$ in $\mathbb{R}^2$.

5.3 Limits of Functions in (X, ρ)

One way of introducing the ideas of limits and continuity of functions defined on an abstract space and taking values in an abstract space is to introduce the idea of distance. If we do this, we restrict our attention to metric spaces and consider the mapping of one metric space into another metric space.

The convergence $\lim_{x \to a} f(x) = L$ means that given $\varepsilon > 0$ there exists $\delta > 0$ such that the distance from $f(x)$ to L is less than ε provided that the distance from x to a is less than $\delta(> 0)$. Now we formulate the corresponding definition for a metric space.

Suppose (X, ρ_1) and (Y, ρ_2) are metric spaces and let $a \in X$. Let f be a function whose range is contained in Y and whose domain contains all $x \in X$ such that $\rho_1(a, x) < r$ for some $r > 0$ except possibly at $x = a$. We also assume that a is a limit (cluster) point of the domain of f.

Definition 5.6 Let f map (X, ρ_1) into (Y, ρ_2). We say that $f(x)$ approaches L (where $L \in Y$) as x approaches to a, if given $\varepsilon > 0$ there exists $\delta > 0$ such that

$$\rho_2(f(x), L) < \varepsilon, \quad \text{when} \quad 0 < \rho_1(x, a) < \delta.$$

We may write in the following ways

$$\lim_{x \to a} f(x) = L \quad \text{or} \quad f(x) \to L \ \text{as} \ x \to a.$$

If $(X, \rho_1) = (Y, \rho_2) = \mathbb{R}^1$, then we have

$$\rho_2(f(x), L) = |f(x) - L| < \varepsilon, \quad \text{when } \rho_1(x, a) = |x - a| < \delta,$$

so that the above definition reduces to the one given for real-valued functions.

Theorem 5.1 *Let (X, ρ) be a metric space and let a be a point in X. Let f and g be real-valued functions whose domains are subsets of X and ranges are sets of real numbers $\mathbb{R}$ with the usual metric. If $\lim_{x \to a} f(x) = L$ and $\lim_{x \to a} g(x) = M$, where $L, M \in \mathbb{R}$, then we have*

(i) $\lim_{x \to a} [f(x) + g(x)] = L + M.$
(ii) $\lim_{x \to a} [f(x) - g(x)] = L - M.$
(iii) $\lim_{x \to a} [f(x)g(x)] = LM.$
(iv) $\lim_{x \to a} \frac{f(x)}{g(x)} = \frac{L}{M},\ M \neq 0.$

Proof (*iii*) Given $\varepsilon > 0$ we must find $\delta > 0$ and we prove that

$$|f(x)g(x) - LM| < \varepsilon, \quad \text{when} \quad 0 < \rho(x, a) < \delta.$$

Since $\lim_{x \to a} g(x) = M$, choose $\varepsilon = 1$ there exists $\delta_1 > 0$ such that

$$|g(x) - M| < 1, \quad \text{when} \quad 0 < \rho(x, a) < \delta_1.$$

Thus

$$|g(x)| < |M| + 1 = Q \text{ (say)}, \quad \text{when} \quad 0 < \rho(x, a) < \delta_1.$$

Since $\lim_{x \to a} f(x) = L$, given $\varepsilon > 0$ there exists $\delta_2 > 0$ such that

$$Q\,|f(x) - L| < \frac{\varepsilon}{2}, \quad \text{when} \quad 0 < \rho(x, a) < \delta_2. \tag{5.24}$$

And there exists $\delta_3 > 0$ such that

$$|L|\,|g(x) - M| < \frac{\varepsilon}{2}, \quad \text{when} \quad 0 < \rho(x, a) < \delta_3. \tag{5.25}$$

Let $\delta = \min(\delta_1, \delta_2, \delta_3)$ and $0 < \rho(x, a) < \delta$. Then from (5.24) and (5.25) we get

$$\begin{aligned} |f(x)g(x) - LM| &= |f(x)g(x) - Lg(x) + Lg(x) - LM| \\ &\leqslant |g(x)|\,|f(x) - L| + |L|\,|g(x) - M| \\ &< Q\,|f(x) - L| + |L|\,|g(x) - M| < \frac{\varepsilon}{2} + \frac{\varepsilon}{2} = \varepsilon. \end{aligned}$$

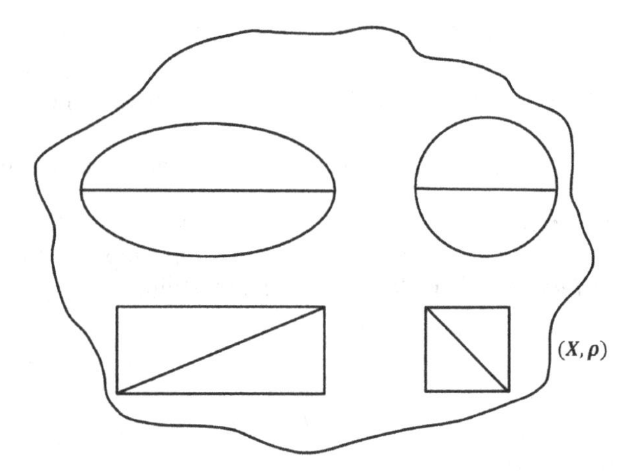

Fig. 5.8 Diameter of set in (X, ρ)

Thus

$$\lim_{x \to a} f(x)g(x) = LM.$$

The proof of other parts of this theorem is exactly along similar lines as the proof given in Chap. 2 (Theorem 2.6) of a sequence of real numbers. ■

5.3.1 The Diameter of a Set in Metric Spaces

The diameter of any non-empty subset A of a metric space (X, ρ), denoted as $\rho(A)$, is defined as follows:

$$\rho(A) = \sup\{\rho(a, b); a, b \in A\},$$

i.e., the diameter of A is the supremum of the distance between pairs of its points. Clearly, the diameter of a circular region, called a circular disc, is the same as its normal diameter. The diameter of an elliptic disc is the same as its major axis. The diameter of a rectangular region is the same as its diagonal (Fig. 5.8).

5.3.2 The Distance Between a Point and a Set in Metric Spaces

The distance between a point p and a set A in the metric space (X, ρ), written as $\rho(p, A)$, is defined as follows:

$$\rho(p, A) = \inf\{\rho(p, a); a \in A\}.$$

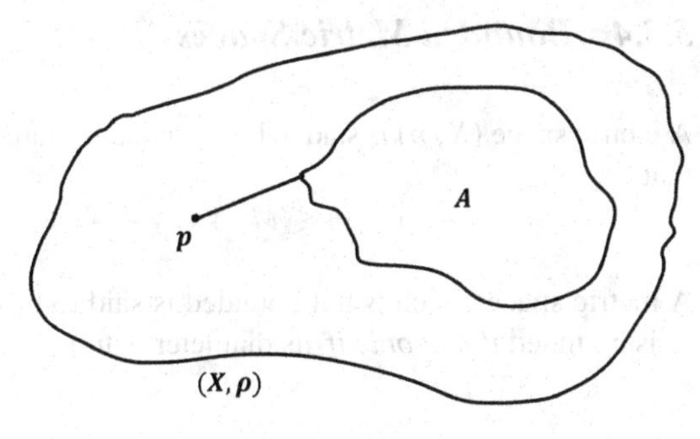

Fig. 5.9 Distance between a point p and a set A in (X, ρ)

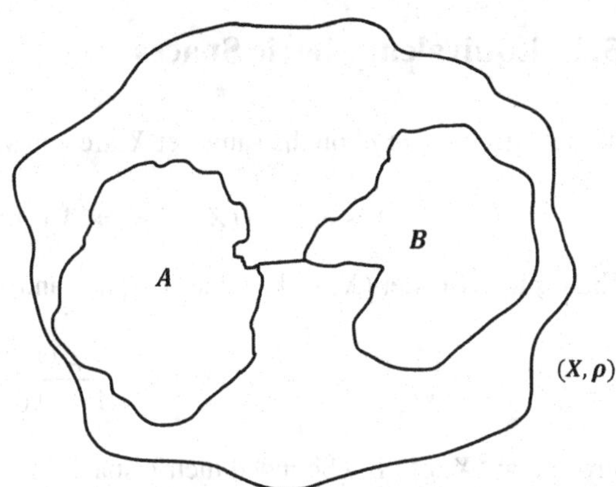

Fig. 5.10 Distance between sets A and B in (X, ρ)

It is the shortest distance between the point p and the points of A, i.e., the distance between the point p and the points of A is the nearest to p (Fig. 5.9).

5.3.3 The Distance Between Two Sets in Metric Spaces

The distance between two non-empty subsets A and B of a metric space (X, ρ), written as $\rho\,(A, B)$, is as

$$\rho(A, B) = \inf\{\rho\,(a, b)\,;\, a \in A, b \in B\}\,.$$

It is clear that the distance between two sets is the shortest distance between the points of one set from those of the other (Fig. 5.10).

5.3.4 *Bounded Metric Spaces*

A metric space (X, ρ) is said to be bounded if there exists a real number $M > 0$ such that

$$\rho(x, y) \leqslant M, \ \forall x, y \in X, \ i.e., \ \rho(X) \leqslant M.$$

A metric space which is not bounded is said to be unbounded. Thus a metric space X is bounded *if and only if* its diameter is finite.

5.4 Equivalent Metric Spaces

Two metrics ρ and ρ' on the same set X are equivalent if for every subset $A \subseteq X$,

$$A \text{ is open in } (X, \rho) \Longleftrightarrow A \text{ is open in } \left(X, \rho'\right). \tag{5.26}$$

Example 5.10 Let (X, ρ) be a metric space and ρ^* be defined

$$\rho^*(x, y) = \frac{\rho(x, y)}{1 + \rho(x, y)}.$$

Prove that (X, ρ^*) is a bounded metric space and ρ^* is equivalent to ρ.

Let G be an open subset in a metric space (X, ρ), and for each $x \in G$ there exists an open sphere with $r > 0$, we have

$$S_r(x) = \{y \in X; \rho(y, x) < r\} \subset G.$$

Let $r^* = \frac{r}{1+r}$. Then $r^* < r$. We see that

$$\rho^*(y, x) < r^* \Longrightarrow \frac{\rho(y, x)}{1 + \rho(y, x)} < \frac{r}{1 + r}.$$

Hence

$$\rho^*(y, x) < r^* \Longrightarrow \rho(y, x) < r,$$

that is,

$$\left\{y \in X; \rho^*(y, x) < r^*\right\} \subset \{y \in X; \rho(y, x) < r\}.$$

Hence every point of G is a center of some open sphere in (X, ρ^*) contained in G. Thus every open set in (X, ρ) is open in (X, ρ^*).

Now let G be any open set in (X, ρ^*). Then there exists an open sphere

$$S_r(x) = \left\{y \in X; \rho^*(y, x) < r\right\} \subset G.$$

Let $0 < r < 1$. Then $\rho^*(y, x) < 1$. We assume $r^{**} = \frac{r}{1-r}$. We see

$$\rho^*(y, x) = \frac{\rho(y, x)}{1 + \rho(y, x)} \Longrightarrow \rho(y, x) = \frac{\rho^*(y, x)}{1 - \rho^*(y, x)}.$$

Therefore

$$\rho(y, x) < r^{**} \Longrightarrow \frac{\rho^*(y, x)}{1 - \rho^*(y, x)} < \frac{r}{1 - r}.$$

Hence,

$$\rho(y, x) < r^{**} \Rightarrow \rho^*(y, x) < r,$$

that is, we see each point of G is a center of some open sphere contained in G, which implies that G is open in (X, ρ). Consequently, the metrics ρ and ρ^* are equivalent.

Lemma 5.1 *Let (X, ρ) and (X, ρ') be two metric spaces. If there exists $0 < C_1 \leqslant C_2$ such that*

$$C_1\rho(x, y) \leqslant \rho'(x, y) \leqslant C_2\rho(x, y), \text{ for every } x, y \in X, \tag{5.27}$$

then (5.26) holds. Thus ρ and ρ' are equivalent on X.

Proof Let $x_0 \in A$ be fixed. Then we define

$$A_{\rho,\delta} = \{y;\ \rho(x_0, y) < \delta\}.$$

Let A be open in (X, ρ). Then there exists $\delta > 0$ such that $A_{\rho,\delta} \subset A$. So we have by (5.27)

$$\left\{y;\ \frac{\rho'(x_0, y)}{C_2} \leqslant \rho(x_0, y) < \delta\right\} \subset A.$$

Hence,

$$A_{\rho',\varepsilon} = \{y;\ \rho'(x_0, y) < C_2\delta = \varepsilon\} \subset A.$$

Therefore, A is open in (X, ρ'). Similarly, we see that if A is open in (X, ρ'), then A is open in (X, ρ). ■

Example 5.11 Let (X, ρ) be a metric space and let $\rho'(x, y) = \min\{1, \rho(x, y)\}$ for all $x, y \in X$. Show that ρ and ρ' are equivalent.

Theorem 5.2 *Let (X, ρ) and (X, ρ') be two metric spaces. Then the following statements are equivalent.*

(*i*) *The metrics ρ and ρ' are equivalent.*
(*ii*) *For each $G \subset X$, G is open in (X, ρ) $\Longleftrightarrow$ G is open in (X, ρ').*

(*iii*) *For each* $F \subset X$, *F is closed in* $(X, \rho) \iff$ *F is closed in* (X, ρ').
(*iv*) *The sequence* $\{x_n\}$ *converges to* x_0 *in* $(X, \rho) \iff$ *it converges to* x_0 *in* (X, ρ').
(*v*) *For each* $A \subset X$, $\overline{A^\rho} = \overline{A^{\rho'}}$, *where* $\overline{A^\rho}$ *and* $\overline{A^{\rho'}}$ *are closures of A in* (X, ρ) *and* (X, ρ') *respectively.*

It is clear that if two metrics ρ and ρ' are equivalent, then the above relations are equivalent.

5.5 Product Metric Spaces

Let (X, ρ_X) and (Y, ρ_Y) be two metric spaces. The product metric $\rho_\prod$ for the Cartesian product space $X \times Y = \{(x, y)\,;\, x \in X, y \in Y\} = Z\ (say)$ is defined as follows:

Let $\rho_\prod : (X \times Y) \times (X \times Y) \to \mathbb{R}$ be a distance function. Then

$$\rho_\prod(z_1, z_2) = \rho_\prod((x_1, y_1), (x_2, y_2)), \tag{5.28}$$

where $z_1 = (x_1, y_1)$ and $z_2 = (x_2, y_2)$ are any pair of points and $z_1, z_2 \in Z$.

The metric space $(X \times Y, \rho_X \times \rho_Y)$, i.e., $(Z, \rho_\prod)$ is called the product metric space of the metric spaces (X, ρ_X) and (Y, ρ_Y).

There is more than one way of combining ρ_X and ρ_Y to form a metric for $X \times Y$, and none of these constructions is "more natural" than the others. We observe that the two metrics $\rho'_\prod$ and $\rho''_\prod$ on Z are defined as

$$\rho'_\prod(z_1, z_2) = \left\{(\rho_X(x_1, x_2))^2 + (\rho_Y(y_1, y_2))^2\right\}^{1/2}, \tag{5.29}$$

and

$$\rho''_\prod(z_1, z_2) = \rho_X(x_1, x_2) + \rho_Y(y_1, y_2). \tag{5.30}$$

All three $\rho_\prod$, $\rho'_\prod$ and $\rho''_\prod$ are metrics on $Z = X \times Y$ and easy to check the axioms $(M1)$ and $(M2)$ of Definition 5.1 in these cases.

Now let us check the triangle inequality for $\rho''_\prod$.

Let

$$\begin{aligned}
\rho''_\Pi((x_1, y_1), (x_2, y_2)) &= \rho_X(x_1, x_2) + \rho_Y(y_1, y_2),\\
\rho''_\Pi((x_1, y_1), (x_3, y_3)) &= \rho_X(x_1, x_3) + \rho_Y(y_1, y_3),\\
\rho''_\Pi((x_2, y_2), (x_3, y_3)) &= \rho_X(x_2, x_3) + \rho_Y(y_2, y_3).
\end{aligned}$$

Then we easily see

$$\begin{aligned}\rho''_{\Pi}((x_1, y_1), (x_2, y_2)) &\leqslant \{\rho_X(x_1, x_3) + \rho_X(x_2, x_3)\} + \{\rho_Y(y_1, y_3) + \rho_Y(y_2, y_3)\} \\ &= \{\rho_X(x_1, x_3) + \rho_Y(y_1, y_3)\} + \{\rho_X(x_2, x_3) + \rho_Y(y_2, y_3)\} \\ &= \rho''_{\Pi}((x_1, y_1), (x_3, y_3)) + \rho''_{\Pi}((x_2, y_2), (x_3, y_3)).\end{aligned}$$

The proofs for other cases ρ_Π and ρ'_Π are also given similarly.

Definition 5.7 Let (X_i, ρ_i) , $i = 1, 2, \cdots, m$ be metric spaces and

$$\rho_{\prod}(\mathbf{x}, \mathbf{y}) = \max\{\rho_i(x_i, y_i)\,;\, 1 \leqslant i \leqslant m\}$$

be the distance function on $X = \prod_{i=1}^{m} X_i$, which is called the product metric spaces $(X_1, \rho_1), (X_2, \rho_2), \cdots (X_n, \rho_n)$, where $\mathbf{x} = (x_1, \cdots, x_m)$, $\mathbf{y} = (y_1, \cdots, y_m)$ and $x_i, y_i \in X_i$, $i = 1, \cdots, m$.

Corollary 5.1 *The three metrics $\rho_{\prod}$, $\rho'_{\prod}$, and $\rho''_{\prod}$ defined on $X \times Y$ are equivalent.*

Proof For all $\mathbf{x}, \mathbf{y} \in X \times Y$ we see that

$$\rho_{\prod}(\mathbf{x}, \mathbf{y}) \leqslant \rho'_{\prod}(\mathbf{x}, \mathbf{y}) \leqslant \sqrt{2}\rho_{\prod}(\mathbf{x}, \mathbf{y}),$$

and

$$\rho_{\prod}(\mathbf{x}, \mathbf{y}) \leqslant \rho''_{\prod}(\mathbf{x}, \mathbf{y}) \leqslant 2\rho_{\prod}(\mathbf{x}, \mathbf{y}).$$

Thus, using Lemma 5.1, we see that ρ_Π, ρ'_Π and ρ''_Π are equivalent. ■

5.6 The Topology of Metric Spaces

In this section, we will discuss the fact that a metric space is a topological space in the usual sense of topology.

Now we turn to generalize metric spaces. The key property which we wish to generalize is that of "open set." For a metric space, the reader should be thinking of the real line, the Euclidean plane, or three-dimensional Euclidean spaces. In there the open sets are easy to find. One can think of them as just "things without a boundary." For example, on the real line, these look like open intervals (a, b) and union of open intervals. In the plane, these would be more like "open balls" with a fixed center. In other words, it would be the interior of a disk.

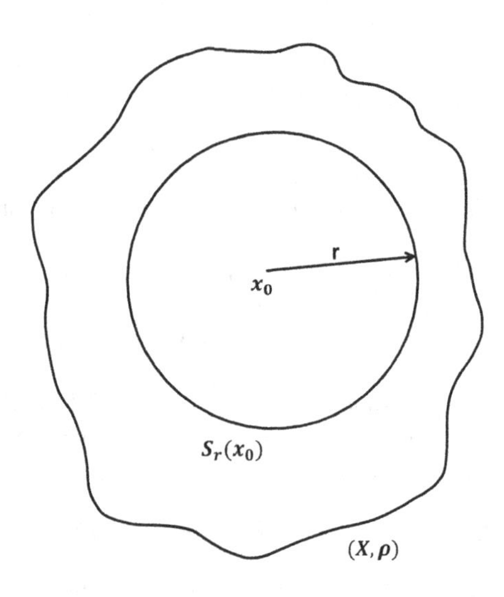

Fig. 5.11 Open sphere in (X, ρ)

5.6.1 Open and Closed Spheres; Balls

The elements of a metric space are usually called the points of the metric space.

Definition 5.8 Let x_0 be a point of a metric space (X, ρ) and a positive real number $r > 0$. The set (Fig. 5.11)

$$S(x_0, r) = S_r(x_0) = \{x \in X; \rho(x, x_0) < r\}$$

is called the open sphere center x_0 and radius r, the set (Fig. 5.12)

$$S[x_0, r] = S_r[x_0] = \{x \in X; \rho(x, x_0) \leqslant r\}$$

is called the closed sphere center x_0 and radius r, and the set

$$B(x_0, r) = B_r(x_0) = \{x \in X; \rho(x, x_0) = r\}$$

is called the ball center x_0 and radius r. Furthermore, the definition immediately implies that

$$B(x_0, r) = S_r[x_0] - S_r(x_0) \neq \phi. \tag{5.31}$$

Now we define the open and closed spheres and balls in a normed linear space.

Definition 5.9 Let $(X, \|\cdot\|)$ be a normed space, and let $r > 0$. Then the set

$$S_r(x_0) = \{x \in X; \|x_0 - x\| < r\}$$

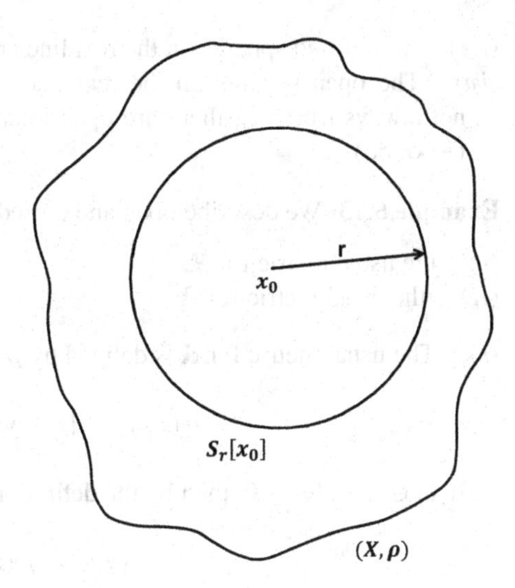

Fig. 5.12 Closed sphere in (X, ρ)

is called the open sphere with the center x_0 and the radius r, the set

$$S_r[x_0] = \{x \in X; \|x_0 - x\| \leqslant r\}$$

is called the closed sphere with the radius r and the set

$$B_r(x_0) = \{x \in X; \|x_0 - x\| = r\}$$

is called the ball with the center x_0 and the radius r.

Example 5.12 If $0 < r_1 < r_2$, then prove that $S_{r_1}(x_0) \subsetneq S_{r_2}(x_0)$.

Let x be an arbitrary element of an open sphere $S_{r_1}(x_0)$, i.e., $x \in S_{r_1}(x_0)$, we have

$$\Longrightarrow \rho(x, x_0) < r_1$$
$$\Longrightarrow \rho(x, x_0) < r_2, (r_1 < r_2)$$
$$\Longrightarrow x \in S_{r_2}(x_0), \textit{ (by the definition of the open sphere).}$$

Hence,

$$S_{r_1}(x_0) \subseteq S_{r_2}(x_0).$$

But we see

$$\phi \neq B\left(x_0, \frac{r_1 + r_2}{2}\right) = S_{r_2} - S_{r_1}.$$

Consequently, we have $S_{r_1}(x_0) \subsetneq S_{r_2}(x_0)$.

Remark 5.3 (i) It is clear that $S_r(x_0) \subset S_r[x_0]$, $\forall x_0 \in X$ and $\forall r > 0$.
(ii) From (5.31) a ball is always non-empty.

(*iii*) The closed spheres on the real line are precisely the closed intervals.
(*iv*) The open spheres on the real line are open intervals, but the converse is not always true, i.e., there are open intervals which are not open spheres, e.g., $(-\infty, \infty)$.

Example 5.13 We describe open and closed spheres for

(*i*) the usual metric for $\mathbb{R}$,
(*ii*) the usual metric for $\mathbb{R}^2$.

(*i*) The usual metric for $\mathbb{R}$ is defined by $\rho : \mathbb{R} \times \mathbb{R} \to \mathbb{R}$ such that

$$\rho(x, y) = |x - y|, \quad \forall\, x, y \in \mathbb{R}.$$

If $x_0 \in \mathbb{R}$ and $r > 0$, then by the definition of an open sphere

$$\begin{aligned} S_r(x_0) &= \{x \in \mathbb{R};\ \rho(x, x_0) < r\} \\ &= \{x \in \mathbb{R};\ |x - x_0| < r\} \\ &= \{x \in \mathbb{R};\ x_0 - r < x < x_0 + r\} \\ &= (x_0 - r, x_0 + r) \\ &=]x_0 - r, x_0 + r[\,. \end{aligned}$$

Hence the open spheres on the real line are open intervals.

$$Forexample, \quad S_{\frac{3}{2}}(1) = \left]1 - \frac{3}{2}, 1 + \frac{3}{2}\right[= \left]-\frac{1}{2}, \frac{5}{2}\right[,$$

$$S_1(-1) =]-1 - 1, -1 + 1[\, =]-2, 0[\,.$$

If $x_0 \in \mathbb{R}$ and $r > 0$, then by the definition of a closed sphere

$$\begin{aligned} S_r[x_0] &= \{x \in \mathbb{R};\ \rho(x, x_0) \leqslant r\} \\ &= \{x \in \mathbb{R};\ |x - x_0| \leqslant r\} \\ &= \{x \in \mathbb{R};\ x_0 - r \leqslant x \leqslant x_0 + r\} \\ &= [x_0 - r, x_0 + r]\,. \end{aligned}$$

Hence the closed spheres on the real line are the closed intervals.

$$Forexample, \quad S\left[1, \frac{5}{2}\right] = \left[1 - \frac{5}{2}, 1 + \frac{5}{2}\right] = \left[-\frac{3}{2}, \frac{7}{2}\right].$$

(*ii*) The usual metric for $\mathbb{R}^2$ is defined by $\rho : \mathbb{R}^2 \times \mathbb{R}^2 \to \mathbb{R}$ such that

$$\rho(z_1, z_2) = \sqrt{(x_1 - x_2)^2 + (y_1 - y_2)^2},$$

where $z_1 = (x_1, y_1)$ and $z_2 = (x_2, y_2)$.
Let $z_0 = (x_0, y_0) \in \mathbb{R}^2$ and $r > 0$. Then by the definition of an open sphere

$$\begin{aligned} S_r(z_0) &= \left\{z = (x, y) \in \mathbb{R}^2; \rho(z, z_0) < r\right\} \\ &= \left\{z = (x, y) \in \mathbb{R}^2; \sqrt{(x - x_0)^2 + (y - y_0)^2} < r\right\} \\ &= \left\{z = (x, y) \in \mathbb{R}^2; (x - x_0)^2 + (y - y_0)^2 < r^2\right\}. \end{aligned}$$

The open sphere consists of all the points of the Cartesian plane, which lies within the circle

$$(x - x_0)^2 + (y - y_0)^2 < r^2.$$

Thus, an open sphere in $\mathbb{R}^2$ is an open disc.
Similarly, by the definition of a closed sphere

$$\begin{aligned} S_r[z_0] &= \left\{z = (x, y) \in \mathbb{R}^2; \rho(z, z_0) \leqslant r\right\} \\ &= \left\{z = (x, y) \in \mathbb{R}^2; \sqrt{(x - x_0)^2 + (y - y_0)^2} \leqslant r\right\}. \end{aligned}$$

The closed sphere consists of all the points of the Cartesian plane, which lies within and on the circumference of the circle:

$$(x - x_0)^2 + (y - y_0)^2 \leqslant r^2.$$

Thus, a closed sphere in $\mathbb{R}^2$ is a closed disc.

Example 5.14 Let (X, ρ) be any discrete metric space. Describe open spheres and closed spheres for ρ. A discrete metric ρ is defined by $\rho : X \times X \to \mathbb{R}$

$$\rho(x, y) = \begin{cases} 0, & \text{when } x = y, \\ 1, & \text{when } x \neq y. \end{cases}$$

Let $x_0 \in X$ and $r > 0$. Then the definition of an **open sphere** is

$$S(x_0, r) = S_r(x_0) = \{x \in X; \rho(x, x_0) < r\}.$$

(i) Let $r \leqslant 1$: If $x = x_0$, then $\rho(x, x_0) = 0$. Thus $x \in S_r(x_0)$, and if $x \neq x_0$, then $\rho(x, x_0) = 1$ $(\geqslant r)$ and

$$x \notin S_r(x_0), \quad \forall x \neq x_0.$$

Therefore,

$$S_r(x_0) = \{x_0\}.$$

(ii) Let $r > 1$: If $x = x_0$, then $\rho(x, x_0) = 0 < r$. We have

$$x \in S_r(x_0),$$

and if $x \neq x_0$, then $\rho(x, x_0) = 1$. Thus $x \in S_r(x_0)$.
Therefore,

$$S_r(x_0) = X.$$

By the definition of a **closed sphere**

$$S[x_0, r] = S_r[x_0] = \{x \in X; \rho(x, x_0) \leqslant r\}.$$

(iii) Let $r < 1$: If $x = x_0$, then $\rho(x, x_0) = 0 < r$. Thus $x \in S_r[x_0]$, and if $x \neq x_0$, then $\rho(x, x_0) = 1\ (> r)$ and

$$x \notin S_r[x_0], \quad \forall x \neq x_0.$$

Therefore,

$$S_r[x_0] = \{x_0\}.$$

(iv) Let $r \geqslant 1$: If $x = x_0$, then $\rho(x, x_0) = 0 < r$. We have

$$x \in S_r[x_0], \tag{5.32}$$

and if $x \neq x_0$, then

$$\rho(x, x_0) = 1 \leqslant r. \tag{5.33}$$

From (5.32) and (5.33), we get $x \in S_r[x_0]$.
Thus,

$$S_r[x_0] = X.$$

$$\begin{aligned}
E.g., \quad & S_{\frac{1}{2}}[x_0] = \{x_0\} && \left(\because r = \frac{1}{2} < 1\right) \\
& S_{\frac{1}{2}}[-1] = \{-1\} && \left(\because r = \frac{1}{2} < 1\right) \\
& S_1[-1] = X && (\because r = 1) \\
& S_2[1] = X && (\because r = 2 > 1) \\
& S_1[2] = X && (\because r = 1).
\end{aligned}$$

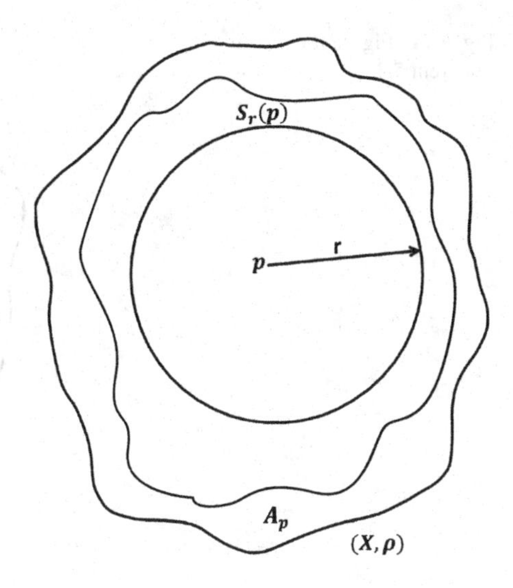

Fig. 5.13 Neighborhoods of a point p in (X, ρ)

5.6.2 *Neighborhoods*

There is an essential switch in going from metric spaces to topological spaces that one must take, and it involves the concepts of neighborhoods, i.e., a neighborhood is one of the basic concepts in a topological space.

Definition 5.10 Let (X, ρ) be a metric space. A subset A_p of X is called a *neighborhood (n.b.d.)* of a point $p \in X$, if there exists an open sphere $S_r(p)$ centered at p and contained in A_p, i.e., (Fig. 5.13)

$$p \in S_r(p) \subseteq A_p, \quad \text{for some } r > 0.$$

Example 5.15 (i) Every open interval (a, b) is a *neighborhood* of each of its points.
(ii) Every closed interval is a *neighborhood* of each of its points except end points.
(iii) The set $\mathbb{R}$ of real numbers is a *neighborhood* of each of its points.
(iv) The set $\mathbb{N}$ of natural numbers, the set $\mathbb{Q}$ of rational numbers and the set $\mathbb{Z}$ of integers are not a *neighborhood* of any of its points.
(v) A subset A of the discrete metric space (X, ρ) is a *neighborhood* of each of its points.

Remark 5.4 (i) In the usual metric space $\mathbb{R}$, (a, b) is a n.b.d. of each of its points.
(ii) An interval $[a, b]$ is a n.b.d. of each of its points except the end points a and b.

Theorem 5.3 *Every open sphere is a neighborhood of each of its points.*

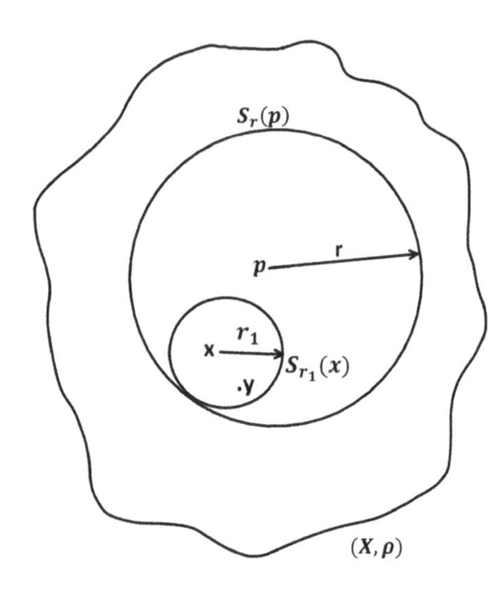

Fig. 5.14 Figure for Theorem 5.3

Proof Let $S_r(p)$ be an open sphere and let $x \in S_r(p)$.

If $x = p$, then $p \in S_r(p) \subseteq S_r(p)$, i.e., $S_r(p)$ is a neighborhood of p (Fig. 5.14).

Let $x \neq p$. To show that $S_r(p)$ is a neighborhood of x, we must show that there exists $r_1 > 0$ such that

$$x \in S_{r_1}(x) \subseteq S_r(p).$$

Now

$$x \in S_r(p) \Longrightarrow \rho(x, p) < r \Longrightarrow r - \rho(x, p) > 0.$$

Take $r_1 = r - \rho(x, p)$, then $y \in S_{r_1}(x) \Longrightarrow \rho(y, x) < r_1$.

By using the triangle inequality, we have

$$\begin{aligned} &\rho(y, p) \leqslant \rho(y, x) + \rho(x, p) < r_1 + \rho(x, p) = r \\ &\Longrightarrow \rho(y, p) < r, \ i.e., \ y \in S_r(p). \end{aligned}$$

Hence

$$S_{r_1}(x) \subseteq S_r(p),$$

that is, $S_r(p)$ is a neighborhood of x. ■

5.7 Continuity of Functions

Definition 5.11 Let (X, ρ_X) and (Y, ρ_Y) be two metric spaces. A function $f : X \to Y$ is said to be continuous at the point a of X if for every $\varepsilon > 0$, and there exists $\delta > 0$ such that (Fig. 5.15)

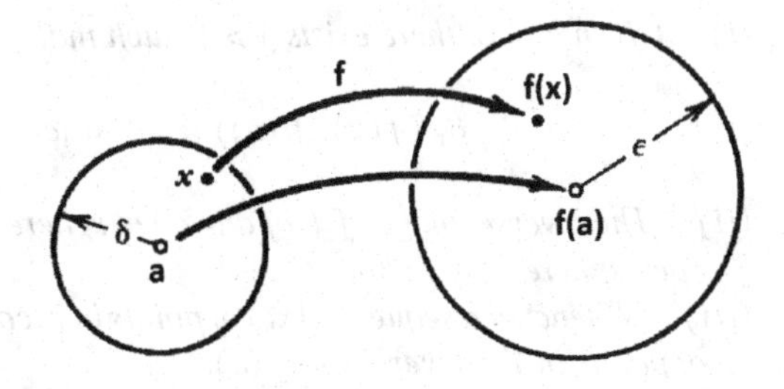

Fig. 5.15 Illustrated definition 5.11 in $\mathbb{R}^n$

$$\rho_Y(f(x), f(a)) < \varepsilon, \quad \text{whenever} \quad \rho_X(x, a) < \delta, \ x \in X,$$

that is

$$f(x) \in S_\varepsilon(f(a)), \quad \text{whenever} \quad x \in S_\delta(a),$$

which means

$$f(S_\delta(a)) \subset S_\varepsilon(f(a)).$$

The inequalities are given in the above definitions illustrated in the following figure in the case of the Euclidean planes $X = \mathbb{R}^2$ and $Y = \mathbb{R}^2$ (Fig. 5.16).

Example 5.16 (*i*) Let $(\mathbb{R}, \rho)$ be the usual metric space (see Example 5.1) and $f : \mathbb{R} \to \mathbb{R}$ be a constant function. Then f is continuous.

(*ii*) The identity function $I : X \to X$ is continuous in a metric space (X, ρ).

Theorem 5.4 *Let* (X, ρ_X) *and* (Y, ρ_Y) *be two metric spaces and a function* $f : X \to Y$. *Then* f *is continuous at* $a \in X$ *if and only if any one of the following conditions holds.*

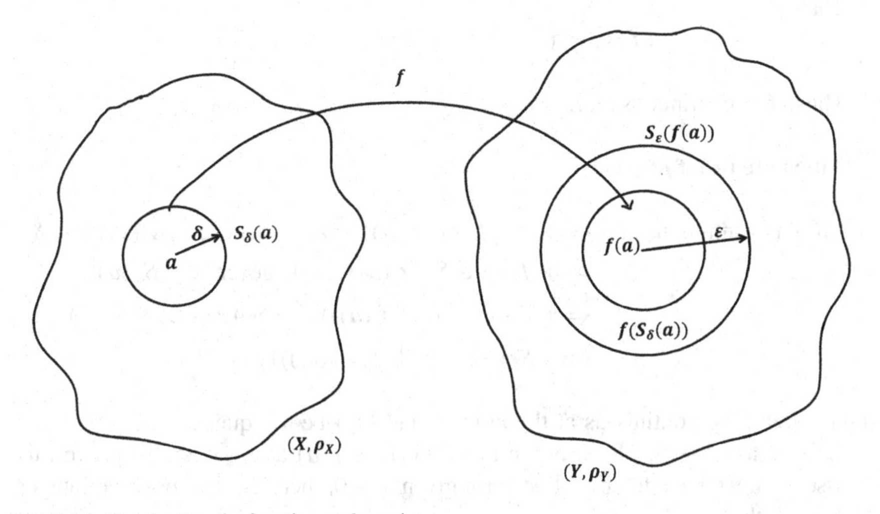

Fig. 5.16 Continuity of a function at the point a

(i) *Given* $\varepsilon > 0$, *there exists* $\delta > 0$ *such that*

$$\rho_Y (f(x), f(a)) < \varepsilon, \quad \textit{whenever} \quad \rho_X (x, a) < \delta.$$

(ii) *The inverse image of* f *of any open sphere* $S_\varepsilon (f(a))$ *about* $f(a)$ *contains an open sphere* $S_\delta (a)$ *about* a.

(iii) *Whenever a sequence* $\{x_n\}$ *of points in* X *converges to* a, *the sequence* $\{f(x_n)\}$ *of points in* Y *converges to* $f(a)$.

Proof (i) This condition is the reformulation of the definition of a continuous function $f : X \to Y$ in the place of the absolute-value function (Fig. 5.17).

(ii) Since f is a continuous function, for a given $\varepsilon > 0$, there exists $\delta > 0$ such that

$$\rho_Y (f(x), f(a)) < \varepsilon, \quad \text{whenever} \quad \rho_X (x, a) < \delta.$$

Therefore, for any $x \in S_\delta(a)$ we get

$$f(x) \in S_\varepsilon (f(a)) \Longrightarrow x \in f^{-1} (S_\varepsilon (f(a))) .$$

Then, we have

$$S_\delta (a) \subset f^{-1} (S_\varepsilon (f(a))) .$$

Thus if f is continuous, then the inverse image of any open sphere $S_\varepsilon (f(a))$ about $f(a)$ contains an open sphere $S_\delta (a)$ about a.

Conversely, if $S_\delta (a) \subset f^{-1} (S_\varepsilon (f(a)))$, then we have $f(S_\delta (a)) \subset S_\varepsilon (f(a))$. This implies that

$$x \in S_\delta (a) \Longrightarrow f(x) \in S_\varepsilon (f(a)) .$$

That is

$$\rho_Y (f(x), f(a)) < \varepsilon, \quad \text{whenever} \quad \rho_X (x, a) < \delta.$$

Thus, f is continuous at a.

Alternate proof of (ii)**.**

$$\begin{aligned}
\text{If } f \text{ is continuous at } a &\Longleftrightarrow \rho_Y (f(x), f(a)) < \varepsilon, \quad \text{whenever} \quad \rho_X (x, a) < \delta \\
&\Longleftrightarrow f(x) \in S_\varepsilon (f(a)), \quad \text{whenever} \quad x \in S_\delta (a) \\
&\Longleftrightarrow x \in f^{-1} (S_\varepsilon (f(a))), \quad \text{whenever} \quad x \in S_\delta (a) \\
&\Longleftrightarrow S_\delta (a) \subset f^{-1} (S_\varepsilon (f(a))) .
\end{aligned}$$

(iii) Let f be continuous at the point a and $\{x_n\}$ be a sequence in X such that $x_n \to a$ as $n \to \infty$. Then prove that $f(x_n) \to f(a)$ as $n \to \infty$. To prove this assertion, we have to show that for a given $\varepsilon > 0$, there exists a positive integer n_0 such that

$$f(x_n) \in S_\varepsilon(f(a)), \quad \text{for all } n \geqslant n_0.$$

Since f is continuous at a, for a given $\varepsilon > 0$, there exists $\delta > 0$ such that

$$f(x) \in S_\varepsilon(f(a)), \quad \text{whenever } x \in S_\delta(a).$$

Therefore,

$$f(S_\delta(a)) \subset S_\varepsilon(f(a)). \tag{5.34}$$

Since $x_n \to a$ as $n \to \infty$, there exists a positive integer n_0 such that

$$x_n \in S_\delta(a), \quad \text{for all } n \geqslant n_0. \tag{5.35}$$

From (5.34) and (5.35), we obtain

$$f(x_n) \in S_\varepsilon(f(a)), \quad \text{for all } n \geqslant n_0.$$

Thus

$$f(x_n) \to f(a) \quad \text{as} \quad n \to \infty.$$

Conversely, let $f(x_n) \to f(a)$, for every sequence $\{x_n\}$ with $x_n \to a$. Assume the contrary, that is, f is not continuous at a. Then there exists an $\varepsilon > 0$ such that there is no open sphere centered at a whose f-image is contained in $S_\varepsilon(f(a))$. Consider the sequence of open spheres

$$S_1(a), S_{\frac{1}{2}}(a), S_{\frac{1}{3}}(a), \cdots, S_{\frac{1}{n}}(a), \cdots.$$

For each positive integer n, there exists $x_n \in S_{\frac{1}{n}}(a)$ such that $f(x_n) \notin S_\varepsilon(f(a))$. Thus we get a sequence $\{x_n\} \subset X$ with $x_n \to a$ such that $f(x_n) \nrightarrow f(a)$. This is a contradiction to the hypothesis. Thus, f is continuous at a.

■

Theorem 5.5 *Let (X, ρ_X), (Y, ρ_Y) and (Z, ρ_Z) be three metric spaces and let functions $f : X \to Y$ and $g : Y \to Z$. If f is continuous at $a \in X$ and g is continuous at $f(a) \in Y$, then $g \circ f$ is continuous at a.*

Proof Suppose that $\{x_n\}$ is a sequence of points in X such that $x_n \to a$ as $n \to \infty$. To prove the theorem, we have to show that

$$\lim_{n\to\infty} g(f(x_n)) = g(f(a)).$$

Since f is continuous at a, we have $\lim_{n\to\infty} f(x_n) = f(a)$.

Let $y_n = f(x_n)$ and $y_n \to f(a)$ as $n \to \infty$ in (Y, ρ_Y). Since g is continuous,

$$\lim_{n\to\infty} g(y_n) = g(f(a)).$$

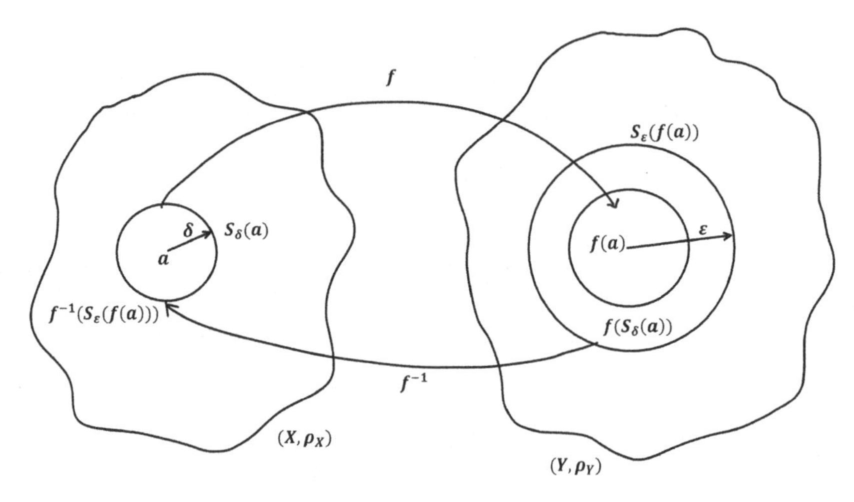

Fig. 5.17 Figure for Theorem 5.4

Substituting the value of y_n, we have

$$\lim_{n\to\infty} g\left(f\left(x_n\right)\right) = g\left(f\left(a\right)\right).$$

Thus, $g \circ f : X \to Z$ is continuous. ∎

5.8 Cluster Points

Let A be any subset of a metric space (X, ρ). A point p of X is called an `adherent point` of A, if every open sphere centered at p (neighborhood of p) contains a point of A.

An adherent point has two types: (i) Isolated points and (ii) Cluster points.

An adherent point p of a subset A of X is called an `isolated point` if there exists an open sphere centered at p which contains no point of A other than p itself.

An adherent point p of a subset A of X is said to be a `cluster point` of A if every open sphere centered at p contains at least one point of A other than p itself.

Definition 5.12 Let A be any subset of a metric space (X, ρ). A point p of X is said to be a cluster point of A if every open sphere centered at p contains at least one point of A other than p itself, i.e., $(S_r(p) - \{p\}) \cap A \neq \phi$.

Remark 5.5 (i) A cluster point $p \in X$ of A is not necessarily in a set A.

(ii) The cluster point is also known as a `limit point` or a `condensation point` or an `accumulation point`.

(iii) If p is a cluster point of A, then every open sphere centered at p contains infinitely many elements of A.

(iv) The set of all cluster points of A is called the `derived set` of A and denoted by A'.

Example 5.17 Let $X = \left\{\frac{1}{n}; n \in \mathbb{N}\right\}$. And consider the usual metric given by the absolute value. Note that 0 is the only limit point of X and all other points are isolated points of X.

5.9 Open Sets

A subset G of a metric space (X, ρ) is called an open set if for each $x \in G$ there exists a positive real number r such that

$$S(x, r) \subseteq G, \text{ i.e., } S_r(x) \subseteq G.$$

That is, a set G is said to be open if it contains a neighborhood of each of its points.

Theorem 5.6 *In any metric space (X, ρ), each open sphere is an open set.*

Proof Let

$$S_r(x_0) = \{x \in X; \rho(x, x_0) < r\} \tag{5.36}$$

be an open sphere in a metric space (X, ρ). Then we will show that $S_r(x_0)$ is an open set.

Let p be an arbitrary point of $S_r(x_0)$, i.e.,

$$\begin{aligned} p \in S_r(x_0) &\Longrightarrow \rho(p, x_0) < r \quad \text{(by the definition of an open sphere)} \\ &\Longrightarrow r - \rho(p, x_0) > 0. \end{aligned}$$

Let

$$r_1 = r - \rho(p, x_0). \tag{5.37}$$

Now we have to show that $S_{r_1}(p) \subseteq S_r(x_0)$.

$$\begin{aligned} \text{Let} \quad x \in S_{r_1}(p) &\Longrightarrow \rho(x, p) < r_1 \quad \text{(by the definition of an open sphere)} \\ &\Longrightarrow \rho(x, p) < r - \rho(p, x_0) \quad \text{(by (5.37))} \\ &\Longrightarrow \rho(x, p) + \rho(p, x_0) < r \quad \text{(by Definition 5.1)} \\ &\Longrightarrow \rho(x, x_0) < r \\ &\Longrightarrow x \in S_r(x_0). \end{aligned}$$

Hence $S_{r_1}(p) \subseteq S_r(x_0)$. Thus, $S_r(x_0)$ is an open set (Fig. 5.18). ■

Theorem 5.7 *In a metric space (X, ρ), the union of an arbitrary collection of open sets is open.*

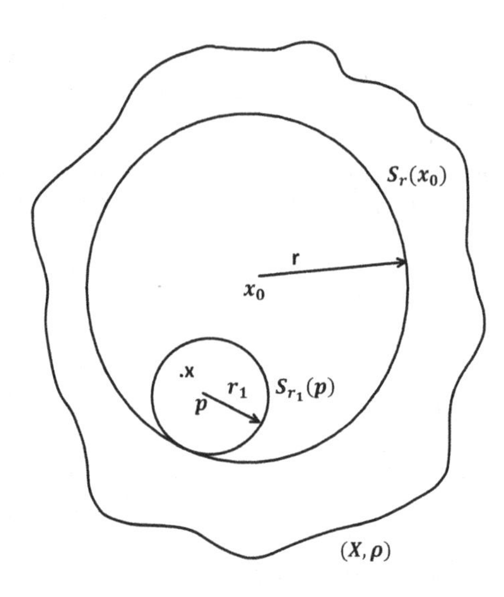

Fig. 5.18 Figure for Theorem 5.6

Proof Let $\{G_\lambda; \lambda \in \Lambda\}$ be an arbitrary collection of non-empty open sets in X, where Λ is any non-empty index set. Then we will show that $G = \cup\{G_\lambda; \lambda \in \Lambda\}$ is an open set (Fig 5.19).

Let x be any element of G. Then

$$x \in G = \cup\{G_\lambda; \lambda \in \Lambda\} \implies x \in G_\lambda, \text{ for some } \lambda \in \Lambda.$$

The set G_λ is open $\exists\, r > 0$ such that

$$\begin{aligned} S_r(x) &\subseteq G_\lambda, \text{ for some } \lambda \in \Lambda \quad \text{(by the definition of an open set)} \\ &\subseteq \cup\{G_\lambda; \lambda \in \Lambda\} = G. \end{aligned}$$

Hence,

$$S_r(x) \subseteq G.$$

Thus,

$$G = \cup\{G_\lambda; \lambda \in \Lambda\} \text{ is an open set in } X.$$

■

Theorem 5.8 *In a metric space* (X, ρ), *the intersection of finite numbers of non-empty open sets is open.*

Proof Let $\{G_i; i = 1, 2, 3, \cdots, n\}$ be a finite collection of open sets of X. Then we will show that $G = \cap\{G_i : i = 1, 2, 3, \cdots, n\}$ is also open.

If $G = \phi$, then G is open (by Theorem 5.12) (Fig. 5.20).

If $G \neq \phi$, let x be any element of G, i.e.,

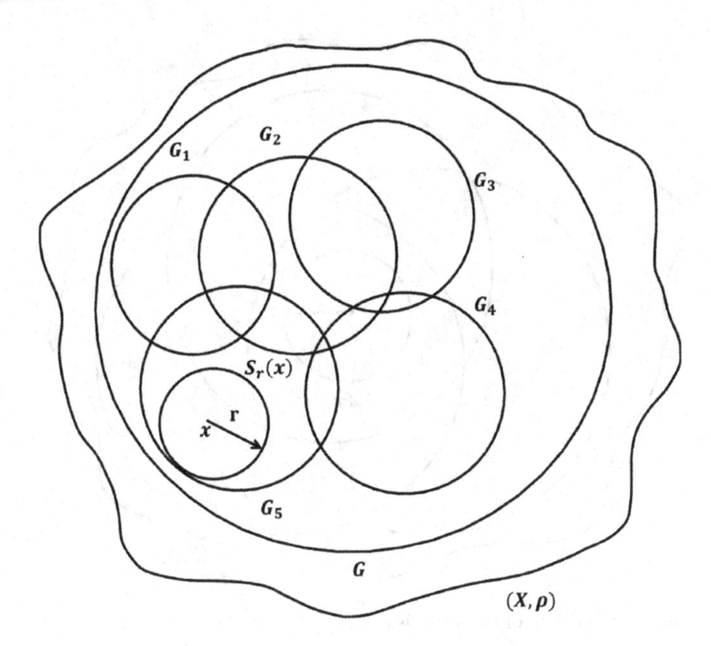

Fig. 5.19 An arbitrary union of open sets is open

$$x \in G = \cap\{G_i; i = 1, 2, 3, \cdots, n\} \implies x \in G_i, \forall i = 1, 2, 3, \cdots, n.$$

Since each $G_i, \forall i = 1, 2, 3, \cdots, n$ is open, there exists $r_i > 0$ such that

$$S_{r_i}(x) \subseteq G_i, \; i = 1, 2, 3, \cdots, n \quad \text{(by the definition of an open set).} \tag{5.38}$$

Let $r = \min\{r_1, r_2, r_3, \cdots r_n\}$. Then

$$\begin{aligned} & S_r(x) \subseteq S_{r_i}(x), \; \forall i = 1, 2, 3, \cdots, n \\ \implies & S_r(x) \subseteq G_i, \; \forall i = 1, 2, 3, \cdots, n \quad \text{(by (5.38))} \\ \implies & S_r(x) \subseteq \cap\{G_i; i = 1, 2, 3, \cdots, n\} = G. \end{aligned}$$

Hence $\forall x \in G, \exists\, r > 0$ such that $S_r(x) \subseteq G$.

Thus

$$G = \cap\{G_i; i = 1, 2, 3, \cdots, n\}$$

is open in X. ■

Theorem 5.9 *Let (X, ρ) be a metric space. A subset G of X is open if and only if it is a union of open spheres.*

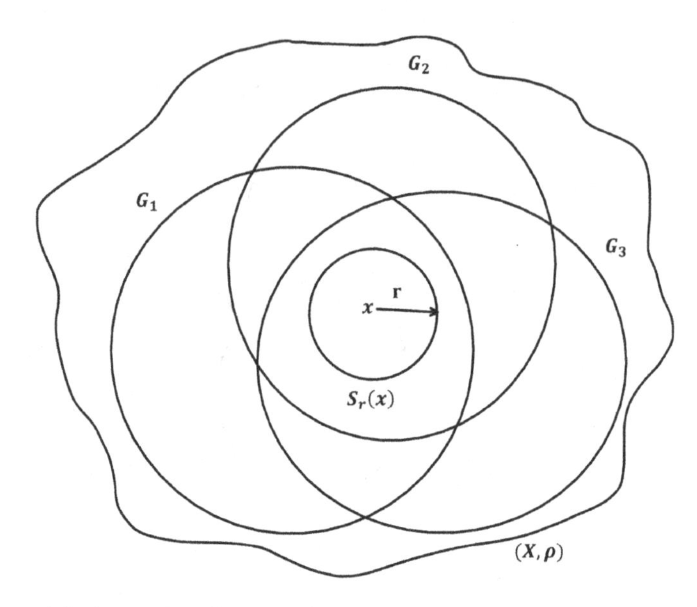

Fig. 5.20 A finite intersection of open sets is open

Proof If G is a non-empty open set, each point of G is the center of some open sphere which is contained in G. Clearly, the union of all such open spheres is precisely the set G.

Conversely, suppose G is the union of open spheres. Then by Theorem 5.6 and Theorem 5.7 we see that G is an open set. ■

Theorem 5.10 *Every non-empty open subset in the usual metric space* $(\mathbb{R}, \rho)$ *is the union of a countable disjoint class of open intervals.*

Proof Let G be a non-empty open subset of the usual metric space $(\mathbb{R}, \rho)$. Let $x \in G$. Since G is open, x is the center of open intervals contained in G. Let I_x be the union of all open intervals which contains x and is contained in G. Here we note that I_x is the maximal open interval which contains x. Then we will show

$$G = \bigcup_{n=1}^{\infty} I_{x_n}, \tag{5.39}$$

where $I_{x_1}, I_{x_2}, \cdots$ are countable open intervals such that $I_{x_m} \cap I_{x_n} = \phi,\ m \neq n$. In fact, if y is another point in G, i.e., $y \in G$, then

$$I_x = I_y \ \text{ or } \ I_x \cap I_y = \phi.$$

If $I_x \neq I_y$ and $I_x \cap I_y \neq \phi$, then $I_x \cup I_y$ would be an open interval contained in G which is larger than I_x. This is a contradiction of the definition of I_x. Thus I_x is a disjoint class of open intervals and G is obviously its union.

Every I_x contains a rational number x_n, then we put $I_{x_n} = I_x$. Since disjoint intervals cannot contain the same rational number x_n, for $I_x \cap I_y = \phi$ we can put $I_{x_m} = I_y$, $m \neq n$. Therefore mutually disjoint intervals I_{x_n} are countable. Hence G is the union (5.39) of a countable disjoint class of open intervals. ■

Theorem 5.11 *Let (X, ρ_1) and (Y, ρ_2) be two metric spaces and $f : X \to Y$. Then f is continuous if and only if $f^{-1}(G)$ is open in X whenever G is open in Y, i.e., "A function f is continuous if and only if the inverse of every open set is open."*

Proof First, assume that f is continuous and $G (\neq \phi)$ is an open set in Y, then we have to show that $f^{-1}(G)$ is open in X.

Let x be a point in $f^{-1}(G)$. Then $f(x) \in G$. Since G is open, there exists an open sphere $S_\varepsilon(f(x))$ centered at $f(x)$ such that

$$S_\varepsilon(f(x)) \subseteq G, \quad \text{for some } \varepsilon > 0.$$

By Definition 5.11 of continuity, for some $\delta > 0$ there exists an open sphere $S_\delta(x)$ such that

$$f(S_\delta(x)) \subseteq S_\varepsilon(f(x)) \subseteq G.$$

Hence

$$S_\delta(x) \subseteq f^{-1}(G), \ i.e., \ f^{-1}(G) \quad \text{is open in } X.$$

Now assume that $f^{-1}(G)$ is open in X, whenever G is open in Y, then we can show that f is continuous.

Let x be an arbitrary point in X. For $\varepsilon > 0$, let $S_\varepsilon(f(x))$ be an open sphere in Y centered at $f(x)$. Therefore, by our assumption its inverse image $f^{-1}(S_\varepsilon(f(x)))$ is an open set in X.

Thus there exists a number $\delta > 0$ such that

$$S_\delta(x) \in f^{-1}(S_\varepsilon(f(x))) \implies f(S_\delta(x)) \subseteq S_\varepsilon(f(x)).$$

Hence, f is continuous at every point of X. ■

5.10 Closed Sets

Definition 5.13 A subset F of a metric space (X, ρ) is said to be closed if it contains each of its cluster points, that is, $F' \subset F$.

When $F \subsetneqq X$, we have the following:

$$F \text{ is closed, that is, } F' \subset F \iff F^c \text{ contains no any cluster point of } F$$
$$\iff X - F = F^c \text{ is open.} \tag{5.40}$$

Specially, $X - \phi = \phi^c = X$ is open, so we see that ϕ is closed. Since X is closed, we may consider that (5.40) is applicable to X, that is, $X - X = X^c$ is open. Hence we conclude that $\phi = X^c$ is open. Thus we have the next theorem.

Theorem 5.12 *In a metric space* (X, ρ)*, both the whole space* X *and the empty set* ϕ *are open as well as closed.*

Example 5.18 We show that every singleton set in $\mathbb{R}$ is closed for the usual metric ρ in $\mathbb{R}$.

Every singleton set in $\mathbb{R}$ is closed for the usual metric ρ in $\mathbb{R}$. Let $\{a\} \subset \mathbb{R}$. Then

$$\mathbb{R} - \{a\} =]-\infty, a[\ \cup \ ]a, \infty[.$$

Since $]-\infty, a[$ and $]a, \infty[$ are open sets, their unions are also open, that is, $\mathbb{R} - \{a\}$ is open. Therefore $\{a\}$ is closed.

Example 5.19 Every closed interval is a closed set for the usual metric on $\mathbb{R}$.

Let $a, b \in \mathbb{R}$, where $a < b$. Now

$$\begin{aligned} \mathbb{R} - [a, b] &= \{x \in \mathbb{R}; x < a \text{ or } x > b\} \\ &= \{x \in \mathbb{R}; x < a\} \cup \{x \in \mathbb{R}; x > b\} \\ &=]-\infty, a[\ \cup \ ]b, \infty[, \end{aligned}$$

that is, $\mathbb{R} - [a, b]$ is open. Hence, $[a, b]$ is closed.

Theorem 5.13 *In a metric space, every closed sphere is a closed set.*

Proof Suppose (X, ρ) is a metric space and let $S_r[x_0]$ be a closed sphere in X, i.e.,

$$S_r[x_0] = \{x \in X; \rho(x, x_0) \leqslant r\} .$$

If $r = 0$, then $S_r[x_0] = \{x_0\}$ is a closed set. Let $r > 0$ and $x \in S_r[x_0]^c$. Then $x \notin S_r[x_0]$ and $\rho(x, x_0) > r$.

Let $d = \rho(x, x_0) - r$ be a positive real number. We take an open sphere $S_d(x)$ whose radius is d and center is x.

Fig. 5.21 Figure for Theorem 5.13

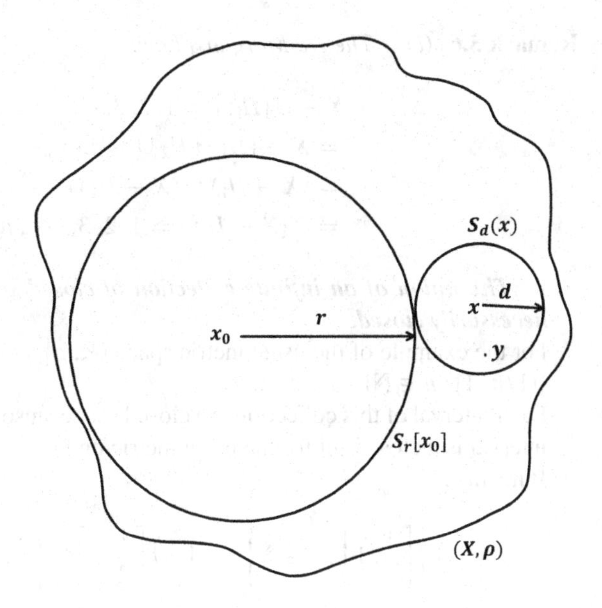

Now we show that $S_d(x) \subset S_r[x_0]^c$ (Fig. 5.21).
Let $y \in S_d(x) \Longrightarrow \rho(x, y) < d$.
Now

$$\rho(x, x_0) \leqslant \rho(x, y) + \rho(y, x_0) < d + \rho(y, x_0)$$
$$\Longrightarrow \rho(y, x_0) > \rho(x, x_0) - d = r.$$

Therefore, $\rho(y, x_0) > r \Longrightarrow y \in S_r\,[x_0]^c$.
Hence, $S_d\,(x) \subset S_r[x_0]^c$.
Thus, $S_r[x_0]^c$ is open $\Longrightarrow$ $S_r\,[x_0]$ is closed. ■

Theorem 5.14 *In a metric space (X, ρ), the finite union of closed sets is closed.*

Proof Let $\{H_i; i = 1, 2, 3, \cdots, n\}$ be a finite collection of closed sets of X. Then prove that $\cup\{H_i; i = 1, 2, 3, \cdots, n\}$ is also a closed set.

Since H_i is closed for $i = 1, 2, 3, \cdots, n$,
$\Longrightarrow (X - H_i)$ is open for $i = 1, 2, 3, \cdots, n$ (by the definition of a closed set)
$\Longrightarrow \cap\{X - H_i; i = 1, 2, 3, \cdots, n\}$ is open (by Theorem 5.8)
$\Longrightarrow X - \cup\{H_i; i = 1, 2, 3, \cdots, n\}$ is open
(by De-Morgan's Law: $A - (B \cup C) = (A - B) \cap (A - C)$)
$\Longrightarrow \cup\{H_i; i = 1, 2, 3, \cdots, n\}$ is closed.

Hence, the finite collection of closed sets is closed. ■

Remark 5.6 (i) *The De-Morgan's Law:*

$$\begin{aligned}&X - \cup\{H_i; i = 1, 2, 3, \cdots, n\}\\&= X - \{H_1 \cup H_2 \cup \cdots \cup H_n\}\\&= (X - H_1) \cap (X - H_2) \cap \cdots (X - H_n)\\&= \cap\{X - H_i; i = 1, 2, 3, \cdots, n\}.\end{aligned}$$

(ii) ***The union of an infinite collection of closed sets in a metric space is not necessarily closed.***
For the example of the usual metric space $[\mathbb{R}, \rho]$, consider the infinite collection $\{[1/n, 1]\,; n \in \mathbb{N}\}$.
Each interval of this collection is a closed set because we know that "every closed interval is a closed set for the usual metric on $\mathbb{R}$."
But, since

$$\cup\left\{\left[\frac{1}{n}, 1\right]; n \in \mathbb{N}\right\} = \{1\} \cup \left[\frac{1}{2}, 1\right] \cup \left[\frac{1}{3}, 1\right] \cup \cdots =]\,0, 1],$$

this infinite collection is not closed.
Thus, the union of an infinite collection of closed sets is not closed.

Theorem 5.15 *In a metric space (X, ρ), the intersection of an arbitrary collection of closed subsets is closed.*

Proof Let $\{H_\lambda; \lambda \in \Lambda\}$ be an arbitrary collection of closed subsets of a metric space (X, ρ). Then prove that $\cap\{H_\lambda; \lambda \in \Lambda\}$ is also a closed set.

$$\begin{aligned}&\text{In fact, } H_\lambda \text{ is closed for } \lambda \in \Lambda\\&\Longrightarrow (X - H_\lambda) \text{ is open } \forall \lambda \in \Lambda\\&\Longrightarrow \cup\{X - H_\lambda;\ \forall \lambda \in \Lambda\} \text{ is open (by Theorem 5.7)}\\&\Longrightarrow X - \cap\{H_\lambda;\ \forall \lambda \in \Lambda\} \text{ is open}\\&\qquad \text{(by De-Morgan's Law; } A - (B \cap C) = (A - B) \cup (A - C))\\&\Longrightarrow \cap\{H_\lambda;\ \forall\, \lambda \in \Lambda\} \text{ is closed.}\end{aligned}$$

Hence, the intersection of an arbitrary collection of closed subsets of X is closed. ■

Theorem 5.16 *A subset of a metric space is closed if and only if its complement is open.*

Proof Let (X, ρ) be a metric space and let $F\ (\subset X)$ be closed.

(i) If $F^c = \phi$, then F^c is open.
(ii) If $F^c \neq \phi$, then let $x \in F^c \Longrightarrow x \notin F$. F is closed $\Longrightarrow$ x is not a cluster point of $F \Longrightarrow$ There exists $r > 0$ such that

$$\begin{aligned} &S_r(x) \cap F = \phi \\ &\Longrightarrow x \in S_r(x) \subseteq F^c. \end{aligned}$$

Hence F^c is open.

Conversely, let F^c be open. And if $x \in F^c$, then there exists $r > 0$ such that

$$S_r(x) \subset F^c, \quad i.e., \quad S_r(x) \cap F = \phi \quad \Longrightarrow \quad x \text{ cannot be a cluster point of } F.$$

Since $x \in F^c$ is an arbitrary point, F does not have any cluster point outside it. Hence F is closed.

■

5.11 The Closure of a Set

Definition 5.14 Let A be any subset of a metric space (X, ρ). The closure of A, denoted by $\overline{A}$, is the set of all adherent points A, i.e.,

$$\overline{A} = A \cup A',$$

that is,

$$\overline{A} = \{x \in X; S_r(x) \cap A \neq \phi, \text{ for all } r > 0\}.$$

Theorem 5.17 *Let A and B be two subsets of a metric space (X, ρ). Then*

(*i*) *$\overline{A}$ is a closed set.*
(*ii*) *A is closed if and only if $A = \overline{A}$. In particular, $\phi = \overline{\phi}$ and $X = \overline{X}$.*
(*iii*) *$A \subseteq B$, then $\overline{A} \subseteq \overline{B}$.*
(*iv*) *If F is a closed subset of X such that $A \subseteq F$, then $\overline{A} \subseteq F$.*
(*v*) *$\overline{A}$ is the intersection of all the closed subsets of X which contain A.*
(*vi*) *$\overline{A} \cup \overline{B} = \overline{A \cup B}$ and $\bar{A} \cap \bar{B} \supseteq \overline{A \cap B}$.*

Proof (*i*) If $x \in X$ and $x \notin \bar{A}$, then x is neither a point of A nor a cluster point of A. Hence the neighborhood of x does not intersect A, i.e., there exists $r > 0$ such that $S_r(x) \cap A = \phi$. Therefore the complement of $\overline{A}$ is open. Hence, $\overline{A}$ is closed.

(*ii*) If A is closed, then every cluster point of A is a point of A. Therefore $\overline{A} = A \cup A'$, $A' \subset A$. Hence, $A = \overline{A}$.

(*iii*) Let $x \in \overline{A}$. Then by the definition of the closure set 5.11, for $r > 0$, we have

$$S_r(x) \cap A \neq \phi \quad \Longrightarrow \quad S_r(x) \cap B \neq \phi \quad \Longrightarrow \quad x \in \overline{B}.$$

Hence, $\overline{A} \subseteq \overline{B}$.

(*iv*) F is closed, therefore F also contains all cluster points of A. Thus, $\overline{A} \subset \overline{F} = F$.

(v) Let H be an intersection of closed sets containing A, i.e.,

$$H = \cap \{B \subset X;\ B \text{ is closed and } B \supset A\}.$$

Then by Theorem 5.15, H is closed. From (ii) and (iii), we have

$$A \subset H \Longrightarrow \overline{A} \subset \overline{H} \Longrightarrow \overline{A} \subset H.$$

On the other hand, we see $A \subset \overline{A}$, so $H \subset \overline{A}$. Thus, we have $\overline{A} = H$.

(vi) A cluster point of A is a cluster point of $A \cup B$, so that $\overline{A} \subseteq \overline{A \cup B}$, similarly $\overline{B} \subseteq \overline{A \cup B}$. Hence, $\overline{A} \cup \overline{B} \subseteq \overline{A \cup B}$.

Now, we have to show that $\overline{A \cup B} \subseteq \bar{A} \cup \bar{B}$. Let $x \in \overline{A \cup B}$ and $x \notin \bar{A} \cup \bar{B}$. Therefore, x is neither a cluster point of A nor B, i.e., there exist open spheres $S_{r_1}(x)$ and $S_{r_2}(x)$ containing no points of A and B, respectively. Let $r = \min\{r_1, r_2\}$. Then the open sphere $S_r(x)$ does not contain any point of A and B. Hence $S_r(x)$ contains no point of $A \cup B$.
Therefore, x cannot be cluster points of $A \cup B$, i.e., $x \notin \overline{A \cup B}$. It is a contradiction. Therefore, $x \in \overline{A \cup B} \implies x \in \bar{A} \cup \bar{B}$. Hence

$$\overline{A \cup B} \subseteq \overline{A} \cup \overline{B}.$$

Thus, we have $\overline{A \cup B} = \overline{A} \cup \overline{B}$.
Next we show that $\overline{A \cap B} \subseteq \overline{A} \cap \overline{B}$. Let $x \notin \overline{A} \cap \overline{B}$. Then there exists $r > 0$ such that

$$S_r(x) \cap (A \cap B) = \phi,$$

that is,

$$x \notin \overline{A \cap B}.$$

Hence, we have

$$\overline{A \cap B} \subseteq \overline{A} \cap \overline{B}.$$

■

5.11.1 *Interior Points*

Definition 5.15 Let A be any subset of a metric space (X, ρ). A point x of A is called an interior point of A, if there exists $r > 0$ such that (Fig. 5.22)

$$x \in S_r(x) \subseteq A.$$

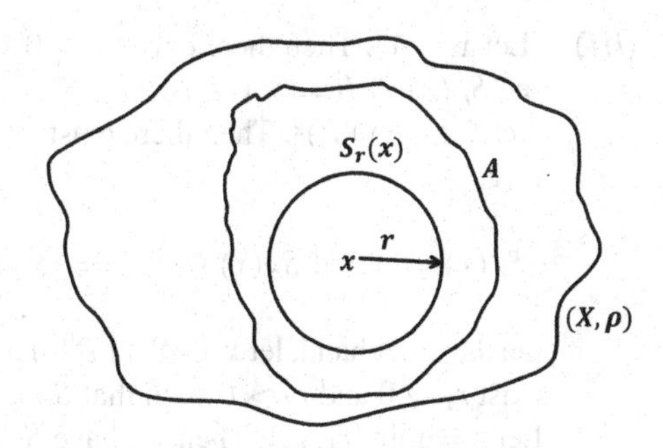

Fig. 5.22 Illustration of interior points

5.11.2 The Interior of a Set

The set of all interior points of A is called the interior of A and is denoted by *int* A or A^0 or A^i. Of course $A^0 \subseteq A$. Symbolically,

$$int\ A = \{x; x \in A \text{ and } S_r(x) \subseteq A, \text{ for some } r > 0\}.$$

Example 5.20 Let A be a subset of the usual metric space $(\mathbb{R}, \rho)$. Then

(i) if $A = [a, b], [a, b), (a, b], (a, b)$, then $A^0 = (a, b)$.
(ii) if $A = \mathbb{N}, \mathbb{Z}, \mathbb{Q}$, then $A^0 = \phi$.
(iii) if $A = \mathbb{R}$, then $A^0 = \mathbb{R}$.

Theorem 5.18 *Let A and B be two subsets of the metric space* (X, ρ). *Then*

(i) *A^0 is open.*
(ii) *A^0 is the largest open subset which is contained in A.*
(iii) *if $A \subseteq B$, then $A^0 \subseteq B^0$.*
(iv) *$A^0 \cap B^0 = (A \cap B)^0$ and $A^0 \cup B^0 \subseteq (A \cup B)^0$.*

Proof (i) Let x be an arbitrary point of A^0. Then by the definition, there exists $r > 0$ such that $S_r(x) \subset A$. Now if we take $y \in S_r(x)$, then there exists $0 < r_1 < r$ such that

$$S_{r_1}(y) \subset S_r(x) \subset A.$$

This means that every point $y \in S_r(x)$ is an interior point.

Hence, we have

$$S_r(x) \subset A^0,$$

that is, A^0 is an open set.

(ii) Let G be an open set contained in A. Let x be an arbitrary point of G. Then, by the definition, there exists $r > 0$ such that $S_r(x) \subset G \subset A$. We have $x \in G \Longrightarrow x \in A^0$. Therefore we have $G \subset A^0 \subset A$ and A^0 is open by (i). Thus A^0 is the largest open subset which is contained in A.

(iii) Let $x \in A^0$. Then there exists $r > 0$ such that $S_r(x) \subseteq A$. Since $A \subseteq B$ we see $S_r(x) \subseteq B \Longrightarrow x \in B^0$.

(iv) Let $x \in (A \cap B)^0$. Then there exists $r > 0$ such that $S_r(x) \subseteq A \cap B$. Therefore,

$$S_r(x) \subseteq A \text{ and } S_r(x) \subseteq B \Longrightarrow x \in A^0 \text{ and } x \in B^0 \Longrightarrow x \in A^0 \cap B^0.$$

On the other hand, let $x \in A^0 \cap B^0$. Therefore $x \in A^0$ and $x \in B^0$. Then there exist $r_1 > 0$ and $r_2 > 0$ such that $S_{r_1}(x) \subseteq A$ and $S_{r_2}(x) \subseteq B$.
Let $r = \min\{r_1, r_2\}$. Then we have $S_r(x) \subseteq A \cap B \Longrightarrow x \in (A \cap B)^0$. Thus $A^0 \cap B^0 = (A \cap B)^0$. Using (iii), we have

$$A \subseteq A \cup B \text{ and } B \subseteq A \cup B \Longrightarrow A^0 \subseteq (A \cup B)^0 \text{ and } B^0 \subseteq (A \cup B)^0.$$

Hence, $A^0 \cup B^0 \subseteq (A \cup B)^0$.

■

Example 5.21 Let $A = [0, 2]$, $B = [2, 4]$ and $A \cup B = [0, 4]$ be sets in the usual metric space $(\mathbb{R}, \rho)$. Then we see that $A^0 = (0, 2)$, $B^0 = (2, 4)$ and $(A \cup B)^0 = (0, 4)$ are sets in the usual metric space $(\mathbb{R}, \rho)$. Show that $A^0 \cup B^0 \subsetneq (A \cup B)^0$.

5.11.3 Exterior Points and Exterior Sets

Definition 5.16 A point $x \in X$ is an exterior point of A, if it is an interior point of the complement of A, i.e., if there exists $r > 0$ such that an open sphere $S_r(x)$ and $S_r(x) \subseteq A^c$ or $S_r(x) \cap A = \phi$. Then we write (Fig. 5.23)

$$ext\ A = \{x;\ x \text{ is an exterior point of } A\}.$$

Definition 5.17 The set of all exterior points of A is called the exterior of A, denoted by $ext\ A$, i.e.,

$$ext\ A = int\ A^c.$$

Theorem 5.19 *Let A be any subset of a metric space (X, ρ). Then we have*

$$ext\ A = \left(\overline{A}\right)^c \quad and \quad int\ A = ext\ A^c = \left(\overline{A^c}\right)^c.$$

Proof We see

$$x \in ext\ A \Longleftrightarrow x \notin \overline{A} \quad \therefore\ ext\ A = \left(\overline{A}\right)^c,$$

and

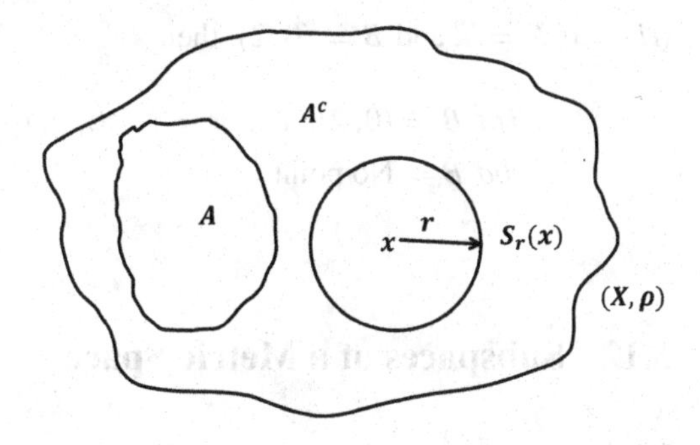

Fig. 5.23 Illustration of exterior points

$$
\begin{aligned}
&x \in int\ A \Longleftrightarrow x \in ext\ A^c, \\
&x \in int\ A \Longleftrightarrow \text{there exists } r > 0 \text{ such that } S_r(x) \subset A \\
&\Longleftrightarrow x \notin \overline{A^c} \Longleftrightarrow x \in \left(\overline{A^c}\right)^c.
\end{aligned}
$$

Therefore, we have the result. ■

5.11.4 Frontier Points and Boundary Points

A point $x \in X$ is said to be a *frontier point (frt)* of $S \subset X$ if it is neither an interior nor an *exterior point (ext)* of S. If the frontier point belongs to S, then it is called a *boundary point (bd)* of S (Fig. 5.24).

Example 5.22 (i) If $X = \mathbb{R}$ and $S = [0, 2)$, then

$$
\begin{aligned}
&int\ S = (0, 2);\ ext\ S = (-\infty, 0) \cup (2, \infty);\ frt\ S = \{0, 2\}; \\
&\quad bd\ S = \{0\}.
\end{aligned}
$$

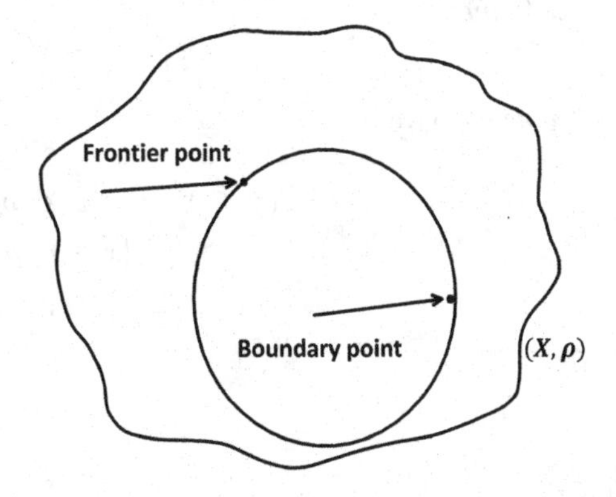

Fig. 5.24 The figure illustrates the frontier and boundary points

(*ii*) If $X = \mathbb{R}$ and $B = (0, 2)$, then

$$int\ B = (0, 2);\ ext\ B = (-\infty, 0) \cup (2, \infty);\ frt\ B = \{0, 2\};$$
$$bd\ B = \text{No point.}$$

5.12 Subspaces of a Metric Space

Let (X, ρ) be a metric space and Y be a non-empty subset in X. The function $\rho_Y : Y \times Y \to \mathbb{R}$, defined by

$$\rho_Y(x, y) = \rho(x, y), \quad \forall x, y \in Y,$$

holds all the properties $(M1)$ through $(M4)$ for the points of the subset Y. The metric ρ_Y is the induced metric on Y. The space (Y, ρ_Y) is called the metric subspace of the metric (X, ρ).

The intervals $[0, 1]$, $(0, 1)$, $[0, 1)$ and the set of rational numbers $\mathbb{Q}$ are the usual metric subspace of $\mathbb{R}$. The real line itself is a subspace of the space $(\mathbb{C}, \rho)$ of complex numbers. The open unit disc, closed unit disc, and the unit circle are subspaces of the space of complex numbers.

Remark 5.7 A subspace of the usual metric space $\mathbb{R}$ in which there is an open subset for the subspace which is not open in $\mathbb{R}$.

Example 5.23 Let Y be a subspace of the usual metric space $(\mathbb{R}, \rho)$.

(*i*) If $Y = [0, 1]$, then the set $\left[0, \frac{1}{3}\right)$ is open in Y but not in $\mathbb{R}$.
(*ii*) If $Y = [0, 1]$, then the set $\left(0, \frac{1}{3}\right]$ is closed in Y but not in $\mathbb{R}$.

Lemma 5.2 *Let (Y, ρ_Y) be a subspace of a metric space (X, ρ_X). If $y \in Y$ and $r > 0$, then*

$$S_r^Y(y) = S_r^X(y) \cap Y.$$

Proof We have

$$\begin{aligned} S_r^X(y) \cap Y &= \{x \in X; \rho_X(x, y) < r\} \cap Y \\ &= \{x \in Y; \rho_Y(x, y) < r\} \\ &= S_r^Y(y) \quad (\because Y \subseteq X). \end{aligned}$$

■

Example 5.24 The set $\mathbb{Q}$ of all the rational numbers can be taken as a subspace of the usual metric space $\mathbb{R}$. It means that if x and y are rational numbers, the distance between x and y is $|x-y|$, just as it is in the usual metric space $\mathbb{R}$. On the other hand, if $\mathbb{Q}$ has the discrete metric, then it will not be a subspace of $\mathbb{R}$ because a different metric is being used on $\mathbb{Q}$.

If Y is a subspace of X, it is important to distinguish between metric spaces (X, ρ_X) and (Y, ρ_Y). For example, if $p \in Y$, and $r > 0$ then open spheres

$$S_r^X(p) = \{x \in X; \rho_X(x, p) < r\} \quad \text{and} \quad S_r^Y(p) = \{x \in Y; \rho_Y(x, p) < r\}$$

are not necessarily the same. The sphere $S_r^Y(p)$ contains only points of Y.

Theorem 5.20 *Let (Y, ρ_Y) be a subspace of the metric space (X, ρ_X) and $A \subseteq Y$. Then*

(i) A is open in Y if and only if there exists a set G which is open in X such that $A = G \cap Y$.
(ii) A is closed in Y if and only if there exists a set F which is closed in X such that $A = F \cap Y$.

Proof Let Y be a subspace in X, and let $A \subseteq Y$.

(i) Let $S_r^X(x)$ and $S_r^Y(x)$ be two open spheres centered at x and radius $r > 0$, respectively, for the metric spaces (X, ρ_X) and (Y, ρ_Y).

Let A be open in Y and let $x \in A$ be arbitrary such that $S_r^Y(x) \subset A$

$$A = \bigcup_{x \in A} S_r^Y(x) = \bigcup_{x \in A} \left(S_r^X(x) \cap Y\right) = G \cap Y, \quad \text{(from Lemma 5.2)}$$

where $G = \bigcup_{x \in A} S_r^X(x)$ is a union of open spheres in X. Here we know that G is an open set in X (Theorem 5.7).

Conversely, let G be open in X and $A = G \cap Y$, and let $x \in A$ be an arbitrary element. Then for $x \in G$, there exists $r > 0$ such that $S_r^X(x) \subset G$. Since $x \in Y$, from Lemma 5.2, we have

$$S_r^Y(x) = S_r^X(x) \cap Y \subset G \cap Y = A \implies S_r^Y(x) \subset A.$$

Thus, A is open in Y.

(ii) We have

$$\begin{aligned}
A \text{ is closed in } Y &\iff Y - A \text{ is open in } Y \\
&\iff Y - A = G \cap Y, \quad \text{where } G \text{ is an open set in } X \\
&\iff A = Y - G \cap Y \quad (\text{from } (i)) \\
&\iff A = X \cap Y - G \cap Y \\
&\iff A = (X - G) \cap Y \\
&\iff A = F \cap Y,
\end{aligned}$$

where $F = X - G$ is a closed set in X.

■

Theorem 5.21 *Let (Y, ρ_Y) be a subspace of the metric space (X, ρ). Then*

(i) *every subset of Y, which is open in Y, is also open in X if and only if Y is open in X.*
(ii) *every subset of Y, which is closed in Y, is also closed in X if and only if Y is closed in X.*

Proof Let Y be a subspace in X.

(i) Suppose every subset of Y, which is open in Y, is also open in X. We will show that Y is open in X. Especially, since Y is an open subset of Y, it must be open in X.

Conversely, suppose Y is open in X. Let A be an open subset of Y. By Theorem 5.20, there exists an open subset G of X such that $A = G \cap Y$. Since G and Y are open subsets of X, from Theorem 5.8, their intersections are also open in X. Therefore, A must be open in X.
(ii) The proof is similar to (i), so we omit the proof.

■

Theorem 5.22 *Let (Y, ρ_Y) be a subspace of a metric space (X, ρ) and $A \subset Y$. Then $x \in Y$ is a cluster point of A in Y if and only if x is a cluster point of A in X.*

Proof Let $x \in Y$ be a cluster point of A in Y. Then for $r > 0$

$$\left(S_r^Y(x) - \{x\}\right) \cap A \neq \phi, \tag{5.41}$$

where $S_r^Y(x)$ is an open sphere in Y. For any given $r > 0$, we have

$$\begin{aligned}
\left(S_r^X(x) - \{x\}\right) \cap A &\supset \left(S_r^X(x) \cap Y - \{x\}\right) \cap A \\
&= \left(S_r^Y(x) - \{x\}\right) \cap A \quad (\text{from Lemma 5.2}) \\
&\neq \phi \quad (\text{from (5.41)}).
\end{aligned}$$

Hence, x is a cluster point of A in X. The converse can be easily established by retracting the above steps. ■

5.12.1 Convergent Sequences

Let (X, ρ) be a metric space. We consider the sequence $\{x_n\} = \{x_1, x_2, x_3, \cdots\}$ of points of X is said to converge to a point x of X.

If for every $\varepsilon > 0$ there exists a positive integer n_0 such that

$$\rho(x_n, x) < \varepsilon, \ \forall n \geqslant n_0, \ i.e., \ \rho(x_n, x) \to 0 \text{ as } n \to \infty,$$

the point x is called the limit of a sequence $\{x_n\}$. Symbolically

$$\lim_{n\to\infty} x_n = x, \ i.e., \ x_n \to x \text{ as } n \to \infty.$$

A sequence which is not convergent is said to be divergent.

Theorem 5.23 *Let (X, ρ) be a metric space. Let $\{x_n\}$ be a convergent sequence in X. Then the limit of $\{x_n\}$ is unique.*

Proof We suppose that

$$x_n \to x, \ x_n \to x' \text{ as } n \to \infty, \ \text{ and } \ \varepsilon = \rho\left(x, x'\right) > 0.$$

Then there is a positive integer N_1 such that

$$\rho(x_n, x) < \varepsilon/2, \ \forall n \geqslant N_1.$$

Similarly, there is a positive integer N_2 such that

$$\rho(x_n, x') < \varepsilon/2, \ \forall n \geqslant N_2.$$

Now let $N = \max\{N_1, N_2\}$, so that

$$\rho(x, x') \leqslant \rho(x, x_n) + \rho(x_n, x') < \frac{\varepsilon}{2} + \frac{\varepsilon}{2} = \varepsilon = \rho(x, x'),$$

which is a contradiction of our hypothesis. Therefore, the limit is unique. We can express this theorem in the following figure (Fig. 5.25). ■

Remark 5.8 If the range of the sequence forms a bounded set, then the sequence is called a *bounded sequence* in a metric space.

Theorem 5.24 *In a metric space, every convergent sequence is bounded.*

Proof Let $\{x_n\}$ be a convergent sequence in a metric space (X, ρ) such that $x_n \to x$, as $n \to \infty$. For each $\varepsilon = 1$, there exists a positive integer N such that

$$\rho(x_n, x) < 1, \ \forall n \geqslant N.$$

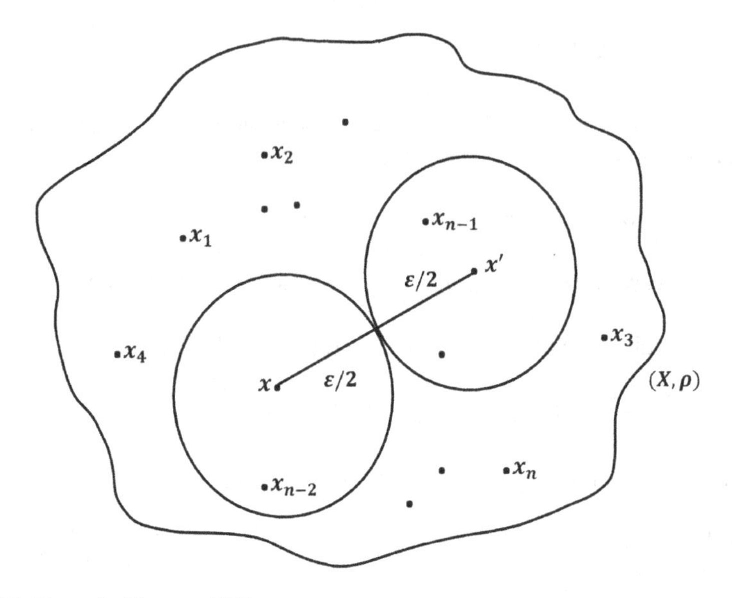

Fig. 5.25 Figure for Theorem 5.23

Let $r = \max\{1, \rho(x_n, x), 1 \leqslant n \leqslant N\}$. Therefore $\rho(x_n, x) \leqslant r, \quad \forall n \in \mathbb{N}$. Now

$$\rho(x_n, x_m) \leqslant \rho(x_n, x) + \rho(x, x_m) \leqslant 2r, \quad \forall n, m \in \mathbb{N}.$$

The diameter of the range of the sequence is bounded by $2r$. ■

Theorem 5.25 *Let A be a subset of a metric space (X, ρ). Then a point x of X is a cluster point of A if and only if there exists a sequence $\{x_n\}$ of points A, none of which equals x, such that $\lim_{n\to\infty} x_n = x$.*

Proof Let $x \in X$ be a cluster point of A. Construct a sequence $\{x_n\}$ such that for each positive integer n, $x_n \in \left(S_{\frac{1}{n}}(x) - \{x\}\right) \cap A$. All the points of the sequence $\{x_n\}$ are in A and none of which equals x. Therefore, for all n

$$x_n \in S_{\frac{1}{n}}(x) \Longrightarrow \rho(x_n, x) < \frac{1}{n} \Longrightarrow \lim_{n\to\infty} x_n = x.$$

Conversely, suppose that $\{x_n\}$ is a sequence in A, and none of which equals x such that $\lim_{n\to\infty} x_n = x$. We will show that $(S_\varepsilon(x) - \{x\}) \cap A \neq \phi$. For given $\varepsilon > 0$, there exists a positive integer N such that

$$x_n \in S_\varepsilon(x), \quad \forall n \geqslant N.$$

Thus $(S_\varepsilon(x) - \{x\}) \cap A \neq \phi$. Hence, x is a cluster point. ■

Corollary 5.2 *Let A be a subset of a metric space X. If $\{x_n\}$ is a sequence of points of A which converges to the point x of X, then $x \in \overline{A}$.*

Corollary 5.3 *A subset A of a metric space X is closed if and only if every convergent sequence of points of A have the cluster point in A.*

Theorem 5.26 *Let (X, ρ_X) and (Y, ρ_Y) be two metric spaces. Let $\{(x_n, y_n)\}$ be a sequence in the product metric space $(X \times Y, \rho)$ with $\rho = \rho_X + \rho_Y$, and let the sequence converge to (x, y). Then equivalently, the sequence $\{x_n\}$ converges to x in X and $\{y_n\}$ converges to y in Y.*

Proof The product metric $\rho : X \times Y \to \mathbb{R}$ is defined as

$$\rho\left((x_1, y_1), (x_2, y_2)\right) = \rho_X\left(x_1, x_2\right) + \rho_Y\left(y_1, y_2\right),$$

and $(X \times Y, \rho)$ is a metric space. Let $x_n \to x$ in X and $y_n \to y$ in Y. Then for each $\varepsilon > 0$, there exists $N_1, N_2 \in \mathbb{N}$ such that

$$\rho_X\left(x_n, x\right) < \frac{\varepsilon}{2}, \ \forall n \geqslant N_1 \quad \text{and} \quad \rho_Y\left(y_n, y\right) < \frac{\varepsilon}{2}, \ \forall n \geqslant N_2.$$

Now, for all $n \geqslant \max\{N_1, N_2\}$, we have

$$\begin{aligned} \rho\left((x_n, y_n), (x, y)\right) &= \rho_X\left(x_n, x\right) + \rho_Y\left(y_n, y\right) \\ &< \frac{\varepsilon}{2} + \frac{\varepsilon}{2} = \varepsilon. \end{aligned}$$

Hence, $(x_n, y_n) \to (x, y)$.

Conversely, $(x_n, y_n) \to (x, y)$ in $X \times Y$. Then, for each $\varepsilon > 0$, there exists $N \in \mathbb{N}$ such that

$$\begin{aligned} &\rho\left((x_n, y_n), (x, y)\right) < \varepsilon, \ \forall n \geqslant N \\ \Longrightarrow\ &\rho_X\left(x_n, x\right) + \rho_Y\left(y_n, y\right) < \varepsilon, \ \forall n \geqslant N \\ \Longrightarrow\ &\rho_X\left(x_n, x\right) < \varepsilon \ \text{ and } \ \rho_Y\left(y_n, y\right) < \varepsilon, \ \forall n \geqslant N \\ \Longrightarrow\ &x_n \to x \ \text{ and } \ y_n \to y. \end{aligned}$$

■

5.13 Cauchy Sequences

A sequence $\{x_n\}$ of points of (X, ρ) is said to be a Cauchy sequence if for each $\varepsilon > 0$ there exists a positive integer n_0 such that

$$\rho\left(x_n, x_m\right) < \varepsilon, \quad \forall n, m \geqslant n_0, \quad i.e., \quad \rho\left(x_n, x_m\right) \to 0 \ \text{as} \ n, m \to \infty.$$

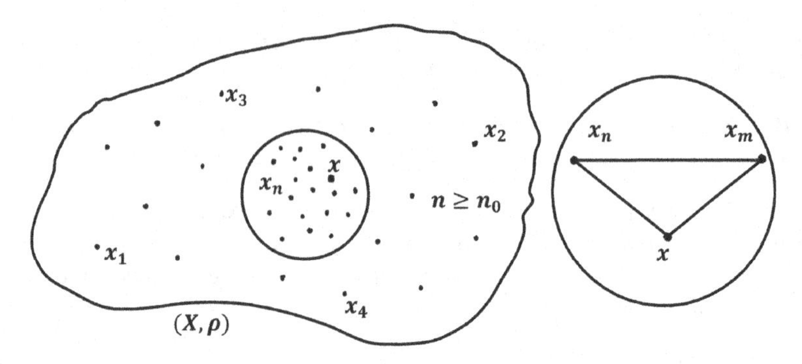

Fig. 5.26 Convergent Cauchy sequence in (X, ρ)

Roughly we can say that all the points of the sequence after some index value are close to each other.

Theorem 5.27 *Every convergent sequence in a metric space (X, ρ) is a Cauchy sequence.*

Proof Let $\{x_n\}$ be a convergent sequence in a metric space (X, ρ) and let x be a limit of x_n (i.e., x_n converges to x). Then we see that for a given $\varepsilon > 0$ there exists $n_0 \in \mathbb{N}$ such that

$$\rho(x_n, x) < \frac{\varepsilon}{2}, \quad \forall n \geqslant n_0.$$

Hence, if $m, n \geqslant n_0$, then

$$\rho(x_m, x) < \frac{\varepsilon}{2} \quad \text{and} \quad \rho(x_n, x) < \frac{\varepsilon}{2}.$$

We have (Fig. 5.26)

$$\begin{aligned}
\rho(x_n, x_m) &\leqslant \rho(x_n, x) + \rho(x, x_m) \quad \text{(by Definition 5.1)} \\
&\leqslant \rho(x_n, x) + \rho(x_m, x) \quad \text{(by Definition 5.1)} \\
&< \frac{\varepsilon}{2} + \frac{\varepsilon}{2} = \varepsilon, \quad \forall m, n \geqslant n_0.
\end{aligned}$$

Hence, $\{x_n\}$ is a Cauchy sequence.

But $\{x_n\}$ is not convergent:

$$x_n = \frac{1}{n} \Longrightarrow \lim_{n \to \infty} x_n = \lim_{n \to \infty} \frac{1}{n} = 0 \notin (0, 1] = X.$$

Since 0 is not a point of X, $\{x_n\}$ is not a convergent sequence in X, but $\{x_n\}$ is a Cauchy sequence.

Thus the converse of the above Theorem 5.27 is not necessarily true. ■

Theorem 5.28 *Every Cauchy sequence in a metric space (X, ρ) is bounded.*

Proof Let $\{x_n\}$ be a Cauchy sequence in a metric space (X, ρ).

By the definition of the Cauchy sequence, for $\varepsilon = 1$ there exists $N \in \mathbb{N}$ such that

$$\rho(x_m, x_n) < 1, \quad \forall m, n \geqslant N.$$

In a particular form, $m = N$

$$\rho(x_N, x_n) < 1, \; \forall n \geqslant N \quad \Longrightarrow \quad x_n \in S_1(x_N), \; \forall n \geqslant N.$$

Let $r = \max\{1, \rho(x_N, x_1), \rho(x_N, x_2), \cdots, \rho(x_N, x_{N-1})\}$.

$$\therefore \; x_n \in S_r(x_N), \quad \text{for all } n \in \mathbb{N}.$$

For $r \geqslant 1$, we have

$$x_n \in S_r(x_N), \text{ for all } n \in \mathbb{N} \quad \Longrightarrow \quad \rho(x_N, x_n) < r.$$

Thus, $\{x_n\}_{n=1}^{\infty}$ is bounded. ■

Theorem 5.29 *Let (X, ρ) be a metric space. If a Cauchy sequence has a convergent subsequence, then the sequence is convergent.*

Proof Let $\{x_n\}$ be a Cauchy sequence in a metric space (X, ρ) and let $\{x_{n_k}\}$ be a convergent subsequence of the sequence $\{x_n\}$, i.e.,

$$\{x_{n_k}\} = \{x_{n_1}, x_{n_2}, x_{n_3}, \cdots\}$$

and for $k = 1, 2, 3, \cdots$, $n_k \in \mathbb{N}$, where $n_1 < n_2 < n_3 < \cdots$.

Let a subsequence $\{x_{n_k}\}$ converge to $l \in X$, i.e.,

$$\lim_{k\to\infty} x_{n_k} = l \Longrightarrow \rho(x_{n_k}, l) \to 0 \text{ as } k \to \infty, \tag{5.42}$$

that is, for a given $\varepsilon > 0$ there exists $N \in \mathbb{N}$ such that

$$\rho(x_{n_k}, l) < \frac{\varepsilon}{2}, \quad \forall n_k \geqslant N.$$

By Theorem 5.27, a sequence $\{x_{n_k}\}$ is a Cauchy sequence.

Now $k \to \infty \Longrightarrow n_k \to \infty$. Hence, by the definition of a Cauchy sequence

$$\rho(x_n, x_{n_k}) \to 0 \,(n, k \to \infty), \tag{5.43}$$

that is, for a given $\varepsilon > 0$ there exists $N \in \mathbb{N}$ such that

$$\rho\left(x_n, x_{n_k}\right) < \frac{\varepsilon}{2}, \quad n, n_k \geqslant N.$$

$$0 \leqslant \rho\left(x_n, l\right) \leqslant \rho\left(x_n, x_{n_k}\right) + \rho\left(x_{n_k}, l\right) < \varepsilon, \quad (n, k \to \infty) \text{ (by } M4). \tag{5.44}$$

From (5.42), (5.43) and (5.44), we get

$$\rho\left(x_n, l\right) \to 0 \text{ as } n \to \infty \implies \lim_{n\to\infty} x_n = l.$$

Hence, the sequence $\{x_n\}$ is convergent. ∎

Theorem 5.30 *If $\{x_n\}$ is a Cauchy sequence in a metric space (X, ρ) and if there is a limit point x of the set $\{x_n; n \in \mathbb{N}\}$, then $\{x_n\}$ converges to x.*

Corollary 5.4 *If a Cauchy sequence has a finite range, then it is convergent.*

Proof Let $\{s_n\}_{n=1}^{\infty} = \{a_1, a_2, \cdots, a_m\}$ for some $m \in \mathbb{N}$. Then we define $\alpha = \min\{|a_i - a_j|;\ i, j = 1, 2, \cdots, m,\ i \neq j\}$. We take $0 < \varepsilon < \alpha$. If $|s_n - s_n'| < \varepsilon$, then we have $s_n = s_n'$, and there exists $a_k = a$ (say) for some $1 \leqslant k \leqslant m$ and $N > 0$ such that $s_n = a_k$ for every $n \geqslant N$. Thus, we conclude $\lim_{n\to\infty} a_n = a$. ∎

Theorem 5.31 *The range of a Cauchy sequence is a bounded set.*

Theorem 5.32 *Let (X, ρ_X) and (Y, ρ_Y) be two metric spaces. If $f : X \to Y$ is a uniformly continuous mapping of metric spaces and if $\{x_n\}$ is a Cauchy sequence in X, then $\{f(x_n)\}$ is a Cauchy sequence in Y.*

Remark 5.9 Theorem 5.32 need not be true if the condition of uniform continuity is replaced by continuity.

Theorem 5.33 *Let (X, ρ_X) and (Y, ρ_Y) be two metric spaces. Let $\{(x_n, y_n)\}$ be a Cauchy sequence in the product metric space $(X \times Y, \rho)$ with $\rho = \rho_X + \rho_Y$ if and only if $\{x_n\}$ and $\{y_n\}$ are Cauchy sequences in X and Y, respectively.*

Proof The product metric $\rho : X \times Y \to \mathbb{R}$ is defined as

$$\rho\left(\left(x_1, y_1\right), \left(x_2, y_2\right)\right) = \rho_X\left(x_1, x_2\right) + \rho_Y\left(y_1, y_2\right),$$

where $(x_1, y_1), (x_2, y_2) \in X \times Y$. Since $\{(x_n, y_n)\}$ is a Cauchy sequence in $(X \times Y, \rho)$, for each $\varepsilon > 0$, there exists a positive integer N such that

$$\begin{aligned}
&\rho\left(\left(x_m, y_m\right), \left(x_n, y_n\right)\right) < \varepsilon, \quad \forall m, n \geqslant N \\
\implies &\rho_X\left(x_m, x_n\right) + \rho_Y\left(y_m, y_n\right) < \varepsilon, \quad \forall m, n \geqslant N \\
\implies &\rho_X\left(x_m, x_n\right) < \varepsilon \text{ and } \rho_Y\left(y_m, y_n\right) < \varepsilon, \quad \forall m, n \geqslant N.
\end{aligned}$$

Hence, $\{x_n\}$ and $\{y_n\}$ are Cauchy sequences in X and Y, respectively.

Conversely, let $\{x_n\}$ and $\{y_n\}$ be Cauchy sequences in X and Y, respectively. Then for each $\varepsilon > 0$, there exists a positive integer N such that

$$\rho_X(x_m, x_n) < \frac{\varepsilon}{2}, \quad \forall m, n \geqslant N_1 \quad \text{and} \quad \rho_Y(y_m, y_n) < \frac{\varepsilon}{2}, \quad \forall m, n \geqslant N_2.$$

Now for $N' = \max\{N_1, N_2\}$ and $\forall m, n \geqslant N'$, we have

$$\begin{aligned}\rho((x_m, y_m), (x_n, y_n)) &= \rho_X(x_m, x_n) + \rho_Y(y_m, y_n)\\ &< \frac{\varepsilon}{2} + \frac{\varepsilon}{2} = \varepsilon.\end{aligned}$$

Thus, $\{(x_n, y_n)\}$ is a Cauchy sequence in a metric space $(X \times Y, \rho)$. ■

If a metric space has the property that all of its Cauchy sequences converge (in the space), then the metric space is called a complete metric space. Examples of complete metric spaces are provided by the Euclidean spaces $\mathbb{R}^n$ with their Euclidean distances.

5.14 Complete Metric Spaces

A metric space (X, ρ) is said to be *complete* if every Cauchy sequence of points of X converges to a point of X. In a complete space it is possible to determine the convergence of a sequence without knowing the limit; merely show that it is a Cauchy sequence.

Theorem 5.34 *If (X, ρ) is a complete metric space and Y is a subspace of X, then Y is complete if and only if Y is closed in X.*

Proof Let Y be a complete subspace of a space X. In order to show that Y is closed, we will prove that each cluster point of Y belongs to Y. Let x be an arbitrary cluster point of Y. Then for $n \in \mathbb{N}$, every open sphere $S_{\frac{1}{n}}(x)$ centered at x contains at least one point x_n of Y other than x, i.e.,

$$\text{for all } n \in \mathbb{N}, \quad x_n \in \left(S_{\frac{1}{n}}(x) - \{x\}\right).$$

Thus we get a sequence $\{x_n\}$ in Y such that

$$\rho(x_n, x) < \frac{1}{n} \quad \Longrightarrow \quad x_n \to x \ as \ n \to \infty.$$

Hence, $\{x_n\}$ is a convergent sequence in Y. By Theorem 5.27 $\{x_n\}$ is a Cauchy sequence in Y. Since Y is a complete metric space, $\{x_n\}$ must converge to a point x in Y. Thus every cluster point of Y is contained in it. Hence, Y is closed, i.e., $\overline{Y} = Y$.

Conversely, let $\{x_n\}$ be a Cauchy sequence of points in $Y(\subset X)$. Therefore it is also a Cauchy sequence in X. Since X is complete, $\{x_n\}$ must converge to a point x in X. Since $\{x_n\} \subset Y$, we see $x \neq Y^c$, because Y is a closed set. Thus, we have $x \in Y$. ■

5.15 Cantor Sets

The *Cantor (ternary)* set was first published in 1883 by German mathematician Georg Cantor. The Cantor set plays a very important role in many branches of mathematics, above all in set theory, chaotic dynamical systems and fractal theory.

The basic *Cantor* set is a subset of the interval [0, 1] and has many definitions and many different constructions. The *Cantor set* is created by repeatedly deleting the open middle thirds of a set of line segments.

To construct the *Cantor set*, first, we set the closed unit interval $F_1 = [0, 1]$.

One starts by deleting the open middle third $\left(\frac{1}{3}, \frac{2}{3}\right)$ from F_1, and remaining two closed intervals denoted by F_2, clearly

$$F_2 = \left[0, \frac{1}{3}\right] \cup \left[\frac{2}{3}, 1\right],$$

where each interval has the length $\frac{1}{3}$.

Next, delete the open middle third intervals $\left(\frac{1}{9}, \frac{2}{9}\right)$ and $\left(\frac{7}{9}, \frac{8}{9}\right)$ from F_2, and remaining closed intervals denoted by F_3, clearly

$$F_3 = \left[0, \frac{1}{9}\right] \cup \left[\frac{2}{9}, \frac{1}{3}\right] \cup \left[\frac{2}{3}, \frac{7}{9}\right] \cup \left[\frac{8}{9}, 1\right],$$

where each interval has the length $\frac{1}{3^2}$ (Fig. 5.27).

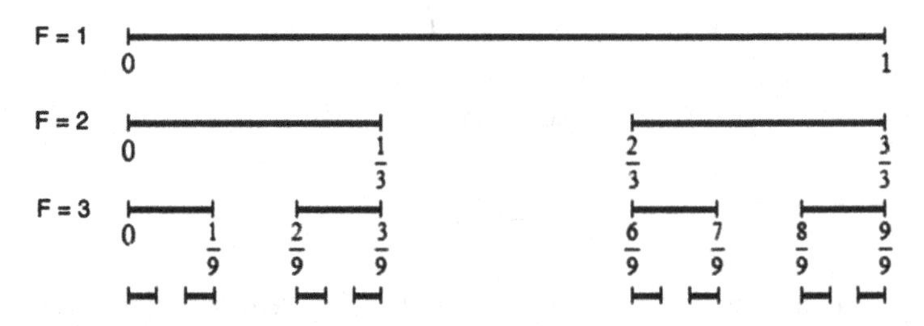

Fig. 5.27 Construction of the Cantor set

We continue this process at each stage, that is, from the previous step, we delete the open middle third interval of each closed interval remaining from the previous step. The Cantor set F is defined by

$$F = \bigcap_{n=1}^{\infty} F_n = \cap\{F_n; n \in \mathbb{N}\}, \quad F_n = \bigcup_{k=1}^{2^{n-1}} F_{n,k}, \tag{5.45}$$

where $F_{n,k}$ $(k = 1, 2, \cdots, 2^{n-1})$ are closed sets which are constructed as above, and they are mutually disjoint. We assume that $F_{n,k} < F_{n,k+1}$ (this means $x \in F_{n,k}$, $y \in F_{n,k+1} \Longrightarrow x < y$). The collection $\{F_n\}$ possesses the following two properties:

(i) $\{F_n\}$ is a descending sequence of closed sets.
(ii) For each n, F_n is the disjoint union of 2^{n-1} closed intervals $F_{n,k}$ $(k = 1, 2, \cdots, 2^{n-1})$, and then each $F_{n,k}$ has the length $1/3^{n-1}$.

Theorem 5.35 *We define F, F_n, $F_{n,k}$ $(k = 1, 2, \cdots, 2^{n-1})$ as (5.45). Then we have $F_{n+1} \subset F_n$ and $diam(F_{n,k}) \to 0$. Then $\bigcap_{n=1}^{\infty} F_n$ contains infinite points with cardinal numbers as* $[0, 1]$.

Proof Each interval $F_{n,k}$ is separated as two disjoint unions,

$$F_{n,k} = F_{n+1,2k-1} \cup F_{n+1,2k}, \text{ and } F_{n+1,2k-1} \cap F_{n+1,2k} = \phi,$$

where

$$diam(F_{n,k}) = \frac{1}{3^{n-1}}, \quad diam(F_{n+1,j}) = \frac{1}{3^{n-2}} \ (j = 2k-1, 2k).$$

Now we consider the sequence of closed sets such that

$$F_1 = F_{1,1} \supset F_{2,k_2} \supset F_{3,k_3} \supset \cdots.$$

Here we define F_{n,k_n}, $k_n = 1, 2, \cdots, 2^{n-1}$ as follows:

$$\begin{aligned}
F_{2,k_2} &= F_{2,1} \text{ or } F_{2,2} \\
F_{3,k_3} &= F_{3,2k_2-1} \text{ or } F_{3,2k_2} \\
&\cdots\cdots\cdots\cdots\cdots\cdots \\
F_{n-1,k_{n-1}} &= F_{n-1,2k_{n-2}-1} \text{ or } F_{n-1,2k_{n-2}} \\
F_{n,k_n} &= F_{n,2k_{n-1}-1} \text{ or } F_{n,2k_{n-1}} \\
&\cdots\cdots\cdots\cdots\cdots\cdots .
\end{aligned}$$

By Theorem 2.10 there exists a unique point $x = x_{k_1,k_2,\cdots} \in [0, 1]$ such that

$$x \in \bigcap_{n=1}^{\infty} F_{n,k_n}, \quad \text{where } k_1 = 1.$$

Now, we define the following set function

$$\psi(F_{n,2k_{n-1}+j}) = j + 1, \quad j = 0, -1, \quad \text{where } \psi(F_{1,1}) = 1.$$

Then we have a one-to-one mapping $a = \{a_n\}_{n=1}^{\infty}$, $a_n = \psi(F_{n,2k_{n-1}+j})$ $(n = 1, 2, 3, \cdots)$, that is,

$$x = x_{k_1,k_2,\cdots} \overset{a}{\Longleftrightarrow} 0.a_1a_2\cdots, \quad \text{where } a_1 = 1.$$

Thus, $\bigcap_{n=1}^{\infty} F_n$ contains infinite points with cardinal numbers as [0, 1]. ■

Remark 5.10 Since each F_n is complement of the union of the removal open intervals and the open intervals $(-\infty, 0)$ and $(1, \infty)$, the Cantor set F is closed.

Theorem 5.36 [*Cantor's Intersection Theorem*] *Let (X, ρ) be a complete metric space and $\{F_n\}$ be a non-empty closed subset of X, for all $n \in \mathbb{N}$ such that $F_{n+1} \subset F_n$ and $diam\,(F_n) \to 0$. Then $\bigcap_{n=1}^{\infty} F_n$ contains exactly one point.*

Proof Let $x_n \in F_n$. For each $n \in \mathbb{N}$, we have a sequence $\{x_n\}$ in X. Since $\{F_n\}$ is a nested sequence and $diam\,(F_n) \to 0$, we have the following.

Let $\varepsilon > 0$ be given. There exists $N \in \mathbb{N}$ such that $diam\,(F_N) < \varepsilon$.
Hence, if $x_m, x_n \in F_n$ with $m, n \geqslant N$, then we see

$$\rho\,(x_m, x_n) \leqslant diam\,(F_n) < \varepsilon.$$

Thus $\{x_n\}$ is a Cauchy sequence. Since (X, ρ) is complete, $\{x_n\}$ converges to some point $x \in X$.

Now we will show that $x \in \bigcap_{n=1}^{\infty} F_n$.

If $N \in \mathbb{N}$ is fixed, then a subsequence $\{x_k\}_{k \geqslant N}$ of $\{x_n\}$ is contained in F_n and it converges to x. Since F_n is a closed subspace of the complete metric space (X, ρ), it is also complete and it follows that $x \in F_n$. Hence $x \in \bigcap_{n=1}^{\infty} F_n$.

Finally, we have to show that x is the only point $\bigcap_{n=1}^{\infty} F_n$. Let $y \in \bigcap_{n=1}^{\infty} F_n$. For all n, both x and y are in F_n. Therefore

$$0 \leqslant \rho\,(x, y) \leqslant diam\,(F_N) \to 0 \;\; as \;\; n \to \infty.$$

Hence,

$$\rho(x, y) = 0 \Longrightarrow x = y.$$

Thus, $\bigcap_{n=1}^{\infty} F_n$ contains exactly one point. ■

In the next Theorem we will discuss the converse of Theorem 5.36.

Theorem 5.37 *Let (X, ρ) be a metric space. Every decreasing sequence F_n of non-empty closed sets with $diam\,(F_n) \to 0$ as $n \to \infty$ has exactly one point in its intersection. Then (X, ρ) is complete.*

Proof Let $\{x_n\}$ be a Cauchy sequence in X. Let $S_n = \{x_n, x_{n+1} \cdots\}$. Since $\{x_n\}$ is a Cauchy sequence, it is easy to verify

$$diam\,(S_n) \to 0 \;\; as \;\; n \to \infty.$$

By Theorem 5.17 we get

$$diam\,\left(\overline{S_n}\right) \to 0 \;\; as \;\; n \to \infty.$$

For our convenience $F_n = \overline{S}_n$. Since $\{F_n\}$ is a decreasing sequence of non-empty closed sets with $diam\,(F_n) \to 0$ as $n \to \infty$, by hypothesis, there exists an $x \in X$ such that

$$x \in \bigcap_{n=1}^{\infty} F_n.$$

Therefore, for all n, we have

$$\rho(x_n, x) \leqslant diam\,(F_n) \to 0 \; as \;\; n \to \infty.$$

Hence, $x_n \to x$ in X. Thus, X is complete. ■

5.15.1 Dense Sets

A subset A of a metric space (X, ρ) is said to be dense in X, if the closure of A is X, i.e., $\overline{A} = X$.

Theorem 5.38 *Let (X, ρ) be a metric space and $A \subset X$. Then the following statements are equivalent:*

(*a*) *A is dense in X.*
(*b*) *The closed set containing A is X.*
(*c*) *The open set which is disjoint from A is ϕ.*

(*d*) *The complement of A has an empty interior, i.e.,* $(A^c)^0 = \phi$.

Proof We see $A \subset X$, then

(*a*) means $\overline{A} = X$.
(*b*) $A \subset F$ (closed) $\Longrightarrow X = \overline{A} \subset F \quad \therefore F = X$.
(*c*) $A \cap G$ (open) $= \phi \Longrightarrow \overline{A} \cap G = \phi \Longrightarrow X \cap G = \phi \Longrightarrow G = \phi$.
(*d*) $A \cap (A^c)^0 = \phi \Longrightarrow \overline{A} \cap (A^c)^0 = \phi \Longrightarrow X \cap (A^c)^0 = \phi \Longrightarrow (A^c)^0 = \phi$.

And we have

$$\left(A^c\right)^0 = \phi \Rightarrow \overline{A} = X.$$

In fact, if $\overline{A} \subsetneqq X$, then we see $A^c \supset \overline{A^c} \neq \phi$. Since $\overline{A^c} \neq \phi$ is open the fact contradictions. ∎

Theorem 5.39 *Let* (X, ρ) *be a metric space. If* $\{G_n\}$ *is a sequence of dense open sets in* X*, then a union of infinite collections of dense open sets is dense in* X*, i.e.,*

$$G = \bigcup_{\lambda \in \Lambda} G_\lambda \ (\Lambda;\ \textit{an index set})\ \textit{is dense in } X.$$

Proof It is trivial that

$$\overline{G} \subset X.$$

From $G_\lambda \subset G$ we see

$$\overline{G_\lambda} \subset \overline{G} \quad \therefore X = \bigcup_{\lambda \in \Lambda} \overline{G_\lambda} \subset \overline{G}.$$

∎

5.15.2 Separable Spaces

A metric space (X, ρ) is called a *separable* space if there exists a countable dense subset A of X. In other words, X is separable if $A(\subseteq X)$ is countable and $\overline{A} = X$.

The set of rational numbers $\mathbb{Q}$ is *separable* in $\mathbb{R}$, since $\mathbb{Q}$ is countable and dense in $\mathbb{R}$.

Theorem 5.40 *Let* (X, ρ) *be a metric space and* $Y \subseteq X$*. If* X *is separable, then* Y *with the induced metric is also separable.*

5.16 The Baire Category Theorem

Definition 5.18 Let X be a metric (topological) space. A subset $A \subset X$ is said to be nowhere-dense in X, if for a given any non-empty open set U, there is a non-empty open subset $V \subset U$ such that $A \cap V = \phi$.

Theorem 5.41 *Let X be a metric (topological) space, and let a subset $A \subset X$ be nowhere-dense in X. Then we have*

$$(\overline{A})^0 = \phi, \quad \overline{ext\ A} = \overline{(\overline{A})^c} = X, \quad and \quad (\overline{(\overline{A})^c})^c = \phi.$$

Proof For a given non-empty open set U there exists a non-empty open subset V such that $V \subset U$ and

$$A \cap V = \phi.$$

Hence, we have

$$\overline{A} \cap V = \phi \Longrightarrow \left(\overline{A}\right)^0 \cap V = \phi. \tag{5.46}$$

If $\left(\overline{A}\right)^0 \neq \phi$, then we can take $x \in \left(\overline{A}\right)^0$. So we have $S_r(x) \subset \left(\overline{A}\right)^0$. Then for any $V(\neq \phi) \subset S_r(x)$ we see $\left(\overline{A}\right)^0 \cap V \neq \phi$, but this is contradictory to (5.46). Thus, we see $(\overline{A})^o = \phi$.

Now, from $ext\ A = \left(\overline{A}\right)^c$ we have

$$\overline{\text{ext } A} = \overline{(\overline{A})^c}.$$

We will show $\overline{\text{ext } A} = X$. If

$$x \in X - \overline{\text{ext } A} = X - \overline{(\overline{A})^c},$$

then

$$x \notin \overline{(\overline{A})^c}.$$

Therefore there exists a sphere $S_r(x)$ such that

$$\overline{(\overline{A})^c} \cap S_r(x) = \phi \quad \therefore\ (\overline{A})^c \cap S_r(x) = \phi,$$

that is,

$$S_r(x) \subset \overline{A}.$$

Then for any non-empty open subset $V \subset S_r(x)$ we see

$$V \subset S_r(x) \subset \overline{A}, \text{ thus } V \cap A \neq \phi.$$

Hence A is not nowhere-dense. Thus

$$X - \overline{(\overline{A})^c} = \phi.$$

Therefore, we have

$$\overline{((\overline{A})^c)^c} = \phi.$$

■

Remark 5.11 The rationals are NOT nowhere-dense. Subsets of nowhere-dense sets are also nowhere-dense. The Cantor set is a nowhere-dense set.

Let C is the Cantor set. We consider the closed sets $\{F_{n,k}\}_{k=1}^{2^{n-1}}$ in (5.45). Let $J_{n,k} = [a_{n,k}, b_{n,k}]$ and $J_{n,k+1} = [a_{n,k+1}, b_{n,k+1}]$. For an open set G, if we take n large enough, then we can find adjacent intervals $J = J_{n,k}$ and $I = (a_{n,k}, b_{n,k+1})$ such that $J, I \subset G$ with $diam J = diam \overline{I} = 3^{n-1}$. Clearly we see $I \cap C = \phi$, so we know that the set C is nowhere-dense.

Remark 5.12 The set $\mathbb{Q}$ of rational numbers is dense in $\mathbb{R}$ and further countable. On the other hand, the Cantor set C is nowhere-dense and further has cardinal numbers as well as [0, 1]. By the way, we note that C has other remarkable properties.

(i) C has a representation as follows:

$$C = \left\{ x; x = \sum_{j=1}^{\infty} \frac{2\varepsilon_j}{3^j}, \ \varepsilon_j = 0, 1 \right\}.$$

(ii) C is a perfect set:

$$C = C'.$$

(iii) Let F_i, $i = 1, 2, 3, \cdots$ be defined to construct the Cantor set. We set

$$O = \bigcup_{i=1}^{\infty} O_i, \text{ where } O_i = F_i^c \bigcap [0, 1].$$

Then we see that the open set O is dense in [0, 1].

The following theorem follows from Definition 5.18 and Theorem 5.41.

Theorem 5.42 *Let A be a subset of a metric space (X, ρ). Then the following statements are equivalent:*

(i) *A is nowhere-dense in X.*
(ii) *$\overline{A}$ does not contain any non-empty open set.*
(iii) *Every non-empty open set has a non-empty open subset which is disjoint from $\overline{A}$.*
(iv) *Every non-empty open set contains a non-empty open subset which is disjoint from A.*
(v) *Every non-empty open set contains an open sphere which is disjoint from A.*

A subset $A \subseteq X$ which can be written as a countable union of nowhere-dense sets is called the **first category**, or **meager**. All subsets of X that cannot be written in this fashion are called the **second category**, or **non-meager**.

Lemma 5.3 *Let A be a nowhere-dense subset of a metric space (X, ρ), and let G be any non-empty open set in X, then there exists an open sphere contained in G which is disjoint from A.*

Proof Since A is a nowhere-dense subset of X, i.e., $\left(\overline{A}\right)^0 = \phi$ and $\left(\overline{A}\right)^0$ is the largest open set which is contained in $\overline{A}$ (by Theorem 5.18 (ii)). Therefore, if G is any non-empty open set, then

$$G \not\subset \overline{A} \quad \left(\because \left(\overline{A}\right)^0 = \phi\right).$$

Then there exists an $x \in G$, therefore $x \notin \overline{A}$. Since G is open, there exists an open sphere $S_r(x)$ for some $r > 0$ such that $S_r(x) \subset G$.

Since $\overline{A}$ is a closed set and $x \notin \overline{A}$, we can choose a positive number $r_1 < r$ such that

$$x \in S_{r_1}(x) \subset S_r(x) \subset G \quad \text{and} \quad S_{r_1}(x) \cap A = \phi.$$

■

The following fundamental theorem was proved by René-Louis Baire in 1899.

Theorem 5.43 *Any complete metric space is of the second category, in other words, if $\{A_n\}$ is a sequence of nowhere-dense sets in a complete metric space (X, ρ), then there exists a point in X which is not a point in any of the A_n's, i.e., $X \neq \bigcup_{n=1}^{\infty} A_n$.*

Proof Since X is open and A_1 is nowhere-dense in X, by Lemma 5.3 there exists an open sphere $S_{r_1}(x_1)$ of radius r_1 with $0 < r_1 < 1$ such that

$$S_{r_1}(x_1) \cap A_1 = \phi.$$

Let F_1 be the concentric closed sphere of radius $\frac{r_1}{2}$, i.e.,

$$F_1 = S_{\frac{r_1}{2}}[x_1] \quad \text{and} \quad F_1^0 \neq \phi.$$

Let $x_2 \in F_1^0$. Since A_2 is nowhere-dense in X, F_1^0 contains an open sphere $S_{r_2}(x_2)$ of radius r_2 with $0 < r_2 < 1$ such that

$$S_{r_2}(x_2) \cap A_2 = \phi.$$

Let F_2 be the concentric closed sphere of radius $\frac{r_2}{2}$, i.e.,

$$F_2 = S_{\frac{r_2}{2}}[x_2] \quad \text{and} \quad F_2^0 \neq \phi.$$

Let $x_3 \in F_2^0$. Since A_3 is nowhere-dense in X, F_2^0 contains an open sphere $S_{r_3}(x_3)$ of radius r_3 with $0 < r_3 < 1$ such that

$$S_{r_3}(x_3) \cap A_3 = \phi.$$

Let F_3 be the concentric closed sphere of radius $\frac{r_3}{2}$. Continuing this process, we obtain a decreasing sequence $\{F_n\}$ of non-empty closed subsets of X, where

$$F_n = S_{\frac{r_n}{2}}[x_n], \quad \text{for every } n \in \mathbb{N},$$

and

$$F_{n+1} = S_{\frac{r_{n+1}}{2}}\left[x_{n+1}\right] \subset S_{r_{n+1}}(x_{n+1}) \subset S_{\frac{r_n}{2}}[x_n] = F_n.$$

Hence

$$F_{n+1} \subset F_n, \quad \text{for all } n \in \mathbb{N}.$$

Also

$$d(F_n) = 2\left(\frac{r_n}{2}\right) < \frac{1}{2^{n-1}}, \quad \text{for all } n \in \mathbb{N} \implies d(F_n) \to 0 \text{ as } n \to \infty.$$

Since (X, ρ) is complete, by Cantor's Intersection Theorem 5.36, it follows that $\bigcap_{n=1}^{\infty} F_n$ is non-empty and contains exactly one point, say $x \in X$.
Then we see

$$x \in F_n, \quad \text{for all } n \in \mathbb{N} \implies x \in S_{\frac{r_n}{2}}[x_n] \subset S_{r_n}(x_n), \quad \text{for all } n \in \mathbb{N},$$

and

$$S_{r_n}(x_n) \cap A_n = \phi \implies x \notin A_n, \text{ for all } n \in \mathbb{N} \implies x \notin \bigcup_{n=1}^{\infty} A_n.$$

Thus

$$\bigcup_{n=1}^{\infty} A_n \neq X.$$

Hence, X cannot be of the first category and therefore X is of the second category. ■

The Baire Category Theorem can be also stated in the following forms.

Theorem 5.44 *If (X, ρ) is a complete metric space, then*

(*i*) *the intersection of every countable collection of dense open subsets of X is non-empty.*

(*ii*) *a complete metric space cannot be a countable union of nowhere-dense closed subsets.*

Proof (*i*) Let $\{O_n\}$ be a countable collection of dense open subsets of X. Then we will show $\bigcap_{n=1}^{\infty} O_n \neq \phi$. Let x_1 be a point of O_1. Then there exists $r_1 > 0$ such that

$$S_{r_1}(x_1) \subset O_1.$$

Since O_2 is dense, we find a point $x_2 \in O_2 \cap S_{r_1}(x_1)$. Since O_2 is open, there exists a sphere S_2 centered at x_2 and contained in O_2, and we may take the radius $r_2 > 0$ of S_2 to be smaller than $r_1/2$ and smaller than $r_1 - \rho(x_1, x_2)$. Then $\overline{S_2} \subset S_1$. Proceeding inductively, we have a sequence $\{S_n\}$ of a sphere such that $\overline{S_n} \subset S_{n-1}$ and $S_n \subset O_n$ whose radius $\{r_n\}$ tends to 0. Let $\{x_n\}$ be the sequence of centers of these spheres. Then for $n,\ m \geqslant N$ we have $x_n \in S_N$ and $x_m \in S_N$. Hence $\rho(x_n, x_m) \leqslant 2r_N$, and $\{x_n\}$ is a Cauchy sequence since $r_n \Longrightarrow 0$. By the completeness of X there is a point x such that $x_n \Longrightarrow x$. Since $x_n \in S_{N+1}$ for $n > N$, we have $x \in \overline{S_{N+1}} \subset S_N \subset O_N$. Hence $x \in \bigcap O_n$.

(*ii*) Let $\{E_n\}$ be a countable collection of nowhere-dense closed sets. Then

$$(\overline{E_n})^c = (E_n)^c$$

is a dense open set, and so there exists a point $x \in \bigcap O_n$, where $O_n = (E_n)^c$. But this means $x \notin \bigcup E_n$.

■

5.17 Uniform Continuity

Definition 5.19 Let (X, ρ_X) and (Y, ρ_Y) be two metric spaces. A mapping $f : X \to Y$ is said to be **uniformly continuous** if for $\varepsilon > 0$ there exists $\delta(\varepsilon) > 0$ such that

$$\rho_X(x, y) < \delta \implies \rho_Y(f(x), f(y)) < \varepsilon, \ \forall x, y \in X.$$

If you are proving that a mapping $f : X \to Y$ is continuous at a point x_0, then you must find the $\delta > 0$, which usually depends on both $\varepsilon > 0$ and x_0. If the mapping is uniformly continuous, the δ depends on ε alone. This distinction might appear to be minor, but it is extremely important in many situations.

Remark 5.13 The difference between the two definitions (continuity and uniformly continuity) is the order of the quantifiers. When we prove the function f is continuous, our proof will have the following form:

Choose $x_0 \in A$. Choose $\varepsilon > 0$. Let $\delta = \delta(x_0, \varepsilon)$ be small enough. Choose $x \in A$. Assume $|x - x_0| < \delta$. Then $|f(x) - f(x_0)| < \varepsilon$.

The expression for $\delta(x_0, \varepsilon)$ can involve both x_0 and ε but must be independent of x. When you prove that f is uniformly continuous, your proof will have the form:

Choose $\varepsilon > 0$. Let $\delta = \delta(\varepsilon)$. Choose $x_0 \in A$. Assume $|x - x_0| < \delta$. Then $|f(x) - f(x_0)| < \varepsilon$.

So, the expression for δ can only involve ε and must not involve either x or x_0.

Remark 5.14 Every function $f : X \to Y$, which is uniformly continuous on X, is necessarily continuous on X, but the converse need not be true.

Example 5.25 Let $f(x) = 2x + 5, \ \ x \in \mathbb{R}$. Then f is uniformly continuous on $\mathbb{R}$.

Choose $\varepsilon > 0$. Let $\delta = \frac{\varepsilon}{2}$. Choose $x_0, x \in \mathbb{R}$. Assume $|x - x_0| < \delta$. Then

$$|f(x) - f(x_0)| = |(2x + 5) - (2x_0 + 5)| = 2|x - x_0| < 2\delta = \varepsilon.$$

Example 5.26 The function $f : \mathbb{R} \to \mathbb{R}$ defined by $f(x) = x^2$ is continuous but not uniformly continuous in the usual metric space $(\mathbb{R}, \rho)$.

First we show f is continuous on $\mathbb{R}$, i.e., for a point $x_0 \in \mathbb{R}$. We have $\forall \varepsilon > 0$, and there exists $\delta(x_0, \varepsilon) > 0$ such that

$$|x - x_0| < \delta \implies \left|x^2 - x_o^2\right| < \varepsilon.$$

Now choose $x_0 \geqslant 0$. Let $a = 1 + x_0$ and $\delta = \min(1, \varepsilon/2a)$. (Note that δ depends on x_0 since a does). Choose $x \in \mathbb{R}$ as $|x - x_0| < \delta$. Then $|x - x_0| < 1$, so $|x| < 1 + x_0 = a$, that is, we have $x, x_0 < a$. Therefore

$$\left|x^2 - x_0^2\right| = (x + x_0)\,|x - x_0| \leqslant 2a\,|x - x_0| < 2a\delta \leqslant 2a\frac{\varepsilon}{2a} = \varepsilon$$

is required.

Now show that f is not uniformly continuous on $\mathbb{R}$, i.e., for a given $\varepsilon > 0$ and $\delta > 0$ there exist $x, x_0 \in \mathbb{R}$ such that

$$|x - x_0| < \delta \quad \text{but} \quad \left|x^2 - x_o^2\right| \geqslant \varepsilon.$$

Let $\varepsilon = 1$. Choose $\delta > 0$. Let $x_0 = \frac{1}{\delta}$ and $x = x_0 + \frac{\delta}{2}$. Then

$$\left|x^2 - x_0^2\right| = \left|\left(\frac{1}{\delta} + \frac{\delta}{2}\right)^2 - \left(\frac{1}{\delta}\right)^2\right| = 1 + \frac{\delta^2}{4} > 1.$$

Example 5.27 Let X be the interval $(0, 1]$ with the subspace metric of $\mathbb{R}$. Define the mapping $f : X \to \mathbb{R}$ by the function $f(x) = |x|^{-1}$. Then we can prove that f is continuous at each point of X, but f is not uniformly continuous on X.

Theorem 5.45 *Let (X, ρ_X), (Y, ρ_Y) and (Z, ρ_Z) be metric spaces and $f : X \to Y$ and $g : Y \to Z$ be uniformly continuous mappings. Then the mapping $g \circ f : X \to Z$ is uniformly continuous.*

Proof Since $f : X \to Y$ is uniformly continuous, corresponding to $\delta > 0$, there exists $\eta > 0$ such that

$$\rho_X(x, y) < \eta \implies \rho_Y(f(x), f(y)) < \delta,$$

for all $x, y \in X$.
Since $g : Y \to Z$ is uniformly continuous, for each $\varepsilon > 0$ there exists $\delta > 0$ such that

$$\rho_Y(f(x), f(y)) < \delta \implies \rho_Z((g \circ f(x)), (g \circ f(y))) < \varepsilon,$$

for all $f(x), f(y) \in Y$.

Hence, for each $\varepsilon > 0$, there exists $\eta > 0$ such that

$$\rho_X(x, y) < \eta \implies \rho_Z((g \circ f(x)), (g \circ f(y))) < \varepsilon,$$

for all $x, y \in X$. Hence $g \circ f$ is uniformly continuous on X. ■

Theorem 5.46 *Let (X, ρ_X) and (Y, ρ_Y) be two metric spaces and $f : X \to Y$ be uniformly continuous. If $\{x_n\}$ is a Cauchy sequence in X, then so is $\{f(x_n)\}$ in Y.*

Proof By the definition of uniform continuity, for every positive real number ε, there exists a positive real number δ such that

$$\rho_Y(f(x), f(y)) < \varepsilon, \quad \text{whenever} \quad \rho_X(x, y) < \delta, \tag{5.47}$$

for all $x, y \in X$.

Since $\{x_n\}$ is a Cauchy sequence, corresponding $\delta > 0$, there exists N such that

$$\rho_X(x_m, x_n) < \delta, \quad \forall m, n \geqslant N. \tag{5.48}$$

From (5.47) and (5.48), we have

$$\rho_Y(f(x_m), f(x_n)) < \varepsilon, \quad \forall m, n \geqslant N.$$

Hence, $\{f(x_n)\}$ is a Cauchy sequence in Y. ■

5.18 Homeomorphisms

Definition 5.20 Let (X, ρ_X) and (Y, ρ_Y) be two metric spaces. And a mapping $f : X \to Y$ is said to be a **homeomorphism** if

(i) f is bijective (1-1 and onto).
(ii) f and f^{-1} are continuous.

In other words, two spaces (X, ρ_X) and (Y, ρ_Y) are homeomorphic when (Y, ρ_Y) can be obtained from (X, ρ_X), and (X, ρ_X) can be obtained from (Y, ρ_Y), continuously and continuously invertible from the other; in effect, when convergence in one space is equivalent to convergence in the other.

Example 5.28 Let $X = \left\{1, \frac{1}{2}, \frac{1}{3}, \cdots\right\}$ with ρ_X is the usual metric on the subset of $\mathbb{R}$ and let $Y = \mathbb{N}$ with the usual metric $\rho_Y (= \rho_X)$. Then $f : X \to Y$ defined by $f\left(\frac{1}{n}\right) = n$ is a homeomorphism from (X, ρ_X) to (Y, ρ_Y).

A function $f : X \to Y$ is called **isometric** when f preserves distances

$$\rho_Y(f(x), f(y)) = \rho_X(x, y), \quad \forall x, y \in X.$$

5.19 The Banach Contraction Mapping Theorem

Definition 5.21 Let (X, ρ) be a metric space and $T : X \to X$ be a mapping. The point $x \in X$ is called a **fixed point** of T, if $Tx = x$, i.e., if x is mapped onto itself.

Remark 5.15 A mapping $T : X \to X$ may have no, one or many fixed points.

Example 5.29 (i) The mapping $T : R^2 \to R^2$ and $c \neq 0, T = c + x$, has no fixed point.

(ii) The mapping $T : R^2 \to R^2$, where $0 < \alpha < 2\pi$,

$$T(\lambda, \mu) = \begin{bmatrix} \cos\alpha & -\sin\alpha \\ \sin\alpha & \cos\alpha \end{bmatrix} \begin{bmatrix} \lambda \\ \mu \end{bmatrix}$$

has only one fixed point (0, 0).

Definition 5.22 Let (X, ρ) be a metric space. A mapping $T : X \to X$ is called a **contraction mapping** of X if there exists a real number α with $0 < \alpha < 1$ such that

$$\rho(Tx, Ty) \leqslant \alpha\rho(x, y), \ \forall x, y \in X.$$

The number α is usually referred to a **Lipschitz constant** of T.

Geometrically, two points x and y have images that are closer together than those points x and y; more precisely, the ratio $\rho(Tx, Ty)/\rho(x, y)$ does not exceed a constant α which is strictly less than 1.

The Banach Contraction Mapping Theorem assures the existence and uniqueness of a fixed point for any contraction mapping on a complete metric space and provides a method to compute it.

Theorem 5.47 (The Banach Contraction Mapping Theorem) *Let (X, ρ) be a complete metric space and let $T : X \to X$ be a contraction on X. Then T has a unique fixed point in X.*

Proof Let $\alpha \in (0, 1)$ be a Lipschitz constant such that

$$\rho(Tx, Ty) \leqslant \alpha\rho(x, y), \ \forall x, y \in X.$$

We prove the theorem in five steps.
Step (i) We construct a sequence $\{x_n\}$ as follows:

Take any point $x_0 \in X$ and inductively construct the sequence $\{x_n\}$ of a point in X as follows:

$$\begin{aligned} x_1 &= Tx_0 \\ x_2 &= Tx_1 = T^2x_0 \\ x_3 &= Tx_2 = T^3x_0 \\ &\cdots\cdots\cdots\cdots \\ &\cdots\cdots\cdots\cdots \\ x_n &= Tx_{n-1} = T^nx_0 \\ &\cdots\cdots\cdots\cdots. \end{aligned}$$

Clearly, $\{x_n\}$ is the sequence of images of x_0 under the repeated application of T.

Step (ii) $\{x_n\}$ is a Cauchy sequence in X.

Let $m > n$. Then

$$\begin{aligned}\rho(x_n, x_m) &= \rho\left(T^n x_0, T^m x_0\right) = \rho(T^n x_0, T^n T^{m-n} x_0) \\ &\leqslant \alpha^n \rho\left(x_0, T^{m-n} x_0\right) = \alpha^n \rho\left(x_0, x_{m-n}\right) \\ &\leqslant \alpha^n \left\{\rho\left(x_0, x_1\right) + \rho\left(x_1, x_2\right) + \cdots + \rho\left(x_{m-n-1}, x_{m-n}\right)\right\} \\ &\leqslant \alpha^n \left\{1 + \alpha + \alpha^2 + \cdots + \alpha^{m-n-1}\right\} \rho(x_0, x_1) \\ &= \alpha^n \frac{1 - \alpha^{m-n}}{1 - \alpha} \rho\left(x_0, x_1\right) \\ &\leqslant \frac{\alpha^n}{1 - \alpha} \rho(x_0, x_1), \quad \left(\text{by } 0 < \alpha < 1, \text{ so } (1 - \alpha^{m-n}) < 1\right) \\ &\to 0 \text{ as } n \to \infty.\end{aligned}$$

This proves that $\{x_n\}$ is a Cauchy sequence.

Step (iii) Since X is complete and $\{x_n\}$ is a Cauchy sequence in X, there exists $x \in X$ such that $x_n \to x$ as $n \to \infty$.

Step (iv) Now we will show that x is a fixed point of T. In fact, for $\varepsilon > 0$ there exists $N > 0$ such that

$$\rho(x_n, x) < \varepsilon, \quad (n \geqslant N).$$

Hence, we see

$$\begin{aligned}\rho\left(x, Tx\right) &\leqslant \rho\left(x, x_n\right) + \rho\left(x_n, Tx\right) \\ &\leqslant \rho\left(x, x_n\right) + \alpha\rho\left(x_{n-1}, x\right) < \varepsilon.\end{aligned}$$

We conclude that $\rho\left(x, Tx\right) = 0$, so that $x = Tx$ (by (M_2) in Definition 5.1). This shows that x is a fixed point of T.

Thus, the existence of a fixed point is established. In the last step, we verify the uniqueness of such a fixed point.

Step (v) x is a unique fixed point of T.
Let, if possible, x and y be two fixed points of T in X. Then $Tx = x$ and $Ty = y$. Now, note that

$$\rho(x, y) = \rho(Tx, Ty) \leqslant \alpha\rho(x, y),$$

which implies $\rho(x, y) = 0$ since $\alpha < 1$. Hence, $x = y$ (by (M_2) in Definition 5.1) and the theorem is proved. ■

Remark 5.16 The above theorem is also referred as the **Banach Fixed Point Theorem**.

5.20 Compactness Arguments

In this section, we discuss the ideas of relative compactness and total boundedness. We relate these ideas in metric spaces, Banach spaces, and function spaces. Before giving a definition of compactness we explain some important preliminaries of compactness.

A family $\{G_\alpha; \alpha \in \Lambda\}$ of subsets of a metric space (X, ρ) is called a **cover** of X if $X = \bigcup\limits_{\alpha \in \Lambda} G_\alpha$. If each G_α is an open subset of X, then the cover is called an **open cover** of X.

If $\Lambda' \subseteq \Lambda$, then $\left\{G_\alpha; \alpha \in \Lambda'\right\}$ is called a **subfamily** of $\{G_\alpha; \alpha \in \Lambda\}$. If a subfamily of a cover of X is still a cover of X, then it is called a **subcover** of X. If Λ' is finite, the subfamily is called a finite subfamily and if it is a subcover of X, it is called a **finite subcover** of X.

The definition of compactness may be conveniently expressed in the language of covers.

Let (X, ρ) be a metric space and $A \subseteq X$. A family $\{G_\alpha; \alpha \in \Lambda\}$ of subsets of X is called a cover for A if $A \subseteq \bigcup\limits_{\alpha \in \Lambda} G_\alpha$.

Definition 5.23 A metric space (X, ρ) is called **compact** if every open cover of X contains a finite subcover.

A subset A of X is called a compact set, if it is compact when considered as a subspace of X, i.e., if every open cover of A has a finite subcover.

Remark 5.17 A metric space (X, ρ) is called **compact** if X itself is compact, i.e., for each family of open subsets $\{G_\alpha\}$ of X for which $\bigcup\limits_{\alpha \in \Lambda} G_\alpha = X$, there exists a finite subfamily $\left\{G_{\alpha_i}; i = 1, 2, \cdots, n\right\}$ such that

$$X = \bigcup_{i=1}^{n} G_{\alpha_i}.$$

Example 5.30 (i) In the usual metric space $(\mathbb{R}, \rho)$ for each integer n, let G_n be the open interval $(n-1, n+1)$. Then a family $\{G_n; n \in \mathbb{Z}\}$, where $\mathbb{Z}$ is the set of all integers, is an open cover of $\mathbb{R}$, but it contains no finite subcover. To see this latter statement, assume there is a finite subcover and let m be the largest subscript such that G_m is in the subcover. Then $m+2$ is not in the union of the sets of the subcover. This is a contradiction. This result shows that $\mathbb{R}$ is not a compact metric space.

(ii) Let $A = [0, 6]$. And consider the open cover

$$G = \{(n-1, n+1)\,;\, n = -\infty, \cdots, \infty\}.$$

Then

$$P = \{(-1, 1)\,, (0, 2)\,, (1, 3)\,, (2, 4)\,, (3, 5)\,, (4, 6)\,, (5, 7)\}$$

is a subcover of A, and happens to be the smallest subcover of G that covers A.
(iii) The open interval (0, 1) is not compact. Suppose that

$$G = \left\{ \left(\frac{1}{n}, 1 \right); n = 2, 3, \cdots, \infty \right\}$$

is an open cover of (0, 1). However, any finite subfamily, say,

$$\left\{ \left(\frac{1}{n_1}, 1 \right), \left(\frac{1}{n_2}, 1 \right), \cdots, \left(\frac{1}{n_r}, 1 \right) \right\}$$

covers only $\left(\frac{1}{N}, 1\right)$, where $N = max \{n_1, n_2, \cdots, n_r\}$. Hence no finite subcollection of these sets will cover (0, 1).

Theorem 5.48 *Every closed subset of a compact metric space is compact.*

Proof Let (X, ρ) be a compact metric space and let F be a closed non-empty subset of X. Let $\{G_\alpha; \alpha \in \Lambda\}$ be a family of an open cover of F. Then

$$\{G_\alpha; \alpha \in \Lambda\} \cup (X - F)$$

is an open cover of X. Since X is compact, $\{G_\alpha; \alpha \in \Lambda\} \cup (X - F)$ has a finite subcover $G_{\alpha_1}, G_{\alpha_2}, \cdots, G_{\alpha_n}, X - F$, say,

$$\left(\bigcup_{i=1}^{n} G_{\alpha_i} \right) \cup (X - F) = X, \quad \text{so that} \quad F \subseteq \bigcup_{i=1}^{n} G_{\alpha_i},$$

i.e., $\bigcup\limits_{i=1}^{n} G_{\alpha_i}$ is a finite cover of F. Hence, F is compact. ■

Theorem 5.49 *A compact subset of a metric space is closed.*

Proof Let (X, ρ) be a metric space and let F be a subset of X. In order to prove that F is closed, we must show that $F^c = X - F$ is open.

Let $x \in F$ and $y \in F^c$. Then $x \neq y$. Let $r_x = \rho(x, y) > 0$. Then

$$S_{\frac{r_x}{2}}(x) \cap S_{\frac{r_x}{2}}(y) = \phi.$$

Clearly,

$$F \subseteq \bigcup_{x \in F} S_{\frac{r_x}{2}}(x).$$

Since F is compact, there exists a finite number of open spheres, say,

$$\left\{ S_{\frac{r_{x_1}}{2}}(x_1), S_{\frac{r_{x_2}}{2}}(x_2), \cdots, S_{\frac{r_{x_n}}{2}}(x_n) \right\}$$

such that

$$F \subseteq \bigcup_{i=1}^{n} S_{\frac{r_{x_i}}{2}}(x_i).$$

Now for each of the x_i, $i = 1, 2, \cdots, n$ there are open spheres $S_{\frac{r_{x_i}}{2}}(x)$ satisfying

$$S_{\frac{r_{x_i}}{2}}(x) \cap S_{\frac{r_{x_i}}{2}}(y) = \phi.$$

Let

$$G = \bigcap_{i=1}^{n} S_{\frac{r_{x_i}}{2}}(y).$$

Since G is an open subset of X containing y $(\in F^c)$,

$$\begin{aligned} S_{\frac{r_{x_i}}{2}}(x) \cap G = \phi, \quad i = 1, 2, \cdots, n \\ \Longrightarrow \bigcup \left\{ S_{\frac{r_{x_i}}{2}}(x_i), i = 1, 2, \cdots, n \right\} \cap G = \phi \\ \Longrightarrow F \cap G = \phi. \end{aligned}$$

Hence,

$$G \subseteq F^c \Longrightarrow y \in G \subseteq F^c,$$

it follows that F^c is a neighborhood of each of its points, and hence F^c is an open set. Thus, its complement F is a closed set. ■

Theorem 5.50 *A compact subset of a metric space is bounded.*

Proof Let G be a compact subset of a metric space (X, ρ). Consider the open spheres, with radius 1, centered at the points of G. Clearly,

$$G \subseteq \bigcup_{x \in G} S_1(x).$$

In view of the compactness of G, there exists $x_1, x_2, \cdots, x_n$ such that

$$G \subseteq \bigcup_{i=1}^{n} S_1(x_i).$$

Let $M = \max \left\{ \rho(x_i, x_j); i, j = 1, 2, \cdots \right\}$, and let $x, y \in G$. Then there exist x_i and x_j such that

$$x \in S_1(x_i) \quad \text{and} \quad y \in S_1(x_j).$$

Therefore,

$$\rho(x, y) \leqslant \rho(x, x_i) + \rho(x_i, x_j) + \rho(x_j, y) < M + 2.$$

Hence, G is bounded. ■

Remark 5.18 Every compact subset of a metric space is closed and bounded but the converse need not be true.

The proof is simple in view of Theorems 5.49 and 5.50.

Conversely, let $X = \left\{\frac{1}{n}; n \in \mathbb{N}\right\}$ and (X, ρ) be a discrete metric space and a metric $\rho : X \times X \to [0, \infty)$ given by

$$\rho(x, y) = \begin{cases} 0 & x = y, \\ 1 & x \neq y. \end{cases}$$

Clearly, X is closed. Now let $x_0 \in X$ and $X \subseteq S_1(x_0)$. And since any sphere of radius greater than 1 necessarily includes the whole set X, X is bounded.

For each $n \in \mathbb{N}$, let $x_n = \left\{\frac{1}{n}\right\}$ and

$$S_{\frac{1}{2}}(x_n) \subseteq X.$$

Moreover,

$$X = \bigcup_{n=1}^{\infty} S_{\frac{1}{2}}(x_n).$$

Therefore, $\{x_n; n \in \mathbb{N}\}$ is an open cover of X. Since $\{x_n; n \in \mathbb{N}\}$ cannot be made smaller, it is an open cover of X without any finite subcover. Thus X is not compact.

Theorem 5.51 [*The Heine-Borel Theorem*]
A subset of $\mathbb{R}$ *is compact if and only if it is closed and bounded.*

Proof Let A be a subset in $\mathbb{R}$. If A is compact, then by Theorems 5.49 and 5.50 we see that A is closed and bounded. If we suppose A is closed and bounded, then there exist a and b such that $A \subset [a, b]$. Now we will show that $[a, b]$ is compact. Then by Theorem 5.48, we know that A is compact. Let $F = \{(c_\lambda, d_\lambda) : \lambda \in \Lambda\}$ be an open cover of $[a, b]$, where Λ is an index set.

It is supposed that $[a, b]$ is not compact. Then there exists no finite subcover of F. Let us write $I_1 = [a, b] = [a_1, b_1]$, where $a = a_1$ and $b = b_1$.

Let us divide I_1 into two equal closed intervals

$$\left[a_1, \frac{a_1+b_1}{2}\right] \text{ and } \left[\frac{a_1+b_1}{2}, b_1\right].$$

In view of our hypothesis, at least one of these two intervals cannot be covered by any finite subfamily of the open cover F. Let that particular interval be denoted by I_2 as follows:

$$I_2 = [a_2, b_2] = \left[a_1, \frac{a_1+b_1}{2}\right] \text{ or } \left[\frac{a_1+b_1}{2}, b_1\right].$$

As above, divide I_2 into two equal closed intervals

$$\left[a_2, \frac{a_2+b_2}{2}\right] \text{ and } \left[\frac{a_2+b_2}{2}, b_2\right],$$

and at least one of these two intervals cannot be covered by any finite subfamily of the open cover F. Let that particular interval be denoted by I_3 as follows:

$$I_3 = [a_3, b_3] = \left[a_2, \frac{a_2+b_2}{2}\right] \text{ or } \left[\frac{a_2+b_2}{2}, b_2\right].$$

Continuing this process an infinite number of times, we obtain a sequence of intervals $I_1, I_2, I_3, \cdots, I_n, \cdots$ satisfying the following conditions:

(i) $I_n \supset I_{n+1}$ $\because n \in \mathbb{N}$.
(ii) for $n \in \mathbb{N}$, $I_n = [a_n, b_n]$ is a closed interval.
(iii) $\lim_{n\to\infty} l(I_n) = 0$, where $l(I_n)$ denotes the length of the interval I_n.
(iv) I_n is not covered by any finite subfamily of F.

In view of the conditions (i), (ii), (iii) and (iv), we find the sequence of closed intervals $\{I_n\}$ that satisfies all the conditions of Theorem 5.36. Hence

$$\bigcap_{n=1}^{\infty} I_n \text{ contains exactly one element, say } x, \text{ i.e.,}$$

$$x \in I_n, \quad \forall n \in \mathbb{N}. \tag{5.49}$$

Now there exists $\lambda \in \Lambda$ such that $x \in (c_\lambda, d_\lambda)$, hence we can find ε_1 as

$$S_{\varepsilon_1}(x) \subset (c_\lambda, d_\lambda).$$

On the other hand, in view of the condition (iii), for a given $\varepsilon > 0$, there exists $N \in \mathbb{N}$ such that $l(I_N) < \varepsilon$. Also, using (5.49), we have $x \in I_N$. Hence,

$$l(I_N) < \varepsilon \Longrightarrow I_N \subset (x - \varepsilon, x + \varepsilon).$$

Now we take $0 < \varepsilon \leqslant \varepsilon_1$, then we see

$$I_N \subset S_\varepsilon(x) \subset S_{\varepsilon_1}(x) \subset (c_\lambda, d_\lambda).$$

But this is contrary to the condition (iv) of the intervals I_n.

Thus we arrive at a contradiction and our hypothesis is wrong. Therefore, every open cover of $[a, b]$ will admit a finite subcover, and so $[a, b]$ is compact. Hence, A is compact. ■

Theorem 5.52 *A continuous image of a compact set is compact.*

Proof Let (X, ρ_X) and (Y, ρ_Y) be two metric spaces. Let $f : X \to Y$ be a continuous function and (X, ρ_X) be compact. Then we need to prove that the image $f(X)$ is a compact subset of Y.

We consider an arbitrary open covering $f(X) \subset \bigcup_{\alpha \in \Lambda} G_\alpha$ of $f(X) = Y_1$ (say), each G_α is open in Y_1. Therefore,

$$X = f^{-1}(Y_1) \subset f^{-1}\left(\bigcup_{\alpha \in \Lambda} G_\alpha\right) = \bigcup_{\alpha \in \Lambda} f^{-1}(G_\alpha).$$

Hence $\left\{f^{-1}(G_\alpha)\right\}$ is an open cover of X. Since X is compact, there exists a finite subcover, say,

$$\left\{f^{-1}(G_{\alpha_1}), f^{-1}(G_{\alpha_2}), \cdots, f^{-1}(G_{\alpha_n})\right\}$$

is an open cover of X such that

$$X = \bigcup_{i=1}^{n} f^{-1}(G_{\alpha_i}). \tag{5.50}$$

Let $y \in Y_1 = f(X)$. Then there exists $x \in X$ such that $y = f(x)$.
From (5.50), we have

$$x \in \bigcup_{i=1}^{n} f^{-1}(G_{\alpha_i}).$$

Therefore,

$$y = f(x) \in G_{\alpha_i}, \quad \text{for some } i.$$

Hence,

$$Y_1 = \bigcup_{i=1}^{n} G_{\alpha_i}.$$

Therefore,

$$\{G_{\alpha_1}, G_{\alpha_2}, \cdots, G_{\alpha_n}\}$$

is a finite subcover of the open cover $\{G_\alpha\}$ of Y_1. ■

Remark 5.19 Let $f : X \xrightarrow{on\ to} Y$ be continuous and X be compact. Then $Y_1 = f(X) = Y$ and therefore Y is compact.

5.20.1 The Finite Intersection Property (FIP)

A family $\{A_\alpha; \alpha \in \Lambda\}$ of sets is said to have the finite intersection property if $\bigcap_{i=1}^{n} A_{\alpha_i} \neq \phi$ holds for every finite set $\{A_{\alpha_i}\}_{i=1}^{n}$.

Example 5.31 (i) The family $\{[-n, n]\,; n \in \mathbb{N}\}$ of closed intervals of $\mathbb{R}$ has a finite intersection property. In fact, let $\{I_{n_i} = [-n_i, n_i]\}_{i=1}^{k}$ be any finite subfamily. Then we see $\bigcap_{i=i}^{k} I_{n_i} = I_{n_m} \neq \phi$, where $I_{n_m} = \min\{n_i; i = 1, 2, \cdots, k\}$, i.e.,

$$[-1, 1] \cap [-2, 2] \cap [-3, 3] \cap \cdots \cap [-1000, 1000] \neq \phi.$$

(ii) The family $\{[-\frac{1}{n}, \frac{1}{n}]\,; n \in \mathbb{N}\}$ of closed intervals of $\mathbb{R}$ has a finite intersection property.

Theorem 5.53 *Let (X, ρ) be a metric space, and let $\{F_\alpha; \alpha \in \Lambda\}$ is a family of closed subsets of X. A subset of K is compact if and only if the family $\{K \cap F_\alpha; \alpha \in \Lambda\}$ has the finite intersection property, then $\bigcap_{\alpha\in\Lambda} K \cap F_\alpha \neq \phi$.*

Proof First, we note that

$$\text{the family } \{K \cap F_\alpha; \alpha \in \Lambda\} \text{ has the finite intersection property} \Longrightarrow \bigcap_{\alpha\in\Lambda} K \cap F_\alpha \neq \phi$$

is equivalent to

$$\bigcap_{\alpha\in\Lambda} K \cap F_\alpha = \phi \Longrightarrow \text{the family } \{K \cap F_\alpha; \alpha \in \Lambda\} \text{ does not have the finite intersection property.} \tag{5.51}$$

Thus we may show that K is compact if and only if (5.51) holds. Let the set K be compact, and let $\{F_\alpha; \alpha \in \Lambda\}$ be a family of closed sets. Assume $\bigcap_{\alpha\in\Lambda} K \cap F_\alpha = \phi$.

Then we see $K \subset \bigcup_{\alpha\in\Lambda} F_\alpha^c$, and so we have an open covering $\bigcup_{\alpha\in\Lambda} F_\alpha^c$ of K. Since K is compact, there exist finite sets $F_{\alpha_1}, F_{\alpha_2}, \cdots, F_{\alpha_n} \in \mathcal{F}$ such that $K \subset \bigcup_{i=1}^{n} F_{\alpha_i}^c$, that is, $\bigcap_{i=1}^{n} K \cap F_{\alpha_i} = \phi$. Hence the collection $\{F_\alpha; \alpha \in \Lambda\}$ does not have the finite intersection property. Conversely, we suppose (5.51). Let $K \subset \bigcup_{\alpha\in\Lambda} O_\alpha$ (O_α; open sets). If we put $F_\alpha := O_\alpha^c$, then F_α is closed, and $\bigcap_{\alpha\in\Lambda} K \cap F_\alpha = \phi$ holds. Then we have an open covering $\bigcup_{\alpha\in\Lambda} F_\alpha^c$ of K. Since $\mathcal{F}_K := \{K \cap F_\alpha; \alpha \in \Lambda\}$ does not have the finite intersection property, there exist finite sets $F_{\alpha_1}, F_{\alpha_2}, \cdots, F_{\alpha_n} \in \mathcal{F}$ such that $K \subset \bigcup_{i=1}^{n} F_{\alpha_i}^c$. Hence we see that K is compact. ■

Corollary 5.5 *Let (X, ρ) be a metric space, and let $\mathcal{F} = \{F_\alpha; \alpha \in \Lambda\}$ is a family of closed subsets of X. The space (X, ρ) is compact if and only if $\mathcal{F}$ has the finite intersection property, then $\bigcap_{\alpha\in\Lambda} F_\alpha \neq \phi$.*

5.20.2 Sequential Compactness

If S is a closed and bounded subset of $\mathbb{R}$, then every infinite subset A of S has a cluster point in S. If a metric space (X, ρ) satisfies that every infinite subspace has a cluster point, then we call this space (X, ρ) the Bolzano-Weierstrass Property (BWP). From this property we are led to the generalization.

Definition 5.24 A metric space (X, ρ) is said to be **sequentially compact** if every sequence $\{x_n\}$ in X has a subsequence $\{x_{n_k}\}$, which converges in X. This is equivalent to saying that every sequence in X has at least one subsequence converging to a point of X.

Example 5.32 Let $(\mathbb{R}, \rho)$ be a usual metric space:

(i) The sequence $\left\{\frac{1}{n}\right\}$ converges to $0 \notin (0, 1]$, so all subsequences of $\left\{\frac{1}{n}\right\}$ converge to 0. Hence $(0, 1]$ is not sequentially compact.
(ii) The sequence $\{n\}, \forall n \in \mathbb{Z}$ is unbounded and all subsequences of $\{n\}$ are unbounded. Therefore these subsequences are not convergent. Hence $\mathbb{Z}$ is not sequentially compact.

Theorem 5.54 *A metric space (X, ρ) is sequentially compact if and only if every infinite subset of X has an accumulation point in X.*

In other ways, a metric space is sequentially compact if and only if it has the Bolzano-Weierstrass Property.

Proof Let A be an infinite subset of a sequentially compact metric space (X, ρ). We shall show that A has a cluster point.

Let $\{x_n\}$ be any sequence of distinct points of A. Since X is sequentially compact, therefore the sequence $\{x_n\}$ contains a convergent subsequence $\{x_{n_k}\}$. Let x be a limit of $\{x_{n_k}\}$, that is,

$$\lim_{k\to\infty} x_{n_k} = x.$$

Consider $S_\varepsilon(x)$ is a neighborhood of x. Then there exist $\varepsilon > 0$ and $N \in \mathbb{N}$ such that

$$\rho\left(x_{n_k}, x\right) < \varepsilon, \quad \forall k \geqslant N \implies x_{n_k} \in S_\varepsilon(x), \quad \forall k \geqslant N.$$

Then the open sphere $S_\varepsilon(x)$ contains infinitely many terms of $\{x_n\}$. Therefore x is a cluster point of A. Hence X has the Bolzano-Weierstrass Property.

Conversely, let us suppose that X has the Bolzano-Weierstrass Property and let $\{x_n\}$ be an arbitrary sequence in X such that the range set $S = \{x_n : n \in \mathbb{N}\}$. There are two possibilities:

(i) Let S be finite. For infinitely many values of n, $x_n = x \in X$, i.e., $x_{n_k} = x,\ k = 1, 2, \cdots$. Then we see $\lim\limits_{k\to\infty} x_{n_k} = x$.

(ii) Let S be infinite. Thus, S has at least one limit point, say, $x_0 \in X$. Therefore the open sphere $S_1(x_0)$ contains infinitely many points of S. Hence, there exists $n_1 \in \mathbb{N}$ such that

$$x_{n_1} \in S_1(x_0) \implies \rho\left(x_{n_1}, x_0\right) < 1.$$

As before, $S_{\frac{1}{2}}(x_0)$ also contains infinitely many points of S. Hence, we have $n_2 \in \mathbb{N}$ such that

$$x_{n_2} \in S_{\frac{1}{2}}(x_0) \implies \rho\left(x_{n_2}, x_0\right) < \frac{1}{2}.$$

Proceeding likewise, we construct a sequence $\{x_{n_k}\}$ such that

$$x_{n_k} \in S_{\frac{1}{k}}(x_0) \implies \rho\left(x_{n_k}, x_0\right) < \frac{1}{k}, \text{ for all } k \in \mathbb{N}.$$

Hence $\{x_{n_k}\}$ is a subsequence of $\{x_n\}$ that converges to x_0, i.e.,

$$\{x_{n_k}\} \to x_0 \ \ as \ \ k \to \infty.$$

Thus $\{x_{n_k}\}$ is a convergent subsequence, and X is sequentially compact.

■

Theorem 5.55 *Every compact metric space is sequentially compact.*

Proof Let (X, ρ) be a compact metric space. In order to show that X is sequentially compact, it is sufficient to show that it has the Bolzano-Weierstrass Property.

Let $\{x_n\}$ be a sequence in (X, ρ) and the associated set $A = \{x_n : n \in \mathbb{N}\}$. If A has a finite range, then there exist a subsequence $\{x_{n_k}\}$ and $K > 0$ such that $x_{n_k} = x_0$ (constant) for $k \geqslant K$. Thus we see $\lim_{k \to \infty} x_{n_k} = x_0$. We consider the case that A is an infinite subset of X. Suppose that A has no limit point. Then corresponding to each $x \in X$, there exists an open sphere $S_{\varepsilon_x}(x)$, $\varepsilon_x > 0$ containing no point of A, other than x, i.e., $(S_{\varepsilon_x}(x) - \{x\}) \cap A = \phi$. We see the family of the open sphere $\left\{S_{\varepsilon_x}(x) : x \in X\right\}$ is an open cover for X. Since X is compact, there exists a finite subcover, i.e., there exists $\{y_i\}_{i=1}^{m} \subset X$ such that

$$A \subset X \subset \bigcup_{i=1}^{m} S_{\varepsilon_{y_i}}(y_i).$$

Then we see that since $S_{\varepsilon_{y_i}}$ contains at most one point of A, the set A contains at most m points. But this contradicts the fact that A is an infinite subset. ■

5.20.3 ε-Net and Totally Bounded

Let A be a subset of a metric space (X, ρ). Given $\varepsilon > 0$, a finite subset $E = \{x_1, x_2, \cdots, x_n\}$ of A is called an ε-net for A if the family $\{S_\varepsilon(x_i) ; 1 \leqslant i \leqslant n\}$ covers A, i.e., $A \subseteq \cup \{S_\varepsilon(x_i) ; 1 \leqslant i \leqslant n\}$. In other words, if for an arbitrary point $a \in A$ at least one point $x \in E$ can be found such that $\rho(a, x) < \varepsilon$.

In other ways, a finite set $E = \{x_1, x_2, \cdots, x_n\}$ of a metric space (X, ρ) is called an ε-net for X if the family $\{S_\varepsilon(x_i) ; 1 \leqslant i \leqslant n\}$ covers X, i.e., $X \subseteq \cup \{S_\varepsilon(x_i) ; 1 \leqslant i \leqslant n\}$.

A metric space is called **totally bounded or precompact** if there is an ε-net for every positive real number ε. Thus, X is totally bounded if and only if for every positive real number ε, X can be covered by a finite union of open spheres of radius ε with centers in E.

Remark 5.20 Every compact space is totally bounded and every subset of a totally bounded space is totally bounded.

5.20.4 Separable Metric Spaces

A metric space contains an everywhere-dense subset which is countable: this theory is very useful in the study of Hilbert spaces.

Definition 5.25 A metric space (X, ρ) is said to be **separable** if it contains an everywhere-dense subset which is countable. In other words, X is separable if $A \subset X$ is countable and $\overline{A} = X$.

Example 5.33 A usual metric space $(\mathbb{R}, \rho)$ is separable since the set of rational numbers $\mathbb{Q}$ is a subset of $\mathbb{R}$ and countable as well as dense in $\mathbb{R}$.

Theorem 5.56 *A totally bounded metric space is separable.*

Proof Let (X, ρ) be a totally bounded metric space. Now for each $n \in \mathbb{N}$ there exists a finite $\varepsilon \left(= \frac{1}{n}\right)$-net, we say $E_n = \{e_1, e_2, \cdots, e_{k_n}\}$. Here we set $E = \bigcup_{n=1}^{\infty} E_n$. Then E is countable. We see

$$X = \bigcup_{i=1}^{k_n} S_{\frac{1}{n}}(e_i).$$

Let $x \in X$. Now we let $x \in S_{\frac{1}{n}}(e_i)$ for some $i = 1, 2, \cdots, k_n \implies \rho(x, e_i) < \frac{1}{n} = \varepsilon$

$$\implies \rho(x, e_i) < \varepsilon \implies S_\varepsilon(x) \cap E \neq \phi \implies x \in \overline{E}.$$

Therefore

$$\overline{E} = X.$$

Hence, E is a dense set in X. Thus, X is separable. ■

Theorem 5.57 *Every sequentially compact metric space is complete and totally bounded.*

Proof Let (X, ρ) be a sequentially compact metric space, and let $\{x_n\}$ be a Cauchy sequence in X. The sequence $\{x_n\}$ has a convergent subsequence, so $\{x_n\}$ is a convergent sequence by Theorems 5.27 and 5.29, i.e., (X, ρ) is complete.

Next, we have to show that X is totally bounded.
Now, we suppose that (X, ρ) is not a totally bounded, then for some $\varepsilon > 0$ there is no ε-net for X. Now we construct a sequence $\{x_n\}$ in X as follows:

Let x_1 be an arbitrary point of X. If $S_\varepsilon(x_1) \neq X$, so choose $x_2 \in X$ such that $x_2 \notin S_\varepsilon(x_1)$ so that $\rho(x_2, x_1) > \varepsilon$. Again

$$S_\varepsilon(x_1) \cup S_\varepsilon(x_2) \neq X.$$

Choose $x_3 \in X$ such that

$$x_3 \notin S_\varepsilon(x_1) \cup S_\varepsilon(x_2) \quad \text{so that} \quad \rho(x_3, x_1) > \varepsilon \text{ and } \rho(x_3, x_2) > \varepsilon.$$

Continuing inductively, we have a sequence $\{x_n\}$ in X, and

$$\rho(x_m, x_n) > \varepsilon, \quad \text{for all } m \neq n.$$

Therefore, $\{x_n\}$ cannot have any convergent subsequence. Hence, this contraction means that X must be totally bounded. ■

Theorem 5.58 *A compact metric space (X, ρ) is complete and totally bounded.*

Proof We see that a metric space (X, ρ) is sequentially compact by Theorems 5.55 and 5.57, and X is complete and totally bounded. ■

Theorem 5.59 *A metric space (X, ρ) is totally bounded if and only if every sequence in X contains a Cauchy subsequence.*

Proof Let $\{x_n\}_{n=1}^{\infty}$ be a sequence of points of a totally bounded metric space X. We will show that $\{x_n\}_{n=1}^{\infty}$ has a Cauchy subsequence.

Since X is totally bounded, given $\varepsilon = 1$, X can be covered by a finite number of subsets $S_1(y_i)$, $i = 1, 2, \cdots, N$, of X such that

$$diam(S_1(y_i)) < \varepsilon, \; (i = 1, 2, \cdots, N) \text{ and } X = \bigcup_{i=1}^{N} S_1(y_i).$$

Then one of these sets A_1 (say) contains x_n for infinitely many values of n. Choose $n_1 \in \mathbb{N}$ such that $x_{n_1} \in A_1$. Clearly, A_1 is totally bounded. Therefore it can be covered by a finite number of subsets of A_1 of diameter less than $1/2$.

One of the sets A_2 (say) contains x_n for infinitely many values of n. Choose $n_2 \in \mathbb{N}$ $(n_2 > n_1)$ such that $x_{n_2} \in A_2$. Since $A_2 \subset A_1$, we see $x_{n_2} \in A_1$. Continuing this process, for $k \in \mathbb{N}$ we obtain a subset A_k $(\subset A_{k-1})$ with $diam\,(A_k) < \frac{1}{k}$, and a term $x_{n_k} \in A_k$ of the sequence $\{x_n\}_{n=1}^{\infty}$. Since

$$x_{n_k}, x_{n_{k+1}}, x_{n_{k+2}}, \cdots$$

all lie in A_k, and since $diam\,(A_k) < \frac{1}{k}$, it follows that $\left\{x_{n_k}\right\}_{k=1}^{\infty}$ is a Cauchy subsequence of $\{x_n\}_{n=1}^{\infty}$.

Conversely, suppose that every sequence $\{x_n\}_{n=1}^{\infty}$ in X contains a Cauchy subsequence. We have to show that X is totally bounded. Suppose X is not totally bounded. Then there exists some $\varepsilon > 0$ such that there does not exist a finite ε-net. Let x_1 be any element of X. Then clearly the open sphere $S_\varepsilon(x_1) \neq X$, otherwise $\{x_1\}$ would be a finite ε−net for X. Again let x_2 be any element of X such that

$$x_2 \notin S_\varepsilon(x_1) \Longrightarrow \rho(x_1, x_2) > \varepsilon.$$

Since $\{x_1, x_2\}$ is not a finite ε-net, we see $S_\varepsilon(x_1) \cup S_\varepsilon(x_2) \neq X$. Let x_3 be any element of X such that

$$x_3 \notin S_\varepsilon(x_1) \cup S_\varepsilon(x_2) \Longrightarrow \rho(x_1, x_3) > \varepsilon \text{ and } \rho(x_2, x_3) > \varepsilon.$$

Continuing this way, we obtain a sequence $\{x_n\}_{n=1}^{\infty}$ of elements in X such that

$$x_n \notin \bigcup_{i=1}^{n-1} S_\varepsilon(x_i) \quad (n = 2, 3, \cdots),$$

so that

$$\rho(x_i, x_n) > \varepsilon \quad (i = 1, 2, \cdots, n-1 \text{ and } n = 1, 2, \cdots) \quad (i \neq n).$$

Consequently, for all m, n, we have

$$\rho(x_n, x_m) > \varepsilon, \quad n \neq m.$$

Thus, $\{x_n\}_{n=1}^{\infty}$ has no Cauchy subsequence. This contradiction shows that X must be totally bounded. ■

Definition 5.26 Let $\mathcal{U} = \{G_\alpha; \alpha \in \Lambda\}$ be any family of subsets of a metric space (X, ρ) with an open cover of X. A **Lebesgue number** for $\mathcal{U}$ is a real number $\varepsilon > 0$ such that for each $x \in X$ there exists $\alpha \in \Lambda$ (dependent on x) for which

$$S_\varepsilon(x) \subseteq G_\alpha,$$

i.e., the sphere $S_\varepsilon(x)$ is contained in some single set from $\mathcal{U}$.

In other ways, a real number $\varepsilon > 0$ is called a **Lebesgue number** for $\mathcal{U}$ if every set $A \subseteq X$ with $\rho(A) < \varepsilon$ is contained in G_α for at least one $\alpha \in \Lambda$.

Remark 5.21 Every open cover of a metric space does not necessarily have a Lebesgue number.

Lemma 5.4 (The Lebesgue Covering Lemma)
Every open cover of a sequentially compact metric space has a Lebesgue number.

Proof Let $\mathcal{U} = \{G_\alpha; \alpha \in \Lambda\}$ be an open cover of X, i.e., $X = \bigcup_{\alpha \in \Lambda} G_\alpha$. Suppose, for the sake of contradiction, that there is no such Lebesgue number ε. In particular, for any $n \in \mathbb{N}$, a number $\frac{1}{n}$ is not a Lebesgue number for $\mathcal{U}$. For a given $n \in \mathbb{N}$ there exist some points $\{x_n\}$ in X such that $S_{\frac{1}{n}}(x_n)$ is not contained in any single set of $\mathcal{U}$.

Since X is a sequential compact metric space, the sequence $\{x_n\}$ has some subsequence $\{x_{n_k}\}$ converging to a point $x \in X$. Now

$$x \in X = \bigcup_{\alpha \in \Lambda} G_\alpha \Longrightarrow x \in G_\alpha, \quad \text{for some } \alpha \in \Lambda.$$

Therefore

$$S_\varepsilon(x) \subseteq G_\alpha, \quad \text{for some } \varepsilon > 0.$$

Since $\{x_{n_k}\}$ converges to x, there exists $K \in \mathbb{N}$ such that

$$x_{n_k} \in S_{\frac{\varepsilon}{2}}(x), \quad \text{for all } k \geqslant K.$$

In particular, we may choose $k \geqslant K$ large enough so that $\frac{1}{n_k} < \frac{\varepsilon}{2}$. It is now sufficient to prove that

$$S_{\frac{1}{n_k}}(x_{n_k}) \subset S_\varepsilon(x). \tag{5.52}$$

Then we have

$$S_{\frac{1}{n_k}}(x_{n_k}) \subset G_\alpha.$$

So the choice of x_{n_k} is contradictory. Suppose $y \in S_{\frac{1}{n_k}}(x_{n_k})$, then

$$\rho(y, x) \leqslant \rho(y, x_{n_k}) + \rho(x_{n_k}, x) < \frac{1}{n_k} + \frac{\varepsilon}{2} < \varepsilon.$$

Hence, $y \in S_\varepsilon(x)$, that is, we have (5.52). ∎

Theorem 5.60 *Every sequentially compact metric space is compact.*

Proof Let (X, ρ) be a sequentially compact metric space and let $\mathcal{U} = \{G_\alpha; \alpha \in \Lambda\}$ be an open cover of X, i.e., $X = \bigcup_{\alpha \in \Lambda} G_\alpha$. By Lemma 5.4 there is a Lebesgue number $\varepsilon > 0$ for $\mathcal{U}$. By Theorem 5.57 X is totally bounded, so there is a finite $\frac{\varepsilon}{3}$-net, say $\{x_1, x_2, \cdots, x_n\}$ for X. Then we see

$$X = \bigcup_{i=1}^{n} S_{\frac{\varepsilon}{3}}(x_i),$$

where

$$\rho\left(S_{\frac{\varepsilon}{3}}(x_i)\right) \leqslant 2 \cdot \frac{\varepsilon}{3} < \varepsilon, \quad (i = 1, 2, \cdots, n).$$

By the definition of a Lebesgue number there exists at least one G_{α_i} such that

$$S_{\frac{\varepsilon}{3}}(x_i) \subseteq G_{\alpha_i}, (i = 1, 2, \cdots, n) \quad \Longrightarrow \quad X \subseteq \bigcup_{i=1}^{n} S_{\frac{\varepsilon}{3}}(x_i) \subseteq \bigcup_{i=1}^{n} G_{\alpha_i},$$

so $\{G_{\alpha_1}, G_{\alpha_2}, \cdots, G_{\alpha_n}\}$ has the finite subcover for X, showing X to be compact. ■

Corollary 5.6 *A metric space is compact if and only if it is sequentially compact.*

Proof By Theorems 5.55 and 5.60, we have the result. ■

5.21 Connectedness

In this section we will give the concept of connectedness. The first is the most basic for the study of continuity. As an application we re-prove the intermediate value theorem.

5.21.1 *Separated Sets*

Two non-empty subsets A and B of a metric space (X, ρ) are said to be separated if

$$A \cap \overline{B} = \phi \text{ and } \overline{A} \cap B = \phi.$$

In other words, non-empty subsets A and B of X are said to be separated if A and B are disjoint and neither A nor B contains any cluster point of the other. If A and B are separated sets, they are disjoint since

$$A \cap B \subseteq A \cap \overline{B} = \phi,$$

but two disjoint sets are not necessarily separated. For example,

$$A = \{x; -\infty < x < 0\} \text{ and } B = \{x; 0 \leq x < \infty\}$$

are disjoint but not separated.

If $A = \{x; -\infty < x < 0\}$ and $B = \{x; 0 < x < \infty\}$, then they are disjoint and separated.

If the union of two separated sets is closed, then separated sets are also closed.

Let A and B be two separated sets, and its union F be closed, i.e., $F = A \cup B$ be closed. Then we see that both A and B are closed. In fact, we have

$$A \cup B = \overline{A \cup B} = \overline{A} \cup \overline{B}.$$

Therefore

$$\begin{aligned}\overline{A} &= \overline{A} \cap \left(\overline{A} \cup \overline{B}\right) \\ &= \overline{A} \cap (A \cup B) \\ &= \left(\overline{A} \cap A\right) \cup \left(\overline{A} \cap B\right) \\ &= A \cup \phi = A,\end{aligned}$$

so that A is closed. Similarly, B is closed.

5.22 Connected and Disconnected Sets

Definition 5.27 A metric space (X, ρ) is said to be connected if it cannot be expressed as the union of two non-empty separated subsets. A metric space, which is not connected, is called disconnected. This means that a metric space X is disconnected if and only if there exist non-empty open subsets A and B such that $A \cap B = \phi$ and $A \cup B = X$ (see Theorem 5.61 (ii)).

You know that ϕ and X are open as well as closed in a metric space (X, ρ). The next theorem shows that in the case of a disconnected metric space these are the only subsets that are both open and closed.

Theorem 5.61 *Let (X, ρ) be a metric space. Then the following statements are equivalent:*

(i) *(X, ρ) is disconnected.*
(ii) *there exist two non-empty disjoint subsets A and B, which are open in (X, ρ) such that $X = A \cup B$.*
(iii) *there exist two non-empty disjoint subsets A and B, which are closed in (X, ρ) such that $X = A \cup B$.*
(iv) *there exists a proper subset of X, which is open as well as closed in (X, ρ).*

Proof $(i) \Longrightarrow (ii)$: Let $X = A \cup B$, $A \cap \overline{B} = \overline{A} \cap B = \phi$. Then $A = \overline{B}^c$, $B = \overline{A}^c$, hence both A and B are open sets.

$(ii) \Longrightarrow (iii)$: Since $A \cap B = \phi$ and both A and B are open sets, $A = B^c$ and $B = A^c$ are closed sets, and $X = A \cup B$.

$(iii) \Longrightarrow (iv)$: Since $A = B^c$, $B = A^c$, we see that both A and B are open sets as well as closed sets.

$(iv) \Longrightarrow (i)$: We see $A \cup B = X$, $A \neq \phi$ and $B = X - A \neq \phi$ are open and closed sets. Then $A \cap B = \phi$ and $X = A \cup B$. Since A is a closed set, we see $A = \overline{A}$, so $\overline{A} \cap B = \phi$. Similarly, we also have $A \cap \overline{B} = \phi$. Hence, X is disconnected and (i) is established. ■

In the next theorem we give the connectedness which is preserved by homeomorphic mappings.

Theorem 5.62 *Let (X, ρ_X) and (Y, ρ_Y) be homeomorphic metric spaces. Then (X, ρ_X) is connected if and only if (Y, ρ_Y) is connected.*

Proof Let $f : (X, \rho_X) \to (Y, \rho_Y)$ be a homeomorphism. Let (Y, ρ_Y) be disconnected and let A_Y is a proper subset of Y, which is simultaneously open and closed in (Y, ρ_Y). Then $f^{-1}(A_Y)$ is also open as well as closed in (X, ρ_X) and a proper subset of (X, ρ_X). Therefore (X, ρ_X) is disconnected.

Similarly, we can show that if (X, ρ_X) is disconnected, then (Y, ρ_Y) is also disconnected.

Thus, (X, ρ_X) is disconnected if and only if (Y, ρ_Y) is disconnected and therefore the theorem follows. ■

Definition 5.28 A non-empty subset Y of a metric space (X, ρ_X) is said to be a **connected (disconnected)** subset of (X, ρ_X) if (Y, ρ_Y) is connected (disconnected).

Let (X, ρ_X) be any metric space and $Y \subseteq X$. If Y contains exactly one point, then it is a connected subset.

Let (X, ρ_X) be any discrete metric space and $Y \subseteq X$. If Y contains exactly more than one point, then it is a disconnected subset.

Theorem 5.63 *Let (X, ρ_X) be a metric space and Y be a subset of X. Then Y is disconnected if and only if there exist non-empty sets A, B such that*

$$Y = A \cup B; \quad \overline{A^X} \cap B = \phi; \quad A \cap \overline{B^X} = \phi, \tag{5.53}$$

where $\overline{A^X}$, $\overline{B^X}$ denote the closure of A or B in (X, ρ_X).

Proof Suppose that Y is a disconnected subset of (X, ρ_X), then by Definition 5.28, there exist non-empty subsets $A \subset Y$ and $B \subset Y$ such that

$$Y = A \cup B; \quad \overline{A^Y} \cap B = \phi; \quad A \cap \overline{B^Y} = \phi, \tag{5.54}$$

where $\overline{A^Y}$, $\overline{B^Y}$ denote the closure of A or B in (Y, ρ_Y). Since $\overline{A^Y} = \overline{A^X} \cap Y$, we have

$$\overline{A^Y} \cap B = \left(\overline{A^X} \cap Y\right) \cap B = \overline{A^X} \cap B \text{ and } \overline{A^X} \cap B = \phi.$$

Similarly, we can obtain

$$A \cap \overline{B^Y} = A \cap \overline{B^X} \text{ and } A \cap \overline{B^X} = \phi.$$

Conversely, assume that there exist non-empty sets A and B, which satisfy the condition (5.53). Since $\overline{A^X} \supseteq \overline{A^Y}$, $\overline{B^X} \supseteq \overline{B^Y}$, from (5.54) we have Y which is a disconnected subset of (X, ρ_X). ■

Theorem 5.64 *A non-empty subset Y of a metric space (X, ρ) is disconnected if and only if there exist open sets G_1 and G_2 in X with the following four properties:*

(i) $G_1 \cap Y \neq \phi$.
(ii) $G_2 \cap Y \neq \phi$.
(iii) $(G_1 \cap G_2) \cap Y = \phi$.
(iv) $Y \subseteq G_1 \cup G_2$.

Furthermore, the word "open" can be replaced by the word "closed."

Proof First, assume that Y is disconnected, i.e., $Y = A \cup B$ and $A \cap B = \phi$, then A is a non-empty proper set of Y, and it is open as well as closed in Y. Therefore $B = Y - A$ is also a non-empty proper subset of Y and it is open as well as closed in Y.

Since A and B are open in Y, there exist open sets G_1 and G_2 in X such that

$$G_1 \cap Y = A \text{ and } G_2 \cap Y = B, \qquad \text{(by Theorem 5.20)}.$$

Clearly,

$$G_1 \cap Y \neq \phi \quad \text{and} \quad G_2 \cap Y \neq \phi,$$

that is, (i), (ii) are satisfied.

Now

$$(G_1 \cap G_2) \cap Y = (G_1 \cap Y) \cap (G_2 \cap Y) = A \cap B = \phi.$$

Hence (iii) is satisfied. Since

$$Y = A \cup B \subseteq G_1 \cup G_2,$$

we have (iv).

Conversely, we suppose $(i) \sim (iv)$. Since G_1 and G_2 are open sets in X, from (iv) we get

$$\begin{aligned} Y = Y \cap (G_1 \cup G_2) &= (Y \cap G_1) \cup (Y \cap G_2) \\ &= A \cup B, \qquad \text{(by Theorem 5.20)}, \end{aligned}$$

where $A \neq \phi$, $B \neq \phi$. We have

$$A \cap B = (G_1 \cap Y) \cap (G_2 \cap Y) = (G_1 \cap G_2) \cap Y = \phi.$$

Thus, Y is disconnected. ■

Theorem 5.65 *If Y is a connected subset of a metric space (X, ρ_X), then*

(i) *if $Y \subseteq A \cup B$, where A and B are separated sets in X, then either $Y \subseteq A$ or $Y \subseteq B$ holds.*
(ii) *if Z is a subset of X such that $Y \subseteq Z \subseteq \overline{Y}$, then Z is connected,*

and in particular, if Y is connected, then $\overline{Y}$ is connected.

Proof (i) If possible, suppose $Y \cap A \neq \phi$ and $Y \cap B \neq \phi$, then it will be shown that Y can be expressed as the union of two non-empty separated sets namely $Y \cap A$ and $Y \cap B$, i.e.,

$$Y = (Y \cap A) \cup (Y \cap B).$$

By Theorem 5.17

$$\overline{Y \cap A} \subseteq \overline{Y} \cap \overline{A}.$$

Therefore,

$$\left(\overline{Y \cap A}\right) \cap (Y \cap B) \subseteq \left(\overline{Y} \cap \overline{A}\right) \cap (Y \cap B) = Y \cap \left(\overline{A} \cap B\right) = \phi,$$

by $\overline{A} \cap B = \phi$. Similarly,

$$(Y \cap A) \cap \left(\overline{Y \cap B}\right) = \phi.$$

Therefore, by Theorem 5.63, Y is disconnected. This is a contradiction.

Alternate Proof of (i).

Since A and B are separated sets and $Y \cap A \subseteq A$, $Y \cap B \subseteq B$, it is obvious $Y \cap A$ and $Y \cap B$ are also separated sets. We see that

$$(Y \cap A) \cap \left(\overline{Y \cap B}\right) = \phi \text{ and } \overline{(Y \cap A)} \cap (Y \cap B) = \phi.$$

Moreover

$$Y = Y \cap (A \cup B) = (Y \cap A) \cup (Y \cap B).$$

Hence, either $Y \subseteq A$ or $Y \subseteq B$ holds.
(ii) Suppose that Z is disconnected, then from Theorem 5.64, we have

$$G_1 \cap Z \neq \phi, \; G_2 \cap Z \neq \phi, \; G_1 \cap G_2 \cap Z = \phi, \; Z \subseteq G_1 \cup G_2,$$

where G_1 and G_2 are two open sets of X.

We have $Y \subset Z$ so that $Y \subset G_1 \cup G_2$. Since Y is connected, and G_1 and G_2 are separated sets, from part (i), we have $Y \subset G_1$ or $Y \subset G_2$.
Now we assume that $Y \subset G_1$, then

$$Y \cap G_2 = \phi \Longrightarrow \overline{Y} \cap G_2 = \phi \ \ (\because G_1 \text{ and } G_2 \text{ are separated sets}).$$

Since $Z \subset \overline{Y}$, we have $Z \cap G_2 = \phi$. This contradicts $Z \cap G_2 \neq \phi$. Thus, Z is connected.

■

Theorem 5.66 *If $\{Y_\alpha; \alpha \in \Lambda\}$ is a family of connected subsets of a metric space (X, ρ_X) and if $\bigcap_{\alpha \in \Lambda} Y_\alpha \neq \phi$, then*

$$Y = \bigcup_{\alpha \in \Lambda} Y_\alpha$$

is also connected.

Proof Suppose that Y is disconnected, i.e.,

$$Y = \bigcup_{\alpha \in \Lambda} Y_\alpha = A \cup B \quad \text{and} \quad A \cap B = \phi,$$

where A and B are separated sets in Y.

There exist $\alpha, \ \beta \in \Lambda$ and $\alpha \neq \beta$ such that

$$Y_\alpha \subset A \quad \text{and} \quad Y_\beta \subset B,$$

by Theorem 5.65 (note $A \neq \phi, \ B \neq \phi$). Since A and B are separated sets, Y_α and Y_β are also separated and such that

$$Y_\alpha \cap Y_\beta = \phi.$$

This is a contradiction $\left(\because \bigcap_{\alpha \in \Lambda} A_\alpha \neq \phi\right)$. Hence, Y is connected. ■

Theorem 5.67 *If $f : X \to Y$ is a continuous mapping of metric spaces and (X, ρ_X) is connected, then $f(X)$ is connected in (Y, ρ_Y).*

Proof If possible, suppose that $f(X)$ is disconnected. Then there exists a non-empty proper subset G of $f(X)$ that is open as well as closed in $f(X)$ (by Theorem 5.61). Therefore, $F = f(X) - G$ is also open as well as closed in $f(X)$.

Since f is continuous, both $f^{-1}(G)$ and $f^{-1}(F)$ are non-empty, open and closed in X. Now we have

$$X = f^{-1}(G) \cup f^{-1}(F),$$

and from $f^{-1}(G) \cap f^{-1}(F) = \phi$ we have

$$f^{-1}(G) \cap \overline{f^{-1}(F)} = \phi \ \text{ and } \ \overline{f^{-1}(G)} \cap f^{-1}(F) = \phi.$$

Therefore, X is disconnected. This contradicts the hypothesis that X is connected.
■

Corollary 5.7 *If $f : X \to Y$ is a continuous mapping of metric spaces and if A is a connected subset of (X, ρ_X), then $f(A)$ is a connected subset of (Y, ρ_Y).*

Theorem 5.68 (The Intermediate Value Theorem) *Let $f : \mathbb{R} \to \mathbb{R}$ be a continuous mapping and $p, q \in \mathbb{R}$ with $p < q$ and $f(p) \neq f(q)$. If ξ is any real number between $f(p)$ and $f(q)$, then there exists a real number s such that $p \leqslant s \leqslant q$ and $f(s) = \xi$.*

Proof Let $I = [p, q]$. Then $f(I)$ is connected. We suppose $\xi \notin f(I)$. Now we set

$$A = \{x; x \in I,\ f(x) < \xi\} \text{ and } B = \{x; x \in I,\ f(x) > \xi\}.$$

Then we have

$$f(I) = A \cup B \text{ and } A \cap \overline{B} = \overline{A} \cap B = \phi.$$

But this contradicts the connectedness of $f(I)$. ∎

Remark 5.22 We can prove this theorem through numerical analysis. For example, suppose $f : \mathbb{R} \to \mathbb{R}$ is continuous and we want to solve the equation $f(x) = 0$. If we can find two finite real numbers p and q such that $f(p)$ and $f(q)$ have opposite signs, then by the Intermediate Value Theorem you know that there is a root of the equation between p and q. To get a better approximation of the root, you merely have to find such numbers p and q that are close together.

5.23 Exercises

1. Choose the correct answer.

   (i) Which of the following define a metric on $\mathbb{R}$, (the set of all real numbers)?
   (a) $\rho(x, y) = y, \forall x, y \in \mathbb{R}$. (b) $\rho(x, y) = x$, if $x = y$ or 1, if $x \neq y$.
   (c) $\rho(x, y) = \max\{x, y\}$. (d) $\rho(x, y) = |x - y| + |x^2 - y^2|$.

   (ii) Let (X, ρ) be a metric space, $A \subset X$ be closed and $B \subset X$ be open. Then
   (a) $A \neq B$. (b) $A \cap B$ is neither open nor closed.
   (c) $B \subset A$. (d) $A \cup B$ is neither open nor closed.

   (iii) If $Q \subset A \subset \mathbb{R}$, which of the following must be true?
   (a) If A is open, then $A = \mathbb{R}$.
   (b) If A is closed, $A = \mathbb{R}$.
   (c) If A is uncountable, then A is closed.
   (d) If A is countable, then A is closed.

   (iv) Which of the following metric spaces (X_i, ρ_i), $1 \leqslant i \leqslant 4$, are complete?
   (a) $X_1 = [0, 1] \subset \mathbb{R}$, ρ_1 defined by $\rho_1(x, y) = \frac{|x-y|}{1+|x-y|}$ for all $x, y \in X$.
   (b) $X_2 = Q$ and ρ_2 is defined by $\rho_2(x, y) = 1$ for all $x, y \in X_2$, $x \neq y$.
   (c) $X_3 = (0, 1) \subset \mathbb{R}$ and ρ_3 is defined by $\rho_3(x, y) = |x - y|$ for all $x, y \in X_3$.
   (d) $X_4 = \mathbb{R}$ and ρ_4 is defined by $\rho_4(x, y) = |e^x - e^y|$ for all $x, y \in X_4$.

2. Let (X, ρ) be a metric space. Prove that for $x, y, z \in X$,

$$|\rho(x, y) - \rho(z, w)| \leqslant \rho(x, z) + \rho(y, z).$$

3. Let X be a non-empty set, and let $B(X, \mathbb{R})$ be the set of all functions $f : \mathbb{R} \to \mathbb{R}$, which are bounded. Show that $D(f, g) = \sup\{d(f(x), g(x)) : x \in X\}$ defines a metric on $B(X, \mathbb{R})$.
4. Let X be a space with a discrete metric. Let $x \in X$. Show that $B[x, 1/2] = B(x, 1/2) = \{x\}$.
5. Prove that the irrational numbers are not closed in $\mathbb{R}$.
6. Given an example of a set in a metric space which is both open and closed.
7. Describe the interior, exterior and frontier of a $A =]0, 1[$ and $B = [0, 1[$ with respect to the usual metric for $\mathbb{R}$.
8. Show that a metric space is discrete if and only if every point of the space is isolated.
9. Prove that if (X, ρ) is a complete space, and each $x \in X$ is a limit point of X, then X is uncountable.
10. Prove or disprove completeness of the following spaces:

    (a) $Q \cap [0, 1]$ as a subspace of $\mathbb{R}$.
    (b) $\{(x, y); x > 0, xy \geqslant 1\}$ as a subspace of $\mathbb{R}^2$.
    (c) A discrete space.

11. Let X be a metric space and let $f : X \to \mathbb{R}$ be a continuous function. Then give a counter example to show that f does not always map Cauchy sequences into Cauchy sequences (example $f(x) = 1/x$ and the sequence $\{1/n\}$).
12. Let $f : \mathbb{R} \to \mathbb{R}$ be continuous and let $f(q) = 0$ for every rational number $q \in Q$. Then, prove that $f(x) = 0$ for every real number $x \in \mathbb{R}$.

Chapter 6
The Riemann Integral

In the eighteenth century, the work on calculus, particularly on integration and its application, was extremely impressive and excellent, but there was no actual "theory" for it.

In calculus, integration was introduced as "finding the area under a curve." While this interpretation is certainly useful, we instead want to think of 'integration' as a more sophisticated form of summation. Geometric considerations will not be so fruitful, whereas the summation interpretation of integration will make many of its properties easy to remember.

In 1854, the Riemann-integral was developed by German mathematician Georg Friedrich Bernhard Riemann and was, when invented, the first rigorous definition of integration applicable to not necessarily continuous functions; then, he described the relationship between integration and differentiation. In other words, we can say "analytical proof of integration." The Lebesgue integral is a more technical and flexible form of the Riemann-integral.
If I is an interval of real numbers, we denote the length of I by $|I|$.

First, as usual, we need to define integration before discussing its properties.

6.1 Partitions

Let $a, b \in \mathbb{R}$. A partition $\mathscr{P}$ is defined as the ordered n-tuple of real numbers $\mathscr{P} = \{x_0, x_1, x_2, \cdots, x_n\}$ such that

$$a = x_0 < x_1 < x_2 < \cdots < x_n = b,$$

and $[x_0, x_1], [x_1, x_2], \cdots, [x_{n-1}, x_n]$ are the sub-intervals of $[a, b]$. We write

$$\Delta x_i = x_i - x_{i-1}, \text{ for } i = 1, 2, \cdots, n,$$

N. Deo and R. Sakai, *Introduction to Mathematical Analysis*,
https://doi.org/10.1007/978-981-97-6568-3_6

so that Δx_i is the length of i-th sub-interval $\left[x_{i-1}, x_i\right]$.

A partition $\mathscr{P} = \{x_0, x_1, x_2, \cdots, x_n\}$ is called a `Uniform Partition` of $[a, b]$, if

$$x_n - x_{n-1} = \cdots = x_i - x_{i-1} = \cdots = x_1 - x_0 = \frac{b-a}{n}.$$

6.2 The Norm or Mesh of the Partition

The maximum difference between any two consecutive points of the partition, i.e., the greatest segments (largest sub-interval), is called the norm or mesh of the partition and denoted as follows:

$$\|\mathscr{P}\| = \max \Delta x_i \ (1 \leqslant i \leqslant n) = \max_{1 \leqslant i \leqslant n} |x_i - x_{i-1}|,$$

then

$$\|\mathscr{P}\| = \sup\{x_i - x_{i-1},\ 1 \leqslant i \leqslant n\}.$$

6.3 Tagged Partitions

A Tagged Partition $\dot{\mathscr{P}}(x, t)$ of an interval $[a, b]$ is a partition together with a finite sequence of numbers $t_1, \cdots, t_n$ subject to the conditions that for each i, $t_i \in [x_{i-1}, x_i]$. In other words, it is a partition together with a distinguished point of every sub-interval.

A Tagged Partition $\dot{\mathscr{P}}(x, t)$ is defined as the set of ordered pairs

$$\dot{\mathscr{P}} = \{([x_{i-1}, x_i], t_i)\}_{i=1}^{n}, \text{ such that } x_{i-1} < t_i < x_i,$$

and the points t_i are called Tags.

6.4 Refinement (Finer)

If $\mathscr{P}$ and $\mathscr{P}^*$ are partitions of $[a, b]$ with $\mathscr{P} \subset \mathscr{P}^*$, then $\mathscr{P}^*$ is said to be a *refinement* of $\mathscr{P}$, i.e., every point of $\mathscr{P}$ is a point of $\mathscr{P}^*$. In other words, $\mathscr{P}^*$ *refines* $\mathscr{P}$ or $\mathscr{P}^*$ is *finer* than $\mathscr{P}$.

If $\mathscr{P}_1$ and $\mathscr{P}_2$ are two partitions, then $\mathscr{P}^*$ is their common refinement if $\mathscr{P}^* = \mathscr{P}_1 \cup \mathscr{P}_2$.

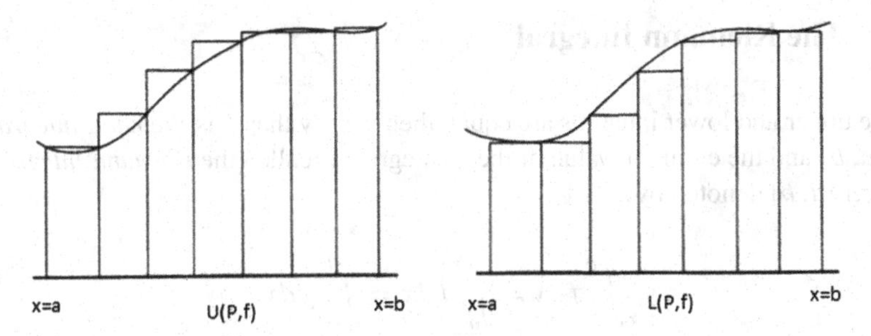

Fig. 6.1 Upper Darboux sum U(P, f) and Lower Darboux sum L(P, f)

6.5 Riemann Sums

Suppose $f : [a, b] \to \mathbb{R}$ is a bounded real function defined on $[a, b]$. Evidently f is a bounded real function defined on each sub-interval of partition $\mathcal{P}$. For each $i = 1, 2, \cdots, n$, we define

$$\begin{aligned} M_i(f) &= \sup\left\{f(x); x \in \left[x_{i-1}, x_i\right]\right\} \\ \text{and} \quad m_i(f) &= \inf\left\{f(x); x \in \left[x_{i-1}, x_i\right]\right\}. \end{aligned}$$

Also we define the two sums

$$\begin{aligned} U(\mathcal{P}, f) &= \sum_{i=1}^{n} M_i \Delta x_i = M_1 \Delta x_1 + M_2 \Delta x_2 + \cdots + M_n \Delta x_n \\ \text{and} \quad L(\mathcal{P}, f) &= \sum_{i=1}^{n} m_i \Delta x_i = m_1 \Delta x_1 + m_2 \Delta x_2 + \cdots + m_n \Delta x_n, \end{aligned}$$

they are respectively called the upper and lower *Riemann (Darboux)* sums (Fig. 6.1).

6.6 Upper and Lower Riemann Integrals

The infimum of the set of upper sums is called the upper integral and the supremum of the set of lower sums is called the lower integral of f. Thus

$$\overline{\int_a^b} f dx = \inf U(\mathcal{P}, f) \text{ and } \underline{\int_a^b} f dx = \sup L(\mathcal{P}, f).$$

These two integrals may or may not be equal.

6.7 The Riemann Integral

If the upper and lower integrals are equal, then we say that f is *Riemann-integrable* on $[a, b]$ and the common value of these integrals is called the *Riemann Integral* of f over $[a, b]$ denoted by

$$\overline{\int_a^b} f dx = \underline{\int_a^b} f dx = \int_a^b f dx.$$

Thus, we write $f \in \mathcal{R}[a, b]$.

Remark 6.1 *(i)* A bounded function is said to be integrable when the upper and lower integrals are equal.

(ii) Since $\mathcal{P} = \{a = x_0, x_1, x_2, \cdots, x_n = b\}$ and $\Delta x_i = x_i - x_{i-1}$,

$$\begin{aligned}\sum_{i=1}^{n} \Delta x_i &= \Delta x_1 + \Delta x_2 + \cdots + \Delta x_n \\ &= (x_1 - x_0) + (x_2 - x_1) + \cdots + (x_n - x_{n-1}) \\ &= x_n - x_0 = b - a.\end{aligned}$$

Thus, $$\sum_{i=1}^{n} \Delta x_i = b - a.$$

Lemma 6.1 *If $f : [a, b] \to \mathbb{R}$ is a bounded function and $\mathcal{P}$ is a partition of $[a, b]$, then $L(\mathcal{P}, f) \leqslant U(\mathcal{P}, f)$.*

Proof Let $\mathcal{P}$ be a partition and f be a bounded function in $[a, b]$. Evidently f is bounded on each sub-interval. Let m_i and M_i be the *infimum* and *supremum* of f in $\left[x_{i-1}, x_i\right]$. Then

$$m_i \leqslant M_i, \quad (i = 1, 2, 3, \cdots, n)$$

$$\implies \quad \sum_{i=1}^{n} m_i \Delta x_i \leqslant \sum_{i=1}^{n} M_i \Delta x_i.$$

Hence,

$$L(P, f) \leqslant U(P, f).$$

■

Lemma 6.2 *If $\mathcal{P}^*$ is a refinement of a partition $\mathcal{P}$, then for a bounded function f,*

$$L\left(\mathcal{P}^*, f\right) \geqslant L(\mathcal{P}, f) \quad \text{and} \quad U\left(\mathcal{P}^*, f\right) \leqslant U(\mathcal{P}, f).$$

Proof If $\mathscr{P}^*$ is a refinement of a partition $\mathscr{P}$, then suppose that $\mathscr{P}^*$ contains just one point more than $\mathscr{P}$. Let ξ be an extra point and $\xi \in (x_{i-1}, x_i)$,
i. e., $x_{i-1} < \xi < x_i$.

The function f is bounded in $[a, b]$, therefore it is bounded in every sub-interval $\left[x_{i-1}, x_i\right]$ $(i = 1, 2, 3, \cdots)$. Let ω_1, ω_2, m_i be the infimum of f in the intervals $\left[x_{i-1}, \xi\right], [\xi, x_i], \left[x_{i-1}, x_i\right]$ respectively, i.e.,

$$\begin{aligned}
\omega_1 &= \inf \{f(x); x_{i-1} \leqslant x \leqslant \xi\}, \\
\omega_2 &= \inf \{f(x); \xi \leqslant x \leqslant x_i\}, \\
m_i &= \inf \{f(x); x_{i-1} \leqslant x \leqslant x_i\},
\end{aligned}$$

clearly, $m_i \leqslant \omega_1, m_i \leqslant \omega_2$.

$$\begin{aligned}
&L\left(\mathscr{P}^*, f\right) - L(\mathscr{P}, f) \\
&= \omega_1 (\xi - x_{i-1}) + \omega_2 (x_i - \xi) - m_i (x_i - x_{i-1}) \\
&= \omega_1 (\xi - x_{i-1}) + \omega_2 (x_i - \xi) - m_i (x_i - \xi) - m_i (\xi - x_{i-1}) \\
&= (\omega_1 - m_i)(\xi - x_{i-1}) + (\omega_2 - m_i)(x_i - \xi) \geqslant 0 \quad (\because \text{each bracket is positive}).
\end{aligned}$$

If $\mathscr{P}^*$ contains k points more than $\mathscr{P}$, we repeat the above arguments k times and arrive at

$$L\left(\mathscr{P}^*, f\right) \geqslant L(\mathscr{P}, f).$$

Let W_1, W_2, M_i be the supremum of f in the intervals $\left[x_{i-1}, \xi\right], [\xi, x_i], \left[x_{i-1}, x_i\right]$ respectively, i.e.,

$$\begin{aligned}
W_1 &= \sup \{f(x); x_{i-1} \leqslant x \leqslant \xi\}, \\
W_2 &= \sup \{f(x); \xi \leqslant x \leqslant x_i\}, \\
M_i &= \sup \{f(x); x_{i-1} \leqslant x \leqslant x_i\},
\end{aligned}$$

clearly, $W_1 \leqslant M_i, W_2 \leqslant M_i$. Now

$$\begin{aligned}
&U(\mathscr{P}, f) - U\left(\mathscr{P}^*, f\right) \\
&= M_i (x_i - x_{i-1}) - \{W_1 (\xi - x_{i-1}) + W_2 (x_i - \xi)\} \\
&= M_i (x_i - \xi) + M_i (\xi - x_{i-1}) - W_1 (\xi - x_{i-1}) - W_2 (x_i - \xi) \\
&= (M_i - W_2)(x_i - \xi) + (M_i - W_1)(\xi - x_{i-1}) \geqslant 0 \quad (\because \text{each bracket is positive}).
\end{aligned}$$

If $\mathscr{P}^*$ contains k points more than $\mathscr{P}$, we repeat the above arguments k times and arrive at

$$U\left(\mathscr{P}^*, f\right) \leqslant U(\mathscr{P}, f).$$

■

Theorem 6.1 *Let* $f : [a, b] \to \mathbb{R}$ *be a bounded function and let* $\mathcal{P}_1$ *and* $\mathcal{P}_2$ *be partitions of* $[a, b]$. *Then*

$$L(\mathcal{P}_1, f) \leqslant U(\mathcal{P}_2, f) \quad and \quad L(\mathcal{P}_2, f) \leqslant U(\mathcal{P}_1, f),$$

i.e., no upper sum can ever be less than any lower sum.

Proof Suppose $\mathcal{P}^*$ is a common refinement of $\mathcal{P}_1, \mathcal{P}_2$, so that

$$P^* = \mathcal{P}_1 \cup \mathcal{P}_2.$$

Since $P_1 \subset P^*$ and $P_2 \subset P^*$, from Lemma 6.2, we have

$$L(\mathcal{P}_1, f) \leqslant L\left(\mathcal{P}^*, f\right) \text{ and } U\left(\mathcal{P}^*, f\right) \leqslant U(\mathcal{P}_2, f).$$

By Lemma 6.1, we have

$$L\left(\mathcal{P}^*, f\right) \leqslant U\left(\mathcal{P}^*, f\right).$$

Using the above three inequalities, we have

$$L(\mathcal{P}_1, f) \leqslant L\left(\mathcal{P}^*, f\right) \leqslant U\left(\mathcal{P}^*, f\right) \leqslant U(\mathcal{P}_2, f).$$

Thus,

$$L(\mathcal{P}_1, f) \leqslant U(\mathcal{P}_2, f).$$

Similarly, by Lemma 6.2, we have

$$L(\mathcal{P}_2, f) \leqslant L\left(\mathcal{P}^*, f\right) \text{ and } U\left(\mathcal{P}^*, f\right) \leqslant U(\mathcal{P}_1, f).$$

By Lemma 6.1

$$L\left(\mathcal{P}^*, f\right) \leqslant U\left(\mathcal{P}^*, f\right).$$

From the above three inequalities, we have

$$L(\mathcal{P}_2, f) \leqslant L\left(\mathcal{P}^*, f\right) \leqslant U\left(\mathcal{P}^*, f\right) \leqslant U(\mathcal{P}_1, f).$$

Thus,

$$L(\mathcal{P}_2, f) \leqslant U(\mathcal{P}_1, f).$$

■

Theorem 6.2 *Let* $f : [a, b] \to \mathbb{R}$ *is a bounded function and let* m *and* M *be lower and upper bounds of* f. *Then*

$$m(b-a) \leqslant L(\mathcal{P}, f) \leqslant U(\mathcal{P}, f) \leqslant M(b-a).$$

In other words, if f is a bounded function on $[a, b]$, then for any partition $\mathcal{P}$ of $[a, b]$, the upper sum $U(\mathcal{P}, f)$ and the lower sum $L(\mathcal{P}, f)$ are bounded.

Proof Let $\mathcal{P} = \{x_0, x_1, x_2, \cdots, x_n\}$ be any partition of $[a, b]$ and let m and M be the infimum and supremum in $[a, b]$ respectively. Again let m_i and M_i be the infimum and supremum in $\left[x_{i-1}, x_i\right]$ respectively, so that

$$
\begin{aligned}
& m \leqslant m_i \leqslant M_i \leqslant M \quad (i = 1, 2, \cdots, n) \\
\Longrightarrow\ & m\Delta x_i \leqslant m_i \Delta x_i \leqslant M_i \Delta x_i \leqslant M \Delta x_i.
\end{aligned}
$$

Putting $i = 1, 2, \cdots, n$ and adding all the inequalities, we get

$$
\begin{aligned}
& \sum_{i=1}^{n} m\Delta x_i \leqslant \sum_{i=1}^{n} m_i \Delta x_i \leqslant \sum_{i=1}^{n} M_i \Delta x_i \leqslant \sum_{i=1}^{n} M \Delta x_i \\
\Longrightarrow\ & m(b-a) \leqslant L(\mathcal{P}, f) \leqslant U(\mathcal{P}, f) \leqslant M(b-a),
\end{aligned}
$$

so that $L(\mathcal{P}, f)$ and $U(\mathcal{P}, f)$ form bounded sets. This shows that the *upper and lower Riemann sums are defined for every bounded function f.* ■

Theorem 6.3 *The lower Riemann-integral cannot exceed the upper Riemann-integral, i.e.,*

$$\underline{\int_a^b} f dx \leqslant \overline{\int_a^b} f dx.$$

Proof If $f : [a, b] \to \mathbb{R}$ is a bounded function and if $\mathcal{P}_1$ and $\mathcal{P}_2$ are two partitions of $[a, b]$, then by Theorem 6.1

$$L(\mathcal{P}_1, f) \leqslant U(\mathcal{P}_2, f).$$

Keeping $\mathcal{P}_2$ fixed and taking the supremum over all partitions $\mathcal{P}_1$, by the above inequality

$$\underline{\int_a^b} f dx = \sup L(\mathcal{P}_1, f) \leqslant U(\mathcal{P}_2, f).$$

Now taking the infimum over all partitions $\mathcal{P}_2$, by the above inequality

$$\underline{\int_a^b} f dx \leqslant \overline{\int_a^b} f dx = \inf U(\mathcal{P}_2, f).$$

Thus,

$$\underline{\int_a^b} f dx \leqslant \overline{\int_a^b} f dx.$$

■

Theorem 6.4 (Riemann's criterion for integrability)
A bounded function $f : [a, b] \to \mathbb{R}$ *on* $[a, b]$ *is integrable if and only if for each* $\varepsilon > 0$ *there exists a partition* $\mathcal{P}$ *of* $[a, b]$ *such that*

$$U(\mathcal{P}, f) - L(\mathcal{P}, f) < \varepsilon.$$

Proof Suppose that f is integrable on a closed interval $[a, b]$, then

$$\overline{\int_a^b} f dx = \underline{\int_a^b} f dx = \int_a^b f dx.$$

Now we take the supremum of $L(\mathcal{P}, f)$ over all partition $\mathcal{P}$ of $[a, b]$. Then

$$\int_a^b f dx = \underline{\int_a^b} f dx = \sup L(\mathcal{P}, f).$$

Therefore, $\int_a^b f dx - \varepsilon/2$ is not an upper bound of a set of all lower sums, hence there is a partition $\mathcal{P}_1$ such that

$$\int_a^b f dx - \frac{\varepsilon}{2} < L(\mathcal{P}_1, f). \tag{6.1}$$

Similarly, $\int_a^b f dx + \varepsilon/2$ is not a lower bound for a set of all upper sums, hence there is a partition $\mathcal{P}_2$ such that

$$U(\mathcal{P}_2, f) < \int_a^b f dx + \frac{\varepsilon}{2}. \tag{6.2}$$

Let $\mathcal{P}$ be a common refinement of $\mathcal{P}_1$ and $\mathcal{P}_2$, i.e., $\mathcal{P} = \mathcal{P}_1 \cup \mathcal{P}_2$. By Lemma 6.2, we have

$$U(\mathcal{P}, f) \leqslant U(\mathcal{P}_2, f) < \int_a^b f dx + \frac{\varepsilon}{2}, \quad (by\ (6.2))$$

and

$$L(\mathcal{P}, f) \geqslant L(\mathcal{P}_1, f) > \int_a^b f dx - \frac{\varepsilon}{2}, \quad (by\ (6.1))$$

which are lower and upper bounds of f. Then

$$U(\mathcal{P}, f) - L(\mathcal{P}, f) < \varepsilon.$$

Suppose $f : [a, b] \to \mathbb{R}$ is bounded on $[a, b]$ and for each $\varepsilon > 0$ there is a partition $\mathcal{P}$ such that

$$U(\mathcal{P}, f) - L(\mathcal{P}, f) < \varepsilon.$$

We have to show that

$$\overline{\int_a^b} f dx = \underline{\int_a^b} f dx,$$

i.e., f is integrable. For any partition $\mathscr{P}$, we know that (Fig. 6.2)

$$\begin{aligned}
L(\mathscr{P}, f) \leqslant \underline{\int_a^b} f dx \leqslant \overline{\int_a^b} f dx \leqslant U(\mathscr{P}, f) \\
\leqslant U(\mathscr{P}, f) - L(\mathscr{P}, f) + L(\mathscr{P}, f) \\
= \varepsilon + L(\mathscr{P}, f).
\end{aligned}$$

Therefore

$$0 \leqslant \overline{\int_a^b} f dx - \underline{\int_a^b} f dx < \varepsilon.$$

Thus,

$$\underline{\int_a^b} f dx = \overline{\int_a^b} f dx, \text{ i.e., } f \text{ is integrable.}$$

■

Example 6.1 Let $f(x) = x^2$ on $[0, k]$, $k > 0$. We show that $f \in \mathscr{R}[0, k]$ and

$$\int_0^k f dx = \frac{k^3}{3}.$$

Given $f(x) = x^2$, $x \in [0, k]$, then we show that $f \in \mathscr{R}[0, k]$.

The interval $[0, k]$ is divided into n equal parts $\Delta x_i = \frac{k}{n}$; $i = 1, 2, \cdots, n$, then we have a partition $\mathscr{P}_n = \{0 = x_0, x_1, x_2, \cdots, x_n = k\}$ of $[0, k]$.

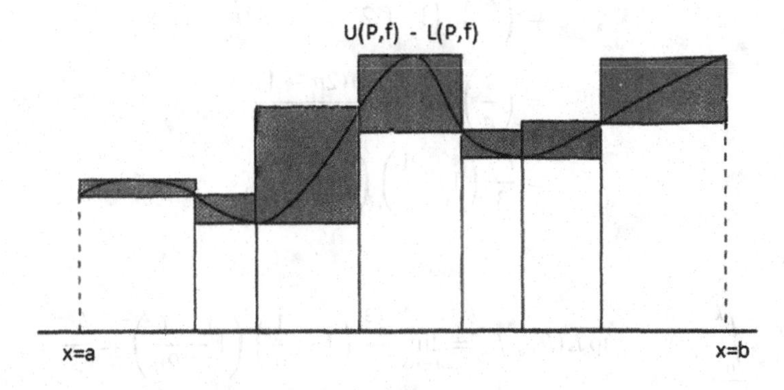

Fig. 6.2 Geometric interpretation of the integrability criteria

Let M_i and m_i be the supremum and the infimum in Δx_i respectively, that is,

$$M_i = \sup\{f(x); x \in \Delta x_i\} = x_i^2 = \left(\frac{ik}{n}\right)^2,$$

$$\text{and} \qquad m_i = \inf\{f(x); x \in \Delta x_i\} = x_{i-1}^2 = \left\{\frac{(i-1)k}{n}\right\}^2.$$

Now

$$\begin{aligned}
U(\mathscr{P}_n, f) &= \sum_{i=0}^{n} M_i \Delta x_i \\
&= M_1 \Delta x_1 + M_2 \Delta x_2 + \cdots + M_n \Delta x_n \\
&= \left(\frac{k}{n}\right)^3 (1^2 + 2^2 + \cdots + n^2) \\
&= \left(\frac{k}{n}\right)^3 \frac{n(n+1)(2n+1)}{6} \\
&= \frac{k^3}{3}\left(1 + \frac{1}{n}\right)\left(1 + \frac{1}{2n}\right).
\end{aligned}$$

Hence

$$\overline{\int_0^k} f\,dx = \inf U(\mathscr{P}_n, f) = \lim_{n\to\infty} \frac{k^3}{3}\left(1 + \frac{1}{n}\right)\left(1 + \frac{1}{2n}\right) = \frac{k^3}{3}.$$

Similarly,

$$\begin{aligned}
L(\mathscr{P}_n, f) &= \sum_{i=0}^{n} m_i \Delta x_i \\
&= m_1 \Delta x_1 + m_2 \Delta x_2 + \cdots + m_n \Delta x_n \\
&= \left(\frac{k}{n}\right)^3 \{1^2 + 2^2 + \cdots + (n-1)^2\} \\
&= \left(\frac{k}{n}\right)^3 \frac{(n-1)n(2n-1)}{6} \\
&= \frac{k^3}{3}\left(1 - \frac{1}{n}\right)\left(1 - \frac{1}{2n}\right).
\end{aligned}$$

Hence

$$\underline{\int_0^k} f\,dx = \sup L(\mathscr{P}_n, f) = \lim_{n\to\infty} \frac{k^3}{3}\left(1 - \frac{1}{n}\right)\left(1 - \frac{1}{2n}\right) = \frac{k^3}{3}.$$

So that

$$\overline{\int_0^k} f dx = \underline{\int_0^k} f dx = \frac{k^3}{3}.$$

Thus,

$$f \in \mathcal{R}[0, k] \text{ and } \int_0^k f dx = \frac{k^3}{3}.$$

Example 6.2 We show that the function defined as follows:

$$f(x) = \begin{cases} 0, & \text{when } x \text{ is rational,} \\ 1, & \text{when } x \text{ is irrational,} \end{cases}$$

is not integrable on any interval, i.e., $f \notin \mathcal{R}[a, b]$.

Given $f(x) = 0, \ x \in \mathbb{Q}$ and $f(x) = 1, \ x \in \mathbb{R} - \mathbb{Q}$, then prove that $f \notin \mathcal{R}[a, b]$.
Let $\mathcal{P} = \{a = x_0, x_1, x_2, \cdots, x_n = b\}$ be any partition of $[a, b]$.
Hence

$$\begin{aligned} M_i &= \sup\{f(x); x \in \Delta x_i\} = \sup\{0, 1\} = 1, \\ m_i &= \inf\{f(x); x \in \Delta x_i\} = \inf\{0, 1\} = 0, \\ \therefore \ M_i &= 1, m_i = 0, \ \forall i = 0, 1, 2, \cdots, n. \end{aligned}$$

Now

$$U(\mathcal{P}, f) = \sum_{i=0}^{n} M_i \Delta x_i = b - a$$

$$\text{and } L(\mathcal{P}, f) = \sum_{i=0}^{n} m_i \Delta x_i = 0.$$

Hence

$$\begin{aligned} &\overline{\int_a^b} f dx = \inf U(\mathcal{P}, f) = \inf(b - a) = b - a \\ &\underline{\int_a^b} f dx = \sup L(\mathcal{P}, f) = \sup\{0\} = 0 \\ \Longrightarrow &\overline{\int_a^b} f dx \neq \underline{\int_a^b} f dx. \end{aligned}$$

Thus, $f \notin \mathcal{R}[a, b]$.

Example 6.3 If the function f is defined on $[a, b]$ by

$$f(x) = c, \; x \in [a, b],$$

where c is a constant, then we show that $f \in \mathcal{R}[a, b]$ and

$$\int_a^b f(x)dx = c(b - a).$$

Given $f(x) = c, \; x \in [a, b]$, we have to show that $f \in \mathcal{R}[a, b]$.

Let $\mathcal{P} = \{a = x_0, x_1, x_2, \cdots, x_n = b\}$ be any partition of $[a, b]$. Let M_i and m_i be the supremum and the infimum in Δx_i respectively $(i = 1, 2, \cdots, n)$. Hence

$$\begin{aligned} M_i &= \sup\{f(x); x \in \Delta x_i\} = \sup\{c\} = c \\ \text{and} \quad m_i &= \inf\{f(x); x \in \Delta x_i\} = \inf\{c\} = c. \\ \therefore M_i &= m_i = c, \quad \forall i = 1, 2, \cdots, n. \end{aligned}$$

Now

$$\begin{aligned} U(\mathcal{P}, f) &= \sum_{i=1}^{n} M_i \Delta x_i = c \sum_{i=1}^{n} \Delta x_i = c(b - a) \\ \text{and} \quad L(\mathcal{P}, f) &= \sum_{i=1}^{n} m_i \Delta x_i = c \sum_{i=1}^{n} \Delta x_i = c(b - a). \end{aligned}$$

Hence

$$\overline{\int_a^b} f dx = \inf U(\mathcal{P}, f) = \inf\{c(b - a)\} = c(b - a),$$

and

$$\underline{\int_a^b} f dx = \sup L(\mathcal{P}, f) = \sup\{c(b - a)\} = c(b - a).$$

Hence

$$\int_a^b f(x)dx = c(b - a).$$

Thus, $f \in \mathcal{R}[a, b]$.

Theorem 6.5 *Every monotonic function f on $[a, b]$ is integrable.*

Proof Suppose f is a monotonic increasing function on $[a, b]$ (the proof is similar to the monotonic decreasing case), therefore it is bounded on $[a, b]$. We see

$$f(a) \leqslant f(x) \leqslant f(b), \quad \forall x \in [a, b].$$

Let $\varepsilon > 0$ be given. If we take a partition $\mathcal{P} = \{a = x_0, x_1, x_2, \cdots, x_n = b\}$ of $[a, b]$ with

$$\max\{\Delta x_i; i = 1, 2, \ldots, n\} < \frac{\varepsilon}{f(b) - f(a)}, \quad M_i = f(x_i) \text{ and } m_i = f(x_{i-1}),$$

then

$$\begin{aligned} U(f, \mathcal{P}) - L(f, \mathcal{P}) &= \sum_{i=1}^{n} (M_i - m_i)\Delta x_i \\ &= \sum_{i=1}^{n} (f(x_i) - f(x_{i-1}))\Delta x_i \\ &< (f(b) - f(a)) \frac{\varepsilon}{f(b) - f(a)} = \varepsilon. \end{aligned}$$

Hence, $f \in \mathcal{R}[a, b]$. ■

Theorem 6.6 *If f is continuous on $[a, b]$, then it is Riemann-integrable. In other words, "Every continuous function is integrable."*

Proof Since f is continuous on the compact (closed and bounded) interval $[a, b]$, f is uniformly continuous on that interval. For every $\varepsilon > 0$ there exists $\delta > 0$ such that

$$\forall x, y \in [a, b] \text{ and } |x - y| < \delta \text{ imply } |f(x) - f(y)| < \frac{\varepsilon}{b - a}.$$

Let $\mathcal{P} = \{a = x_0, x_1, x_2, \cdots, x_n = b\}$ be any partition of $[a, b]$ such that

$$\max\{x_i - x_{i-1}; i = 1, 2, \cdots, n\} < \delta.$$

Since f acquires its maximum and minimum on each sub-intervals $[x_{i-1}, x_i]$, for $x, y \in \left[x_{i-1}, x_i\right]$ we have

$$M_i = \max_{x \in [x_{i-1}, x_i]} f(x) \text{ and } m_i = \min_{y \in [x_{i-1}, x_i]} f(y).$$

Therefore

$$M_i - m_i < \frac{\varepsilon}{b - a}, \quad \forall i = 1, 2, \cdots, n.$$

Hence

$$U(f, \mathcal{P}) - L(f, \mathcal{P}) = \sum_{i=1}^{n} (M_i - m_i)(x_i - x_{i-1}) < \sum_{i=1}^{n} \frac{\varepsilon}{b - a}(x_i - x_{i-1}) = \varepsilon.$$

Thus, f is integrable, i.e., $f \in \mathcal{R}[a, b]$. ■

6.8 Properties of the Riemann Integral

Theorem 6.7 (Scaling) *If $f : [a, b] \to \mathbb{R}$ is an integrable function on $[a, b]$ and λ is any real number, then $\lambda f \in \mathcal{R}[a, b]$ and*

$$\int_a^b \lambda f = \lambda \int_a^b f. \tag{6.3}$$

Proof (i) If $\lambda = 0$, then the function $\lambda f = 0$ and (6.3) is obvious.
(ii) Let $\lambda > 0$, and let $f \in \mathcal{R}[a, b]$. Then for a given $\varepsilon > 0$ there exists a partition $\mathcal{P} = \{a = x_0, x_1, x_2, \cdots, x_n = b\}$ of $[a, b]$ such that

$$U(\mathcal{P}, f) - L(\mathcal{P}, f) < \frac{\varepsilon}{\lambda}.$$

Therefore f is bounded in $[a, b]$ (by Theorem 6.4). Hence $|\lambda f| = \lambda |f|$ is also bounded in $[a, b]$.

Let M_i and m_i be the supremum and the infimum on $[x_{i-1}, x_i]$ respectively. Then λM_i and λm_i are the supremum and the infimum on $\Delta x_i = [x_{i-1}, x_i]$, hence we have

$$U(\mathcal{P}, \lambda f) = \sum_{i=1}^{n} (\lambda M_i)\Delta x_i = \lambda \sum_{i=1}^{n} M_i \Delta x_i = \lambda U(\mathcal{P}, f).$$

Similarly,

$$L(\mathcal{P}, \lambda f) = \lambda L(\mathcal{P}, f).$$

Therefore

$$U(\mathcal{P}, \lambda f) - L(\mathcal{P}, \lambda f) = \lambda [U(\mathcal{P}, f) - L(\mathcal{P}, f)] < \lambda \cdot \frac{\varepsilon}{\lambda} = \varepsilon.$$

Hence,

$$\lambda f \in \mathcal{R}[a, b], \quad i.e., \quad \overline{\int_a^b} \lambda f = \underline{\int_a^b} \lambda f = \int_a^b \lambda f.$$

Thus,

$$\begin{aligned} \int_a^b \lambda f = \overline{\int_a^b} \lambda f &= \inf U(\mathcal{P}, \lambda f) = \lambda \inf U(\mathcal{P}, f) \\ &= \lambda \overline{\int_a^b} f = \lambda \int_a^b f, \end{aligned}$$

or

$$\int_a^b \lambda f = \underline{\int_a^b} \lambda f = \sup L(\mathcal{P}, \lambda f) = \lambda \sup L(\mathcal{P}, f)$$
$$= \lambda \underline{\int_a^b} f = \lambda \int_a^b f. \tag{6.4}$$

(*iii*) Let $\lambda < 0$. If $\lambda = -1$, then

$$U(\mathcal{P}, -f) = -L(\mathcal{P}, f), \text{ for all partitions } \mathcal{P} \text{ of } [a, b].$$

Hence we have

$$\begin{aligned}\overline{\int_a^b}(-f) &= \inf\{U(\mathcal{P}, -f); \mathcal{P}[a, b]\}\\ &= \inf\{-L(\mathcal{P}, f); \mathcal{P}[a, b]\} \quad [\because U(\mathcal{P}, -f) = -L(\mathcal{P}, f)]\\ &= -\sup\{L(\mathcal{P}, f); \mathcal{P}[a, b]\} = -\underline{\int_a^b} f.\end{aligned}$$

Similarly,

$$\underline{\int_a^b}(-f) = -\overline{\int_a^b} f.$$

Since f is integrable, we have

$$\overline{\int_a^b}(-f) = -\underline{\int_a^b} f = -\overline{\int_a^b} f = \underline{\int_a^b}(-f).$$

Thus, $-f$ is integrable and

$$\int_a^b (-f) = -\int_a^b f. \tag{6.5}$$

For the case of $\lambda < 0$, using (6.5) and (6.4) to $-\lambda$, we obtain

$$\int_a^b \lambda f = -\int_a^b (-\lambda) f = -(-\lambda)\int_a^b f = \lambda \int_a^b f.$$

■

Theorem 6.8 (Splitting integrals) *Let a, c, b be real numbers with $a < c < b$. And consider any function $f : [a, b] \to \mathbb{R}$ which is integrable over $[a, c]$ and over $[c, b]$. Then $f \in \mathcal{R}[a, b]$, and*

$$\int_a^b f = \int_a^c f + \int_c^b f.$$

Proof Let $f \in \mathcal{R}[a, c]$ and $f \in \mathcal{R}[c, b]$. Choose $\varepsilon > 0$, then $\mathcal{P}_1$ and $\mathcal{P}_2$ are partitions of $[a, c]$ and $[c, b]$ respectively such that

$$\begin{aligned} U(\mathcal{P}_1, f) - L(\mathcal{P}_1, f) &< \frac{\varepsilon}{2} \\ \text{and} \quad U(\mathcal{P}_2, f) - L(\mathcal{P}_2, f) &< \frac{\varepsilon}{2}. \end{aligned}$$

Now put together the points of $\mathcal{P}_1$ and $\mathcal{P}_2$, then $\mathcal{P}$ is the combined partition of interval $[a, b]$, i.e., $\mathcal{P} = \mathcal{P}_1 \cup \mathcal{P}_2$. Then

$$U(\mathcal{P}, f) = U(\mathcal{P}_1, f) + U(\mathcal{P}_2, f) \geqslant \int_a^c f + \int_c^b f \tag{6.6}$$

$$\text{and} \quad L(\mathcal{P}, f) = L(\mathcal{P}_1, f) + L(\mathcal{P}_2, f) \leqslant \int_a^c f + \int_c^b f. \tag{6.7}$$

Consequently,

$$\begin{aligned} U(\mathcal{P}, f) - L(\mathcal{P}, f) &= [U(\mathcal{P}_1, f) - L(\mathcal{P}_1, f)] + [U(\mathcal{P}_2, f) - L(\mathcal{P}_2, f)] \\ &< \frac{\varepsilon}{2} + \frac{\varepsilon}{2} = \varepsilon. \end{aligned}$$

Therefore, $f \in \mathcal{R}[a, b]$. Now we have

$$\begin{aligned} \int_a^b f \leqslant U(\mathcal{P}, f) = U(\mathcal{P}_1, f) + U(\mathcal{P}_2, f) &< L(\mathcal{P}_1, f) + L(\mathcal{P}_2, f) + \varepsilon \\ &\leqslant \int_a^c f + \int_c^b f + \varepsilon, \end{aligned}$$

and similarly

$$\int_a^b f \geqslant \int_a^c f + \int_c^b f - \varepsilon.$$

Thus,

$$\int_a^b f = \int_a^c f + \int_c^b f.$$

■

Theorem 6.9 (Triangle inequalities) *If a function* $f \in \mathcal{R}[a, b]$, *then taking* $|f| \in \mathcal{R}[a, b]$,

$$\left| \int_a^b f \right| \leqslant \int_a^b |f|.$$

Proof Since $f \in \mathcal{R}[a, b]$, by definition, for $\varepsilon > 0$ there exists a partition $\mathcal{P} = \{a = x_0, x_1, x_2, \cdots, x_n = b\}$ of $[a, b]$ such that

$$U(\mathcal{P}, f) - L(\mathcal{P}, f) < \varepsilon. \tag{6.8}$$

Let M_i and m_i respectively be the supremum and the infimum of $f(x)$ in the interval $[x_{i-1}, x_i]$ and also let M_i' and m_i' respectively be the supremum and the infimum of $|f(x)|$ in the same interval. For $\forall x, y \in [x_{i-1}, x_i]$, we have

$$|f(x)| - |f(y)| \leqslant |f(x) - f(y)|, \tag{6.9}$$

and

$$|f(x) - f(y)| \leqslant M_i - m_i. \tag{6.10}$$

From (6.9) and (6.10) we get

$$|f(x)| - |f(y)| \leqslant M_i - m_i. \tag{6.11}$$

Now, taking the supremum over x in (6.11), we obtain

$$M_i' - |f(y)| \leqslant M_i - m_i. \tag{6.12}$$

Again, taking the supremum over y in (6.12), we get

$$\begin{aligned} & M'_i + \sup(-|f(y)|) \leqslant M_i - m_i \\ \Longrightarrow\ & M'_i - inf|f(y)| \leqslant M_i - m_i \\ \Longrightarrow\ & \qquad M'_i - m_i' \leqslant M_i - m_i. \end{aligned} \tag{6.13}$$

Using (6.13) and (6.8), we have

$$\begin{aligned} U(\mathcal{P}, |f|) - L(\mathcal{P}, |f|) &= \sum_{i=0}^{n} (M' - m'_i)\Delta x_i \\ &\leqslant \sum_{i=0}^{n} (M_i - m_i)\Delta x_i = U(\mathcal{P}, f) - L(\mathcal{P}, f) < \varepsilon. \end{aligned}$$

Hence, prove that $|f| \in R[a, b]$.
Now we have to show that

$$\left|\int_a^b f\right| \leqslant \int_a^b |f|.$$

Since

$$- f(x) \leqslant |f(x)| \Longrightarrow -|f(x)| \leqslant f(x) \text{ and } f(x) \leqslant |f(x)|,$$

we have

$$-|f(x)| \leqslant f(x) \leqslant |f(x)|.$$

Hence

$$-\int_a^b |f(x)|dx \leqslant \int_a^b f(x)dx \leqslant \int_a^b |f(x)|dx.$$

Finally, we have

$$\left|\int_a^b f\right| \leqslant \int_a^b |f|.$$

■

6.9 The Fundamental Theorem of Calculus

The fundamental theorem of calculus has two versions and both the versions express the same thing, that is, differentiation and integration are inverse operations.

Theorem 6.10 (Part-I Integral of a Derivative) *If* $f : [a, b] \to \mathbb{R}$ *is differentiable on* $[a, b]$, *and* $f' \in \mathcal{R}[a, b]$, *then*

$$\int_a^b f'(x)dx = f(b) - f(a). \tag{6.14}$$

Proof Let $\mathcal{P} = \{a = x_0, x_1, x_2, \cdots, x_n = b\}$ be a partition of $[a, b]$. Let $\varepsilon > 0$. Then

$$U\left(\mathcal{P}, f'\right) - L\left(\mathcal{P}, f'\right) < \varepsilon. \tag{6.15}$$

Applying the Mean-Value Theorem, for f defined on $\left[x_{i-1}, x_i\right]$ there exists the point $t_i \in \left[x_{i-1}, x_i\right]$ such that

$$f\left(x_i\right) - f\left(x_{i-1}\right) = f'(t_i)\left(x_i - x_{i-1}\right).$$

Hence

$$f(b) - f(a) = \sum_{i=1}^{n}\left[f\left(x_i\right) - f\left(x_{i-1}\right)\right] = \sum_{i=1}^{n} f'(t_i)\left(x_i - x_{i-1}\right).$$

By the definition of the Riemann-integral

$$L\left(\mathcal{P}, f'\right) \leqslant f(b) - f(a) \leqslant U\left(\mathcal{P}, f'\right). \tag{6.16}$$

On the other hand,

$$L(\mathcal{P}, f') \leqslant \int_a^b f' \leqslant U(\mathcal{P}, f').$$

From (6.15) and (6.16), we get

$$\left| \int_a^b f' - [f(b) - f(a)] \right| < \varepsilon.$$

Since $\varepsilon > 0$ is arbitrary, (6.14) holds. ■

Theorem 6.11 (Part-II Derivative of an Integral) *If a function $f : [a, b] \to \mathbb{R}$ is bounded and $f \in \mathcal{R}[a, b]$, let the function ϕ be defined by*

$$\phi(x) = \int_a^x f(t)dt, \quad a \leqslant x \leqslant b.$$

Then ϕ is continuous on $[a, b]$. If f is continuous at x_0 in (a, b), then ϕ is differentiable at x_0 and $\phi'(x_0) = f(x_0)$.

Proof Since f is bounded and Riemann-integrable on $[a, b]$, there exists a number $M > 0$ such that

$$|f(x)| \leqslant M, \quad \text{for all } x \in [a, b].$$

If x and y are two points of $[a, b]$, such that $a \leqslant x < y \leqslant b$, then

$$\begin{aligned} |\phi(y) - \phi(x)| &= \left| \int_a^y f(t)dt - \int_a^x f(t)dt \right| \\ &= \left| \int_x^y f(t)dt \right| \leqslant |y - x|\, M. \end{aligned}$$

Choose $\varepsilon > 0$ and let $\delta = \varepsilon/M$. If $|y - x| < \delta$, then $|\phi(y) - \phi(x)| < \varepsilon$. Thus ϕ is continuous on $[a, b]$.

Let $\varepsilon > 0$. Since f is continuous at $x_0 \in (a, b)$, there exists $\delta > 0$, such that

$$t \in (a, b) \text{ and } |t - x_0| < \delta \text{ imply } |f(t) - f(x_0)| < \varepsilon,$$

$$\begin{aligned} \left| \frac{\phi(x) - \phi(x_0)}{x - x_0} - f(x_0) \right| &= \left| \frac{1}{x - x_0} \int_{x_0}^x f(t)dt - f(x_0) \right| \\ &= \frac{1}{|x - x_0|} \left| \int_{x_0}^x [f(t) - f(x_0)]dt \right| \\ &< \frac{1}{|x - x_0|} |x - x_0|\, \varepsilon < \varepsilon. \end{aligned}$$

Hence,

$$\lim_{x \to x_0} \frac{\phi(x) - \phi(x_0)}{x - x_0} = f(x_0).$$

In other words, $\phi'(x_0) = f(x_0)$. ■

6.9.1 Mean-Value and Change-of-Variable Theorems

An antiderivative is often required for applying the fundamental theorem of calculus, for this reason integration by substitution is an important tool of calculus. It is the counterpart to the chain rule of differentiation and the last two theorems are in a sense analogs of the Mean-Value Theorem for derivatives.

Theorem 6.12 (The Change-of-Variable) *Let $[a, b] \subseteq \mathbb{R}$ be an interval and let $g : [a, b] \to \mathbb{R}$ be a continuously differentiable function. If $g([a, b]) \subset [c, d]$ and $f : [c, d] \to \mathbb{R}$ is a continuous function on $g([a, b])$, then*

$$\int_{g(a)}^{g(b)} f(x)dx = \int_a^b f(g(t))g'(t)dt, \tag{6.17}$$

the substitution $x = g(t)$ and thus $dx = g'(t)\,dt$.

Proof Since f is continuous on the closed and bounded interval $[a, b]$, it possesses an antiderivative F, i.e., $F' = f$. Since F and g are differentiable, the chain rule gives

$$(F \circ g)'(t) = F'(g(t))g'(t) = f(g(t))g'(t).$$

Applying the second fundamental theorem of calculus,

$$\begin{aligned}\int_a^b f(g(t))g'(t)dt &= (F \circ g)(b) - (F \circ g)(a)\\ &= F(g(b)) - F(g(a))\\ &= \int_{g(a)}^{g(b)} f(x)dx.\end{aligned}$$

■

Theorem 6.13 (First Mean-Value Theorem) *If $f : [a, b] \to \mathbb{R}$ is a continuous function, then there is $\xi \in [a, b]$ such that*

$$\int_a^b f(x)dx = f(\xi)(b - a).$$

Proof Let $m = inf\{f(x); x \in [a, b]\}$ and $M = \sup\{f(x); x \in [a, b]\}$. Then

$$m(b-a) \leqslant \int_a^b f(x)dx \leqslant M(b-a), \quad \text{if } b \geqslant a.$$

Hence, there is a number $\mu \in [m, M]$ such that

$$\int_a^b f(x)dx = \mu(b-a).$$

Since f is continuous on $[a, b]$, it attains every value between its bounds m and M. Therefore, there exists a number $\xi \in [a, b]$ such that $f(\xi) = \mu$. Thus,

$$\int_a^b f(x)dx = f(\xi)(b-a).$$

■

Theorem 6.14 (The Second Mean-Value Theorem) *If $f : [a, b] \to \mathbb{R}$ is a monotonic function, then there is $\xi \in [a, b]$ such that*

$$\int_a^b f(x)dx = f(a)(\xi - a) + f(b)(b - \xi).$$

Proof Let f be a decreasing function on $[a, b]$. Then $f \in \mathcal{R}[a, b]$. Let $g : [a, b] \to \mathbb{R}$ be a continuous function defined by $g(x) = f(a)(x-a) + f(b)(b-x)$. Since f is a decreasing function,

$$g(a) = f(b)(b-a) \leqslant \int_a^b f(x)dx \leqslant f(a)(b-a) = g(b).$$

By the continuity of g, there is $\xi \in [a, b]$ such that

$$g(\xi) = \int_a^b f(x)dx.$$

Thus

$$\int_a^b f(x)dx = g(\xi) = f(a)(\xi - a) + f(b)(b - \xi).$$

Similarly, we obtain the result for an increasing function f. ■

Example 6.4 From calculus $u = \log x$

$$\int_2^4 \frac{(\log x)^4}{x} dx = \int_{\log 2}^{\log 4} u^4 du = \frac{(\log 4)^5 - (\log 2)^5}{5}.$$

By the substitution formula and by integrating the chain rule, we get the integration by parts formula and by integration of the product rule.

6.10 Exercises

1. Prove that the function f, defined on $[a, b]$ by
$$f(x) = \begin{cases} x & \text{when } x \in \mathbb{Q}, \\ -x & \text{when } x \notin \mathbb{Q}, \end{cases}$$
is not integrable over $[a, b]$, whereas $|f| \in \mathcal{R}[a, b]$.
2. If f is defined on $[0, 1]$ by $f(x) = 0$ when x is irrational or zero and by $f(x) = 1/q^{1/3}$ when x is any non-zero rational number p/q with least $p, q \in \mathbb{N}$, prove that $f \in \mathcal{R}[0, 1]$ and
$$\int_0^1 f = 0.$$
3. If $f \in \mathcal{R}[a, b]$ and $\int_a^b f^2(x)dx = 0$, then prove that $f(x) = 0$ at its points of continuity.
4. Distinguish between an integral and a primitive function.
5. Give a different example of a discontinuous function which is primitive but not the integral on a closed interval.
6. If $f(x) = x[1/x]$ when $0 < x \leqslant 1$ and $f(0) = 1$, prove that $f \in \mathcal{R}[0, 1]$ and $\int_0^1 f = \frac{\pi^2}{12}$, where $[1/x]$ denotes an integral part of $1/x$.
7. If f is bounded, defined on $[0, 1]$ and $f(x) = (-1)^{n-1}$ when $1/(n+1) < x < 1/n$; $n \in \mathbb{N}$, then prove that $f \in \mathcal{R}[0, 1]$ and
$$\int_0^1 f = 2\log 2 - 1.$$
8. If f is continuous on $[a, b]$, if
$$f(x) \geqslant 0 \quad (a \leqslant x \leqslant b),$$
and if
$$\int_a^b f(x)dx = 0,$$
prove that f is identically zero on $[a, b]$.
9. Let
$$f(x) = \begin{cases} x & \text{if } x \in \mathbb{Q} \\ \sin x & \text{if } x \notin \mathbb{Q}. \end{cases}$$
Prove that $f'(0) = 1$.

10. True or false? If f is a function on $[a, b]$, if $c \in [a, b]$, and if $f'(c) > 0$, then f is strictly increasing on some open sub-interval of $[a, b]$ containing c.
11. If f is a real-valued function on $[a, b]$ and if f has a right-hand derivative at $c \in [a, b]$, prove that f is continuous from the right at c.

Chapter 7
Sequences and Series of Functions

In the present chapter, we confine our attention to sequences and series of functions. Sequences and series of functions are especially useful in obtaining approximations of a given function and defining new functions from known ones. We will consider sequences whose terms are functions and pay attention to the general properties that are associated with the uniform convergence of sequences and series of functions.

Throughout this chapter we consider a non-empty subset $\mathbf{E} \subset \mathbb{R}$, and then we study real-valued functions $f, \ f_n$ on $\mathbf{E}$.

7.1 Sequences of Functions

Definition 7.1 Let $\mathbf{E}$ be a non-empty subset of $\mathbb{R}$. And suppose that for each $n \in \mathbb{N}$ there is a function $f_n : \mathbf{E} \to \mathbb{R}$, then $\{f_n\}$ is called a sequence of functions from $\mathbf{E}$ to $\mathbb{R}$.

For example,

$$\{x, x^2, x^3, \cdots\}, \qquad x \in [0, 1],$$

$$\{\tan^{-1} x, \tan^{-1} x^2, \tan^{-1} x^3, \cdots\}, \quad x \in \left[0, \frac{\pi}{2}\right].$$

7.2 Series of Functions

A function series is a series, where the summands are not just real or complex numbers but functions.

N. Deo and R. Sakai, *Introduction to Mathematical Analysis*,
https://doi.org/10.1007/978-981-97-6568-3_7

Definition 7.2 Suppose $\{f_n\}$ is a sequence of functions defined on a subset $\mathbf{E}$ of $\mathbb{R}$ into $\mathbb{R}$, i.e., $f_n : \mathbf{E} \to \mathbb{R}$, then a series of functions in $\mathbb{R}$ is a formal sum

$$\sum_{n=1}^{\infty} f_n(x) = f_1(x) + f_2(x) + f_3(x) + \cdots, \quad x \in \mathbf{E}.$$

Each function $f_n(x), n \in \mathbb{N}$ is called the term of the series $\sum_{n=1}^{\infty} f_n(x)$.

For example,

$$\sum_{n=1}^{\infty} \frac{\sin nx}{n^2} = \frac{\sin x}{1^2} + \frac{\sin 2x}{2^2} + \frac{\sin 3x}{3^2} + \cdots, \qquad x \in [0, \pi],$$

$$\sum_{n=1}^{\infty} \tan^{-1} nx = \tan^{-1} x + \tan^{-1} 2x + tan^{-1} 3x + \cdots, \qquad x \in \left[0, \frac{\pi}{2}\right].$$

7.3 Pointwise Convergence

Definition 7.3 A sequence $\{f_n\}$ of functions ($f_n : \mathbf{E} \to \mathbb{R}$ and $\mathbf{E} \subset \mathbb{R}$) converges pointwise to $f : \mathbf{E} \to \mathbb{R}$ on $\mathbf{E}$, if the sequence $\{f_n\}_{n=1}^{\infty}$ converges at each point x of $\mathbf{E}$, i.e., for each $\varepsilon > 0$ and for every point $x \in \mathbf{E}$ there is $m \in \mathbb{N}$ (depends on both $\varepsilon > 0$ and x) such that

$$|f_n(x) - f(x)| < \varepsilon, \quad \text{whenever } n > m.$$

Symbolically,

$$\lim_{n\to\infty} f_n(x) = f(x), \qquad \text{for each } x \in \mathbf{E}.$$

Thus the function f is defined as the limit (pointwise limit) of the sequence $\{f_n\}_{n=1}^{\infty}$ on $\mathbf{E}$.

If the series $\sum_{n=1}^{\infty} f_n(x)$ converges pointwise to a function $f(x)$ on $\mathbf{E}$ for every $x \in \mathbf{E}$, and if we define

$$\sum_{n=1}^{\infty} f_n(x) = f(x), \quad \forall x \in \mathbf{E},$$

the function f is called the sum of the series $\sum_{n=1}^{\infty} f_n$.

For a geometric interpretation of Pointwise Convergence, first we fix a value x_0, then we choose an arbitrary neighborhood around $f(x_0)$, which corresponds

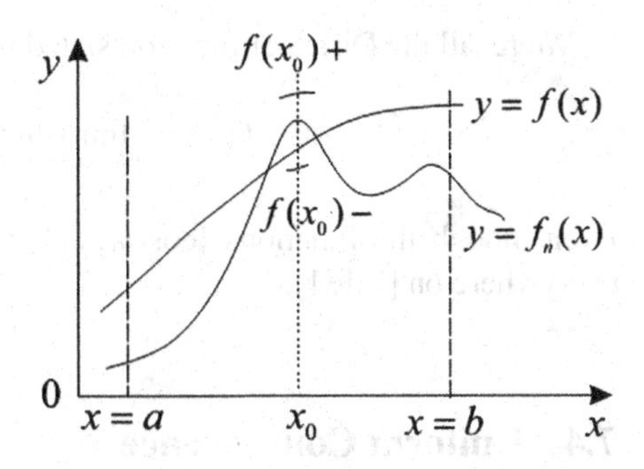

Fig. 7.1 Pointwise convergence

to a vertical interval centered at $f(x_0)$. Then finally we pick $m \in \mathbb{N}$ so that $f_n(x_0)$ intersects the vertical line $x = x_0$ inside the interval $(f(x_0) - \varepsilon, f(x_0) + \varepsilon)$ (Fig. 7.1).

Example 7.1 Let $f_n(x) = x^n$, $x \in [0, 1] \subset \mathbb{R}$, $n \in \mathbb{N}$.

$$\begin{aligned}\lim_{n\to\infty} f_n(x) &= \begin{cases} 0, & if\ 0 \leqslant x < 1, \\ 1, & if\ x = 1, \end{cases} \\ &= f(x).\end{aligned}$$

It follows that $f_n \to f$ converges pointwise on $\mathbb{R}$.

Clearly, this function will fail for uniform convergence (see Definition 7.4, below), because of the difference between $f(1)$ and $\lim_{n\to\infty} f_n(x) = 0$ for each $0 \leqslant x < 1$.

If the limit function $f(x)$ exists at every point of $[0, 1]$, then we say that the sequence of the function $\{f_n(x)\}$ is pointwise convergence in $[0, 1]$. The limit function is not always continuous, although each f_n is continuous.

Example 7.2 If $f_n(x) = x^n$, $\forall x \in [1, 2] \subset R$, $n \in \mathbb{N}$,

$$\lim_{n\to\infty} f_n(x) = \lim_{n\to\infty} x^n = \begin{cases} \infty, & \text{if } 1 < x \leqslant 2, \\ 1, & \text{if } x = 1. \end{cases}$$

Hence the limit function $f(x)$ does not exist at every point of $(1, 2]$.

Example 7.3 If $f_n(x) = 1/(1 + nx)$ $(x \geqslant 0)$, $n \in \mathbb{N}$,

$$\begin{aligned}\lim_{n\to\infty} f_n(x) &= \begin{cases} 0, & \text{if } x > 0, \\ 1, & \text{if } x = 0, \end{cases} \\ &= f(x).\end{aligned}$$

Remark 7.1 Even though for each $n \in \mathbb{N}$, f_n is differentiable or continuous, the limit function f is neither differentiable nor continuous.

We recall the Dirichlet function stated in Example 4.5, that is,

$$f(x) = \lim_{m\to\infty}\left(\lim_{n\to\infty}\{\cos(m!\pi x)\}^{2n}\right).$$

Even though the functions $\{\cos(m!\pi x)\}^{2n}$ are continuous, $f(x)$ is discontinuous everywhere on $[-1, 1]$.

7.4 Uniform Convergence

Uniform convergence is a type of convergence stronger than pointwise convergence. A sequence $\{f_n\}$ of functions converges uniformly to a limiting function f if the speed of convergence of $f_n(x)$ to $f(x)$ does not depend on x.

Definition 7.4 A sequence $\{f_n\}$ of functions defined on a set $\mathbf{E}$ is said to converge uniformly on $\mathbf{E}$ if a given $\varepsilon > 0$, there exists an $m \in \mathbb{N}$ such that for all $x \in \mathbf{E}$ and all $n \geqslant m$ we have

$$|f_n(x) - f(x)| < \varepsilon.$$

Symbolically,

$$\lim_{n\to\infty} f_n = f,$$

i.e., the sequence $\{f_n\}$ of functions converges to f.

"Every uniformly convergent sequence is pointwise convergent but its converse need not be true" (see Example 7.1).

We say that the series $\sum_{n=1}^{\infty} f_n(x)$ converges uniformly on $\mathbf{E}$ if the sequence $\{s_n\}$ of partial sums defined by

$$\sum_{k=1}^{n} f_k(x) = s_n(x)$$

converges uniformly on $\mathbf{E}$.

The geometric meaning of uniform convergence is illustrated in Fig. 7.2. If $f_n \to f$ uniformly on $\mathbf{E} = [a, b]$, for each $\varepsilon > 0$ there is $m \in \mathbb{N}$ (depends on $\varepsilon > 0$ but not on $x \in \mathbf{E}$) such that

$$|f_n(x) - f(x)| < \varepsilon, \quad a \leqslant x \leqslant b, \text{ if } n \geqslant m,$$

$$i.e., \quad f(x) - \varepsilon < f_n(x) < f(x) + \varepsilon, \quad a \leqslant x \leqslant b, \text{ if } n \geqslant m.$$

In other words, for $n \geqslant m$ the graph of

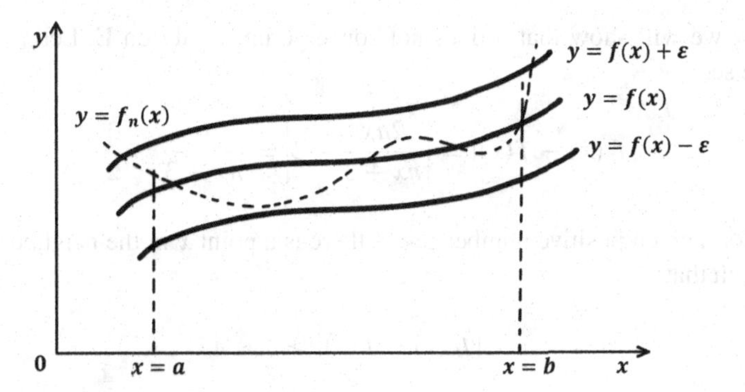

Fig. 7.2 Uniform convergence

$$y = f_n(x), \quad a \leqslant x \leqslant b$$

lies in the strip between the graphs of $y = f(x) - \varepsilon$ and $y = f(x) + \varepsilon$.

Example 7.4 Suppose $f_n(x) = (\sin nx)/n,\ x \in \mathbb{R}$, then $f_n \to 0$ converges pointwise on $\mathbb{R}$. In fact, $f_n \to 0$ converges uniformly on $\mathbb{R}$. Let $\varepsilon > 0$ and $m = 1/\varepsilon$. Then for $n > m$ and *for all* $x \in \mathbb{R}$ we have

$$|f_n(x) - 0| = \left|\frac{\sin nx}{n}\right| \leqslant \frac{1}{n} < \frac{1}{m} = \varepsilon.$$

7.5 Comparing Uniform with Pointwise Convergence

- For pointwise convergence we could first fix a value for x, and then choose m. Consequently, m depends on both ε and x.
- For uniform convergence $f_n(x)$ must be uniformly close to $f(x)$ for all x in the domain. Thus m only depends on ε but not on x.

Example 7.5 The sequence $\{2nx/(nx + 3)\}$ converges pointwise on $\mathbf{E} = (0, 2]$ but it does not converge uniformly on $\mathbf{E}$.

Let $f_n(x) = 2nx/(nx + 3)$. Then

$$\lim_{n\to\infty} f_n(x) = \lim_{n\to\infty} \frac{2x}{x + (3/n)} = 2.$$

Hence, $\{f_n(x)\}$ converges to

$$f(x) = 2$$

at each $x \in \mathbf{E}$, therefore it converges pointwise on $\mathbf{E}$.

Now we will show that it does not converge uniformly on **E**. Let $x \in [0, \frac{1}{n}]$. Then we see

$$|f_n(x) - f(x)| = \left|\frac{2nx}{nx+3} - 2\right| = \frac{6}{nx+3} \geqslant \frac{3}{2}.$$

Hence, for a given positive number $\varepsilon < \frac{3}{2}$ there is a point x in the neighborhood of $x = 0$ such that

$$|f_n(x) - f(x)| > \varepsilon,$$

that is, the sequence $\{f_n\}$ is not uniformly convergent.

Example 7.6 The sequence $\{f_n(x)\} = \{x^n\}$ converges uniformly on $[0, k],\ k < 1$ and only converges pointwise on $[0, 1]$.

Let $f_n(x) = x^n$, $x \in [0, k],\ k < 1,\ n \in \mathbb{N}$, and let us give $\varepsilon > 0$. We set $m = \log(\varepsilon/2)/\log k$, then for $n \geqslant m$ we see

$$|x^n - 0| \leqslant k^n \leqslant k^m = (e^{\log k})^{\log(\varepsilon/2)/\log k} = e^{\log(\varepsilon/2)} = \varepsilon/2 < \varepsilon.$$

Therefore the sequence $\{f_n\}$ converges uniformly on $[0, k]$. For $x \in [0, 1]$ we see

$$f(x) = \lim_{n\to\infty} f_n(x) = \begin{cases} 0, \text{ if } 0 \leqslant x < 1, \\ 1, \quad\quad \text{if } x = 1. \end{cases}$$

Thus, the sequence $\{f_n\}$ converges pointwise. For $x > 1$ we see

$$\lim_{n\to\infty} f_n(x) = \infty.$$

Example 7.7 The sequence

$$\{f_n(x)\} = \left\{\frac{1}{x+n}\right\}$$

is uniformly convergent in the interval $[0, b],\ b > 0$.

The limit function is

$$f(x) = \lim_{n\to\infty} f_n(x) = \lim_{n\to\infty} \frac{1}{x+n} = 0,\ \forall x \in [0, b].$$

So that the sequence $\{f_n(x)\}$ also converges pointwise to 0. Now we show that $\{f_n\}$ is uniformly convergent. For a given $\varepsilon > 0$ we take $m \in \mathbb{N}$ as

$$\frac{1}{m} < \varepsilon < \frac{1}{m-1}.$$

Then for $n \geqslant m,\ x \in [0, b]$, we have

$$|f_n(x) - 0| = \frac{1}{x+n} \leqslant \frac{1}{x+m} \leqslant \frac{1}{m} < \varepsilon.$$

Thus, the sequence $\{f_n\}$ is uniformly convergent in the interval $[0, b],\ b > 0$.

The next theorem enables us to test a sequence for uniform convergence without guessing what the limit function might be. It is analogous to Cauchy's convergence criterion for sequences of constants.

Theorem 7.1 (Cauchy's Criterion for Uniform Convergence) *A sequence of functions $\{f_n\}$ defined on* **E** *converges uniformly on a set* **E** *if and only if for every $\varepsilon > 0,\ x \in \mathbf{E}$ there is an integer n_0 such that*

$$|f_n(x) - f_m(x)| < \varepsilon,\quad m, n \geqslant n_0. \tag{7.1}$$

Proof `A Necessary condition:` Suppose that $\{f_n\}$ converges uniformly to the limit function f on **E**. Then, for a given $\varepsilon > 0$, there is an integer n_0 such that

$$|f_n(x) - f(x)| < \frac{\varepsilon}{2},\ n \geqslant n_0,\ x \in \mathbf{E},$$

so that for $m, n \geqslant n_0$ and $x \in \mathbf{E}$

$$|f_n(x) - f_m(x)| < |f_n(x) - f(x)| + |f(x) - f_m(x)| < \frac{\varepsilon}{2} + \frac{\varepsilon}{2} = \varepsilon.$$

`A Sufficient condition:` Since the Cauchy condition (7.1) holds, for each fixed $x \in \mathbf{E}$, the sequence of real numbers $\{f_n(x)\}$ is a Cauchy sequence. By Theorem 2.15 the sequence $\{f_n(x)\}$ converges to a limit $f(x)$ on **E**, i.e.,

$$f(x) = \lim_{n\to\infty} f_n(x), \quad \text{for all } x \in \mathbf{E}.$$

Keeping m fixed in (7.1) and letting $n \to \infty$, we have

$$|f_m(x) - f(x)| < \varepsilon, \quad m \geqslant n_0 \text{ and for all } x \in \mathbf{E}.$$

This shows that $f_m \to f$ converges uniformly on **E**. ■

In 1880, one of the most convenient and useful tests for uniform convergence of an infinite series of functions was published by Weierstrass. It is analogous to the comparison test for determining the convergence of series of real and complex numbers. It is known as the Weierstrass M-test.

Theorem 7.2 (The Weierstrass M-test) *Let $\sum\limits_{k=1}^{\infty} f_k(x)$ be a series of real-valued functions on* **E** *which converges uniformly on* **E**, *and let $\{M_k\}$ be a sequence of* *non-negative real numbers with* $\sum_{k=1}^{\infty} M_k < \infty$ *such that*

$$|f_k(x)| \leqslant |M_k|, \quad \textit{for all } x \in \mathbf{E} \textit{ and for } k \in \mathbb{N}.$$

Then $\sum_{k=1}^{\infty} f_k(x)$ *converges uniformly on* $\mathbf{E}$, *if* $\sum_{k=1}^{\infty} M_k$ *converges.*

Proof Let $s_n(x) = \sum_{k=1}^{n} f_k(x)$ and $t_n = \sum_{k=1}^{n} M_k$ be partial sums. Then, for all $x \in \mathbf{E}$ and $m > n \geqslant n_0$,

$$\begin{aligned} |s_m(x) - s_n(x)| = \left| \sum_{k=n+1}^{m} f_k(x) \right| &\leqslant \sum_{k=n+1}^{m} |f_k(x)| \\ &\leqslant \sum_{k=n+1}^{m} M_k = t_m - t_n. \end{aligned} \tag{7.2}$$

Since $\sum_{k=1}^{\infty} M_k < \infty$, $\{t_n\}$ is a convergent sequence, and so it is a Cauchy sequence (by Theorem 2.14). Hence we can choose $\varepsilon > 0$, and there exists $n_1 > n_0$, such that

$$|t_m - t_n| < \varepsilon, \quad m, n \geqslant n_1.$$

From (7.2), we get

$$|s_m(x) - s_n(x)| < \varepsilon, \quad m, n \geqslant n_1 \text{ and for all } x \in \mathbf{E}.$$

Therefore, the sequence $\{s_n\}$ converges uniformly on $\mathbf{E}$. This means the series $\sum_{k=1}^{\infty} f_k(x)$ converges uniformly on $\mathbf{E}$. ■

Example 7.8 The series

$$\sin x + \frac{\sin 2x}{2^2} + \frac{\sin 3x}{3^2} + \cdots$$

converges uniformly on $\mathbb{R}$.

Since

$$\left| \frac{\sin nx}{n^2} \right| \leqslant \frac{1}{n^2}, \quad \forall x \in \mathbb{R},$$

we know that the series $\sum (1/n^2)$ is convergent.

Thus, by the Weierstrass M-test, the given series converges uniformly on $\mathbb{R}$.

Example 7.9 The series

$$\sum_{n=1}^{\infty} r^n \cos nx \text{ and } \sum_{n=1}^{\infty} r^n \sin nx, \quad 0 < r < 1,$$

converge uniformly on $\mathbb{R}$.

Given

$$f_n(x) = r^n \cos nx, \quad \forall x \in \mathbb{R},\ 0 < r < 1,$$

then we see

$$|f_n(x)| \leqslant |r^n|\,|\cos nx| \leqslant r^n.$$

Hence,

$$|f_n(x)| \leqslant M_n = r^n.$$

Therefore,

$$\sum_{n=1}^{\infty} M_n = \sum_{n=1}^{\infty} r^n.$$

The series $\sum_{n=1}^{\infty} r^n$ is convergent because of $0 < r < 1$, thus by the Weierstrass M-test, the given series $\sum_{n=1}^{\infty} r^n \cos nx$ converges uniformly on $\mathbb{R}$.

Similarly, we see that the series $\sum_{n=1}^{\infty} r^n \sin nx$ converges uniformly on $\mathbb{R}$.

Example 7.10 The series

$$f(x) = \sum_{n=1}^{\infty} \frac{\sin nx}{n^p}, \quad p > 1$$

converges uniformly on $\mathbb{R}$.

If we set

$$f_n(x) = \frac{\sin nx}{n^p}, \quad \forall x \in \mathbb{R},$$

then we have

$$|f_n(x)| = \left|\frac{\sin nx}{n^p}\right| \leqslant \frac{1}{n^p}.$$

Since $\sum_{n=1}^{\infty} \frac{1}{n^p} < \infty$, by the Weierstrass M-test, the given series converges uniformly on $\mathbb{R}$.

Example 7.11 The series

$$\sum_{n=1}^{\infty} f_n(x) = \sum_{n=1}^{\infty} (\pm 1)^n \frac{x^{2n}}{n^p \left(1 + x^{2n}\right)}, \quad p > 1$$

converges uniformly on $\mathbb{R}$.

If we set

$$f_n(x) = (\pm 1)^n \frac{x^{2n}}{n^p (1 + x^{2n})}, \quad x \in \mathbb{R},$$

then we see

$$|f_n(x)| = \frac{x^{2n}}{n^p (1 + x^{2n})} \leqslant \frac{1}{n^p}, \quad x \in \mathbb{R}.$$

Since

$$\sum_{n=1}^{\infty} \frac{1}{n^p} < \infty,$$

by the Weierstrass M-test, the given series converges uniformly on $\mathbb{R}$.

Example 7.12 The series

$$\sum_{n=1}^{\infty} \frac{a_n x^n}{1 + x^{2n}}$$

converges uniformly on $[a, b]$, if $\sum_{n=1}^{\infty} a_n$ is absolutely convergent.

If we set

$$f_n(x) = \frac{a_n x^n}{1 + x^{2n}}, \quad x \in [a, b],$$

then we see

$$|f_n(x)| = \left| \frac{a_n x^n}{1 + x^{2n}} \right| = |a_n| \left| \frac{x^n}{1 + x^{2n}} \right| \leqslant C_n |a_n| \leqslant |a_n|,$$

where

$$C_n = \max \left\{ \frac{|a|^n}{1 + |a|^{2n}}, \frac{|b|^n}{1 + |b|^{2n}} \right\} \leqslant 1.$$

Since $\sum_{n=1}^{\infty} |a_n| < \infty$, by the Weierstrass M-test, the given series converges uniformly on $[a, b]$.

Example 7.13 The series $\sum_{k=1}^{\infty} (\sin kx)/k$ converges uniformly on the interval $[\pi/2, 3\pi/2]$. However, it does not converge uniformly on $[-\pi/2, \pi/2]$.

We remark that the series is the Fourier expansion of the function $(\pi - x)/2$ in $(0, 2\pi)$, that is,

$$\sum_{k=1}^{\infty} \frac{\sin kx}{k} = \frac{\pi - x}{2}.$$

If $x = 0$ or 2π, then we see that the series equals 0, trivially, so the series is not continuous on $[0, 2\pi]$. However, if $0 < a < b < \pi$, the series converges uniformly to $(\pi - x)/2$ on $[a, b]$.

7.6 Uniform Convergence and Continuity

As we know that the pointwise limit of a sequence of continuous functions need not be continuous, now we show that the uniform limit of a sequence of continuous functions must be continuous.

Theorem 7.3 *Let $\{f_n\}$ be a sequence of functions, which converges uniformly to f on $\mathbf{E}$. If each of the functions $f_n (n = 1, 2, 3, \cdots)$ is continuous at $x_0 \in \mathbf{E}$, then f is continuous at x_o.*

Proof Since the sequence $\{f_n\}$ of functions converges uniformly to f on $\mathbf{E}$, for $\varepsilon > 0$ there is a real number $m(\varepsilon) \in \mathbb{N}$ independent of x such that

$$|f_n(x) - f(x)| < \frac{\varepsilon}{3}, \quad \forall n \geqslant m \text{ and } x \in \mathbf{E}.$$

Let $x_0 \in \mathbf{E}$. We have

$$|f_n(x_0) - f(x_0)| < \frac{\varepsilon}{3}, \quad \forall n \geqslant m.$$

Again since $\{f_n\}$ is continuous at $x_0 \in \mathbf{E}$, by the definition of continuity, for every $\varepsilon > 0$ there exists $\delta(\varepsilon) > 0$ such that

$$|f_n(x) - f_n(x_0)| < \frac{\varepsilon}{3}, \quad \text{whenever } |x - x_0| < \delta.$$

Now if $|x - x_0| < \delta$, we have

$$\begin{aligned}|f(x) - f(x_0)| &= |f(x) - f_n(x) + f_n(x) - f_n(x_0) + f_n(x_0) - f(x_0)| \\ &\leqslant |f(x) - f_n(x)| + |f_n(x) - f_n(x_0)| + |f_n(x_0) - f(x_0)| \\ &< \frac{\varepsilon}{3} + \frac{\varepsilon}{3} + \frac{\varepsilon}{3} = \varepsilon.\end{aligned}$$

Hence

$$|f(x) - f(x_0)| < \varepsilon, \quad \text{when} \quad |x - x_0| < \delta.$$

Consequently, f is continuous at x_0. ■

In the proof of Theorem 7.3, we can take $\varepsilon > 0$ which is arbitrary, then this establishes the continuity of f at an arbitrary point $x_0 \in \mathbf{E}$ (see Fig. 7.3).

Theorems for sequences of functions translate easily into theorems for series of functions. Here is an example.

Theorem 7.4 *Let a series $\sum\limits_{k=1}^{\infty} g_k$ of functions uniformly converges to f on* **E**, *and let us suppose that each g_k $(k = 1, 2, 3, \ldots)$ is continuous on* **E**. *Then f is continuous on* **E**.

Proof Since each g_n $(n = 1, 2, 3, \cdots)$ is continuous on **E**, each partial sum

$$f_n = g_1 + g_2 + g_3 + \cdots + g_n$$

is continuous, and then the sequence $\{f_n\}$ converges uniformly to f on **E**. Choose $\varepsilon > 0$, then there is a real number $m(\varepsilon) \in \mathbb{N}$ independent of x such that

$$|f_n(x) - f(x)| < \frac{\varepsilon}{3}, \quad \forall n \geqslant m \text{ and } x \in \mathbf{E}.$$

For a particular value m, we have

$$|f_m(x) - f(x)| < \frac{\varepsilon}{3}, \quad x \in \mathbf{E}.$$

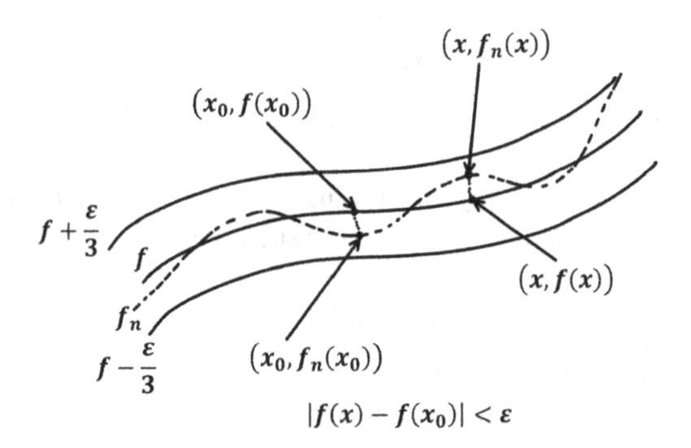

Fig. 7.3 Uniformly convergence of f_n

Since $m \in \mathbb{N}$ is fixed, f_m is continuous on $\mathbf{E}$. By the definition of continuity, for every $\varepsilon > 0$ and $x_0 \in \mathbf{E}$ there exists $\delta(\varepsilon) > 0$ such that

$$|f_m(x) - f_m(x_0)| < \frac{\varepsilon}{3}, \quad \text{whenever } |x - x_0| < \delta.$$

Now if $|x - x_0| < \delta$, we have

$$\begin{aligned} |f(x) - f(x_0)| &= |f(x) - f_m(x) + f_m(x) - f_m(x_0) + f_m(x_0) - f(x_0)| \\ &\leqslant |f(x) - f_m(x)| + |f_m(x) - f_m(x_0)| + |f_m(x_0) - f(x_0)| \\ &< \frac{\varepsilon}{3} + \frac{\varepsilon}{3} + \frac{\varepsilon}{3} = \varepsilon. \end{aligned}$$

Hence, the limit function is continuous. ■

We gave an example earlier to convince the reader that the limit of a sequence of continuous functions need not be Riemann-integrable yielded a limit function that was unbounded. The added hypothesis of uniform convergence will eliminate such occurrences.

Theorem 7.5 *If a sequence $\{f_n\}$ of functions converges uniformly to f on* $\mathbf{E}$, *and f_n is bounded on* $\mathbf{E}$ *for each positive integer $n \in \mathbb{N}$, then f is bounded on* $\mathbf{E}$.

Proof For $\varepsilon = 1$ there is a real number N such that

$$|f_n(x) - f(x)| < 1, \quad n \geqslant N \text{ and } x \in \mathbf{E}.$$

Since $\{f_n\}$ is bounded, there is a real number M such that

$$|f_N(x)| \leqslant M, \quad \text{for all } x \in \mathbf{E}.$$

Hence, for all $x \in \mathbf{E}$,

$$|f(x)| \leqslant |f(x) - f_N(x)| + |f_N(x)| \leqslant 1 + M.$$

Thus, f is bounded on $\mathbf{E}$. ■

7.7 Uniform Convergence and Integration

One should now hope that, if the sequence $\{f_n\}$ of functions converges uniformly to f on $[a, b]$ and for each positive integer n, f_n is Riemann-integrable on $[a, b]$, then f is Riemann-integrable on $[a, b]$. Our next theorem shows that a uniform limit of integrable functions must be integrable.

Theorem 7.6 (Term-by-term integration) *If a sequence* $\{f_n\}$ *of functions is Riemann-integrable on an interval* $[a, b]$ *and if* $f_n \to f$ *converges uniformly on* $[a, b]$*, then* f *is Riemann-integrable on* $[a, b]$ *and*

$$\lim_{n\to\infty} \int_a^b f_n(x)dx = \int_a^b f(x)dx.$$

Proof Since each f_n is bounded on $[a, b]$, f is also bounded on $[a, b]$. Choose $\varepsilon > 0$, then there is a real number $m(\varepsilon) \in M$ such that

$$|f_n(x) - f(x)| < \frac{\varepsilon}{b-a}, \quad n \geqslant m \text{ for all } x \in [a, b].$$

Using Theorem 6.9, we obtain

$$\begin{aligned}\left|\int_a^b f_n(x)dx - \int_a^b f(x)dx\right| &= \left|\int_a^b [f_n(x) - f(x)]\, dx\right| \\ &\leqslant \int_a^b |f_n(x) - f(x)|dx \leqslant \int_a^b \frac{\varepsilon}{b-a}dx = \varepsilon, \ n \geqslant m.\end{aligned}$$

This shows that the sequence $\left\{\int_a^b f_n(x)dx\right\}$ converges to $\int_a^b f(x)dx$, i.e.,

$$\lim_{n\to\infty} \int_a^b f_n(x)dx = \int_a^b f(x)dx.$$

■

Corollary 7.1 *Suppose* $\{f_n\}$ *is a sequence of continuous functions on an interval* $[a, b]$ *and* $f_n \to f$ *uniformly on* $[a, b]$*, then*

$$\lim_{n\to\infty} \int_a^b f_n(x)dx = \int_a^b f(x)dx.$$

Corollary 7.2 *Suppose* $f_n \in \mathcal{R}[a, b]$*, and if the series*

$$f(x) = \sum_{n=1}^{\infty} f_n(x)dx, \quad x \in [a, b]$$

converges uniformly on $[a, b]$*, then*

$$\int_a^b f(x)dx = \sum_{n=1}^{\infty} \int_a^b f_n(x)dx.$$

In other words, the series may be integrated term by term.

7.8 Uniform Convergence and Differentiation

Theorem 7.7 (Term-by-term differentiation) *Suppose $\{f_n\}$ is a sequence of functions differentiable on $[a, b]$. Further suppose that the sequence $\{f_n(x_0)\}$ converges for some $x_0 \in [a, b]$ and $\{f_n'\}$ converges uniformly to g on $[a, b]$, then the sequence $\{f_n\}$ converges uniformly to f on $[a, b]$ and f is differentiable. Then we have*

$$f'(x) = \lim_{n\to\infty} f_n'(x), \quad a \leqslant x \leqslant b.$$

That is,

$$g(x) = f'(x), \quad a \leqslant x \leqslant b.$$

Proof By hypothesis the sequence $\{f_n(x_0)\}$ is convergent for some $x_0 \in [a, b]$ and $\{f_n'\}$ is uniformly convergent. Choose $\varepsilon > 0$, then there is a real number N independent of x such that

$$|f_n(x_0) - f_m(x_0)| < \frac{\varepsilon}{2}, \quad m, n \geqslant N, \tag{7.3}$$

and

$$\left|f_n'(t) - f_m'(t)\right| < \frac{\varepsilon}{2(b-a)}, \quad m, n \geqslant N, \quad t \in [a, b]. \tag{7.4}$$

Applying the mean-value theorem (differential) to the function $(f_n - f_m)$, we get

$$[f_n(x) - f_m(x)] - [f_n(t) - f_m(t)] = (x - t)\left[f_n'(\xi) - f_m'(\xi)\right], \; t < \xi < x. \tag{7.5}$$

Therefore, for all positive integers $m, n \geqslant N$, (7.3), (7.4) and (7.5) imply that

$$\begin{aligned}
&|f_n(x) - f_m(x)| \\
&= |f_n(x) - f_m(x) - f_n(x_0) + f_m(x_0) + f_n(x_0) - f_m(x_0)| \\
&\leqslant |f_n(x) - f_m(x) - f_n(x_0) + f_m(x_0)| + |f_n(x_0) - f_m(x_0)| \\
&< \frac{\varepsilon\,|x - x_0|}{2(b-a)} + \frac{\varepsilon}{2} < \frac{\varepsilon}{2} + \frac{\varepsilon}{2} = \varepsilon \quad (\text{since } |x - x_0| \leqslant b - a),
\end{aligned} \tag{7.6}$$

for all $x \in [a, b]$. Consequently, $\{f_n\}$ converges uniformly to a function f on $[a, b]$, i.e.,

$$f(x) = \lim_{n\to\infty} f_n(x), \quad x \in [a, b].$$

Now let us fix a point $x \in [a, b]$ and $n \in \mathbb{N}$. And we define

$$F_n(y) = \frac{f_n(y) - f_n(x)}{y - x}, \quad y \in [a, b] \setminus \{x\} \tag{7.7}$$

and

$$F(y) = \frac{f(y) - f(x)}{y - x}, \quad y \in [a, b] \setminus \{x\}. \tag{7.8}$$

Then

$$\lim_{y \to x} F_n(y) = \lim_{y \to x} \frac{f_n(y) - f_n(x)}{y - x} = f'_n(x).$$

Now for all $m, n \geqslant N$ and $x \in [a, b]$ and from the inequality in (7.6)

$$|F_n(y) - F_m(y)| = \frac{|f_n(y) - f_n(x) - f_m(y) + f_m(x)|}{|y - x|} < \frac{\varepsilon}{2(b - a)}$$

so that $\{F_n\}$ converges uniformly, for all $y \in [a, b] \setminus \{x\}$. Since $\{f_n\}$ converges uniformly to f on $[a, b]$, we conclude from (7.7) and (7.8)

$$\lim_{n \to \infty} F_n(y) = \lim_{n \to \infty} \frac{f_n(y) - f_n(x)}{|y - x|} = \frac{f(y) - f(x)}{|y - x|} = F(y).$$

Consequently, $\{F_n\}$ converges uniformly to F on $[a, b]$, for all $y \in [a, b] \setminus \{x\}$.

Now we proceed to show that

$$g(x) = f'(x), \quad a \leqslant x \leqslant b.$$

Since $\{f'_n\}$ converges uniformly to g on $[a, b]$, by Theorem 7.6 (Term-by-term integration), for any $y \in [a, b]$ we have

$$\lim_{n \to \infty} \int_a^y f'_n(x)dx = \int_a^y g(x)dx.$$

Thus by Theorem 6.11, we get

$$\lim_{n \to \infty} [f_n(y) - f_n(a)] = \int_a^y g(x)dx.$$

By hypothesis, $\{f_n\}$ converges uniformly to f on $[a, b]$, therefore

$$\lim_{n \to \infty} f_n(y) = f(y) \text{ and } \lim_{n \to \infty} f_n(a) = f(a).$$

Hence

$$f(y) - f(a) = \int_a^y g(x)dx, \quad a \leqslant y \leqslant b.$$

By Theorem 6.10, we obtain

$$f'(y) = g(y), \quad a \leqslant y \leqslant b,$$

and the theorem is proved. ■

7.9 The Weierstrass Approximation Theorem

All the functions which are infinitely differentiable admit the Taylor series expansion, but the convergence of the series may be extremely slow for computation. But we can still approximate a continuous function by a polynomial function or polynomial for short in other ways. Hence we seek an effective way to generate a polynomial approximation using a finite set of points on an interval, say $[a, b]$. The polynomial so obtained should converge to the given function as the number of points in $[a, b]$ increased. In other words, the polynomial obtained is the best approximation to the function on $[a, b]$.

The polynomial approximation is an extremely useful and powerful technique in digital computation. The starting point of finding the best polynomial approximation is the Weierstrass Approximation Theorem. In 1885, the Weierstrass Approximation Theorem was originally stated by a German mathematician *Karl Weierstrass*.

Theorem 7.8 *If $f : [a, b] \to \mathbb{R}$ is a continuous function, then for each $\varepsilon > 0$, there exists a polynomial $P_n : [a, b] \to \mathbb{R}$ such that*

$$|f(x) - P_n(x)| < \varepsilon, \quad a \leqslant x \leqslant b. \tag{7.9}$$

In other words, any continuous function defined on a closed interval is the uniform limit of a sequence of polynomials.

Over the next twenty-five or so years numerous alternative proofs were given to the Weierstrass Approximation Theorem, but unfortunately, all of them are rather difficult results. One of the most elementary proofs is based on the following theorem which is given by Serge Bernstein.

Let $f : [0, 1] \to \mathbb{R}$ be a continuous function. Then Bernstein constructed the following sequence of polynomials $B_n(f)$ (called the Bernstein polynomials):

$$(B_n f)(x) = \sum_{k=0}^{n} \binom{n}{k} x^k (1-x)^{n-k} f\left(\frac{k}{n}\right), \quad x \in [0, 1]. \tag{7.10}$$

This is a polynomial of degree at most n and its coefficients depend on the values of the function f at the $n + 1$ equally spaced points $0, 1/n, 2/n, \cdots, k/n, \cdots, 1$, and on the binomial coefficients

$$\binom{n}{k} = \frac{n!}{k!(n-k)!} = \frac{n(n-1)\cdots(n-k+1)}{1, 2, \cdots, k}.$$

The formula (7.10) is also called a linear Bernstein operator B_n, for each n. The operator $B_n(f)$ is linear, that is,

$$B_n\,(af + bg) = aB_n f + bB_n g. \tag{7.11}$$

Since the inequality

$$f \geqslant g \Longrightarrow B_n f \geqslant B_n g \tag{7.12}$$

is satisfied by the operator B_n as follows:

$$f \geqslant g \Longrightarrow f(x) \geqslant g(x), \quad \text{for all } x \in [0, 1],$$

the operator B_n is a monotone operator.

Lemma 7.1 *For $x \in \mathbb{R}$ and $n \geq 0$, we have*

$$\sum_{k=0}^{n} (nx - k)^2 \binom{n}{k} x^k (1 - x)^{n-k} = nx(1 - x) \leqslant \frac{n}{4}.$$

Proof We know that by the binomial theorem

$$(p + q)^2 = \sum_{k=0}^{n} \binom{n}{k} p^k q^{n-k} = 1 \qquad (\because p + q = 1)\,.$$

For every $x \in \mathbb{R}$ and $n \geq 0$, we have

$$\sum_{k=0}^{n} \binom{n}{k} x^k (1 - x)^{n-k} = 1. \tag{7.13}$$

From Bernstein polynomials (7.10), we have

$$(B_n 1)\,(x) = \sum_{k=0}^{n} \binom{n}{k} x^k (1 - x)^{n-k} = [x + (1 - x)]^n = 1.$$

When the function $f(x) = x$, we have

$$\begin{aligned}(B_n f)\,(x) &= \sum_{k=0}^{n} \frac{k}{n} \binom{n}{k} x^k (1 - x)^{n-k} \\ &= \sum_{k=1}^{n} \frac{(n-1)!}{(k-1)!\,(n-k)!} x^k (1 - x)^{n-k}\end{aligned}$$

$$= x \sum_{k=0}^{n-1} \frac{(n-1)!}{k!\,(n-k-1)!} x^k (1-x)^{n-k-1}$$
$$= x\,(B_{n-1}1)\,(x) = x.$$

We can write in the following way

$$\sum_{k=0}^{n} k \binom{n}{k} x^k (1-x)^{n-k} = nx. \tag{7.14}$$

When the function $f(x) = x^2$, we have

$$\begin{aligned}
(B_n f)\,(x) &= \sum_{k=0}^{n} \binom{n}{k} x^k (1-x)^{n-k} \left(\frac{k}{n}\right)^2 \\
&= \sum_{k=0}^{n} \left(\frac{k\,(k-1)+k}{n^2}\right) \binom{n}{k} x^k (1-x)^{n-k} \\
&= \frac{1}{n^2} \left[\sum_{k=2}^{n} \frac{n!}{(k-2)!\,(n-k)!} x^k (1-x)^{n-k} \right. \\
&\quad \left. + \sum_{k=1}^{n} \frac{n!}{(k-1)!\,(n-k)!} x^k (1-x)^{n-k} \right] \\
&= \frac{1}{n^2} \left[\sum_{k=0}^{n-2} \frac{n!}{k!\,(n-k-2)!} x^{k+2} (1-x)^{n-k-2} \right. \\
&\quad \left. + \sum_{k=0}^{n-1} \frac{n!}{(k)!\,(n-k-1)!} x^{k+1} (1-x)^{n-k-1} \right]
\end{aligned}$$

$$\begin{aligned}
&= \frac{1}{n^2} \left[n(n-1)x^2 \sum_{k=0}^{n-2} \frac{(n-2)!}{k!\,(n-k-2)!} x^k (1-x)^{n-k-2} \right. \\
&\quad \left. + nx \sum_{k=0}^{n-1} \frac{(n-1)!}{(k)!\,(n-k-1)!} x^k (1-x)^{n-k-1} \right] \\
&= \frac{1}{n^2} \left[n(n-1)x^2\,(B_{n-2}1)\,(x) + nx\,(B_{n-1}1)\,(x) \right] \\
&= \frac{1}{n^2} \left[n(n-1)x^2 + nx \right] = \frac{n-1}{n} x^2 + \frac{1}{n} x.
\end{aligned}$$

Therefore, we see

$$\sum_{k=0}^{n} k^2 \binom{n}{k} x^k (1-x)^{n-k} = n(n-1)x^2 + nx. \tag{7.15}$$

From (7.14) and (7.15), we have

$$\begin{aligned}
&\sum_{k=0}^{n} (nx-k)^2 \binom{n}{k} x^k (1-x)^{n-k} \\
&= \sum_{k=0}^{n} \left(n^2x^2 - 2nkx + k^2\right) \binom{n}{k} x^k (1-x)^{n-k} \\
&= n^2x^2 - 2nx(nx) + \left\{n^2x^2 + nx(1-x)\right\} = nx(1-x).
\end{aligned}$$

Since

$$0 \leqslant (2x-1)^2 = 4x^2 - 4x + 1 = -4x(1-x) + 1,$$

we have

$$\sum_{k=0}^{n} (nx-k)^2 \binom{n}{k} x^k (1-x)^{n-k} = nx(1-x) \leqslant \frac{n}{4}. \tag{7.16}$$

■

Here is Bernstein's version of the Weierstrass Approximation Theorem.

Theorem 7.9 (Bernstein's Approximation Theorem) *Let $f : [0, 1] \to \mathbb{R}$ be a continuous function, and let $B_n(x)$ ($\in \mathcal{P}$) be the Bernstein function. Then for each $\varepsilon > 0$, there exists $n_0 = n_0(\varepsilon) \in \mathbb{N}$ such that*

$$|f(x) - B_n(x)| < \varepsilon, \quad 0 \leqslant x \leqslant 1 \text{ and } n \geqslant n_0,$$

i.e., for every continuous function f on $[0, 1]$, we have

$$B_n f \to f \text{ uniformly on } [0, 1].$$

Proof Let $\varepsilon > 0$ be given. Now, on the compact interval $[0, 1]$, f is uniformly continuous, so $\delta > 0$ such that, for all $x, y \in [0, 1]$,

$$|x - y| < \delta \implies |f(x) - f(y)| \leqslant \frac{\varepsilon}{2}. \tag{7.17}$$

Let

$$M = \sup \{|f(x)| : x \in [0, 1]\}.$$

Let $N = M/(\varepsilon\delta^2)$. From (7.13), we have

$$f(x) = \sum_{k=0}^{n} f(x)\binom{n}{k} x^k(1-x)^{n-k}. \tag{7.18}$$

Therefore

$$|B_n f(x) - f(x)| \leqslant \sum_{k=0}^{n} \left| f\left(\frac{k}{n}\right) - f(x) \right| \binom{n}{k} x^k(1-x)^{n-k}. \tag{7.19}$$

To estimate the sum on the right-hand side, let us divide the set $\{0, 1, 2, \cdots, n\}$ into two sets;

$$k \in A \text{ if } \left|\frac{k}{n} - x\right| < \delta \text{ and } k \in B \text{ if } \left|\frac{k}{n} - x\right| \geqslant \delta.$$

With the help of (7.17), for $k \in A$, we have $\left|f\left(\frac{k}{n}\right) - f(x)\right| < \frac{\varepsilon}{2}$. Therefore from (7.13), we have

$$\begin{aligned} \sum_{k\in A} \left| f\left(\frac{k}{n}\right) - f(x) \right| \binom{n}{k} x^k(1-x)^{n-k} &\leqslant \sum_{k\in A} \frac{\varepsilon}{2} \binom{n}{k} x^k(1-x)^{n-k} \\ &\leqslant \frac{\varepsilon}{2}. \end{aligned} \tag{7.20}$$

Again, using (7.17), for $k \in B$, we have $\left|\frac{k}{n} - x\right| \geqslant \delta \Longrightarrow (k - nx)^2 \geqslant n^2\delta^2$. With the help of Lemma 7.1, we have

$$\begin{aligned} &\sum_{k\in B} \left| f\left(\frac{k}{n}\right) - f(x) \right| \binom{n}{k} x^k(1-x)^{n-k} \\ &\leqslant 2M \sum_{k\in B} \binom{n}{k} x^k(1-x)^{n-k}, \text{ where } |f(x)| \leqslant M \text{ for any } x \in [0,1] \qquad (7.21) \\ &\leqslant \frac{2M}{n^2\delta^2} \sum_{k\in B} (k-nx)^2 \binom{n}{k} x^k(1-x)^{n-k} \\ &\leqslant \frac{2M}{n^2\delta^2} \cdot \frac{n}{4} = \frac{M}{2n\delta^2} < \frac{M}{2N\delta^2} = \frac{\varepsilon}{2}. \quad \text{(by (7.16)).} \end{aligned} \tag{7.22}$$

Finally, from (7.19), (7.20) and (7.21), we get

$$|B_n f(x) - f(x)| \leqslant \varepsilon.$$

■

7.10 Exercises

1. Show that the sequence $\{\frac{x}{n}\}$ does not converge uniformly on $\mathbb{R}$, but converges uniformly on every bounded subset of $\mathbb{R}$.
2. Show that 0 is a point of non-uniform convergence of the sequence $\{f_n(x)\}$ where

$$f_n(x) = \tan^{-1} nx, \text{ for } x \geqslant 0.$$

3. Show that 0 is a point of non-uniform convergence of the series

$$1 + (x - 1) + (x^2 - x) + (x^3 - x^2) + \cdots.$$

4. Test the sequence $\{x^{\frac{1}{2n-1}}\}$ for uniform convergence on $[-1, 1]$.
5. Show that the series $\sum_{n=1}^{\infty} \frac{x}{n(n+1)}$ is uniformly convergent on $[0, k]$, where k is a positive number, but it is non-uniformly convergent on $[0, \infty)$.
6. Show that the series $\frac{x}{x+1} + \frac{x}{(x+1)(2x+1)} + \frac{x}{(2x+1)(3x+1)} + \cdots$ is uniformly convergent on $[k, \infty[$, where k is a positive number. Show that the series is non-uniformly convergent near the point 0.
7. Show that the following series converges uniformly on the intervals indicated:

   (a) $\sum_{n=1}^{\infty} \frac{1}{n^2+x^2} \quad (0 \leqslant x < \infty)$.

   (b) $\sum_{n=1}^{\infty} e^{-nx} x^n \quad (0 \leqslant x \leqslant 100)$.

8. Determine the intervals on which the series $\sum_{n=1}^{\infty} \frac{1}{1+n^2x^2}$ converges uniformly.
9. Prove that the series $\sum_{n=1}^{\infty} (-1)^n \frac{x^2+n}{n^2}$ converges uniformly in any bounded intervals.
10. Show that the series $a^x = 1 + \frac{\log a}{1!} x + \frac{(\log a)^2}{2!} x^2 + \cdots + \frac{(\log a)^{n-1}}{(n-1)!} x^{n-1} + \cdots$ can be integrated and differentiated term by term.
11. Examine, for term-by-term integration, the series $\sum x^{n-1}(1 - 2x^n)$ in the interval $0 \leqslant x \leqslant 1$.
12. Test the series $\sum \left(\frac{n^2x}{1+n^4x^2} - \frac{(n-1)^2x}{1+(n-1)^4x^2}\right)$ which converges on the interval $[0, 1]$ uniformly. Can the series be integrated term by term?

Chapter 8
The Lebesgue Integral

In this chapter, we will define and study the Lebesgue integral. The concepts of an upper function and its Lebesgue integral are introduced first. This leads to a discussion of outer measure, measure, and measurable functions. The definition and derivation of the important properties of the Lebesgue integral on the real line, including the Monotone and Dominated Convergence Theorems, are then carried out. The Lebesgue integral on n−dimensional Euclidean space is discussed, and versions of the Fubini and Tonelli Theorems on the equality of multiple and iterated integrals are established.

8.1 Outer Measure and Measurable Sets

8.1.1 Algebra and σ-Algebra

We denote the class of all subsets of $\mathbb{R}$ by $\mathcal{S}(\mathbb{R})$.

Definition 8.1 A collection $\mathcal{A}$ of subsets of $\mathbb{R}$ is called an algebra of sets if

(i) $A - B = A \cap B^c$ is in $\mathcal{A}$ whenever $A,\ B \in \mathcal{A}$,
(ii) $A \cup B$ is in $\mathcal{A}$ whenever $A,\ B \in \mathcal{A}$, and
(iii) $\mathbb{R} \in \mathcal{A}$.
If an algebra $\mathcal{A}$ satisfies
(iv) $\bigcup_{j=1}^{\infty} A_j \in \mathcal{A}$ for $A_i \in \mathcal{A}$ $(i = 1, 2, \cdots)$,

then $\mathcal{A}$ is called a σ-algebra.

N. Deo and R. Sakai, *Introduction to Mathematical Analysis*,
https://doi.org/10.1007/978-981-97-6568-3_8

Let $\mathcal{A}$ be an algebra, and let $A,\ B,\ A_i \in \mathcal{A},\ i = 1, 2, \cdots, n$. Then we have

(a) $A^c = \mathbb{R} - A \in \mathcal{A}$,
(b) $A \cap B = (A^c \cup B^c)^c \in \mathcal{A}$,
(c) $A \cup A_2 \cup \cdots \cup A_n,\ \ A_1 \cap A_2 \cap \cdots \cap A_n \in \mathcal{A}$.

Let us define

$$\mathcal{R}(\mathbb{R}) = \left\{ E = \bigcup_{i=1}^{n} [a_i, b_i);\ a_i, b_i \in \mathbb{R} \cup \{\pm\infty\}\ (i = 1, 2, \cdots, n),\ n \geqslant 1 \right\},$$

where if $a = -\infty$, then we set $[a, b) = (-\infty, b)$. $\mathcal{R}(\mathbb{R})$ satisfies $(i),\ (ii)$ and (iii) in Definition 8.1.

For $E = \bigcup_{i=1}^{n} [a_i, b_i) =: \bigcup_{i=1}^{n} I_i \in \mathcal{R}(\mathbb{R}),\ I_i \cap I_j = \phi$ we define the measure $m(E)$ of E by

$$m(E) = \sum_{i=1}^{n} (b_i - a_i) = \sum_{i=1}^{n} m(I_i),$$

where if $I = (-\infty, b)$ or $= [a, +\infty)$, then $m(I) = \infty$. Especially, if $E \cap F = \phi,\ E, F \in \mathcal{R}(\mathbb{R})$, then we have

$$m(E \cup F) = m(E) + m(F). \tag{8.1}$$

Definition 8.2 Let μ be a set function which is defined on a σ-algebra $\mathcal{A}$ with values in $\mathbb{R} \cup \{\pm\infty\}$. Then μ satisfies the following.

(i) $\mu(\phi) = 0$.
(ii) if $A, B \in \mathcal{A},\ \ A \cap B = \phi$, then $\mu(A \cup B) = \mu(A) + \mu(B)$.
(iii) if $A_j \in \mathcal{A}\ (j = 1, 2, \cdots),\ \ \bigcup_{j=1}^{\infty} A_j \in \mathcal{A}$ and $A_i \cap A_j = \phi\ (i \neq j)$.

Then

$$\mu\left(\bigcup_{j=1}^{\infty} A_j\right) = \sum_{j=1}^{\infty} \mu(A_j),$$

where μ is said a measure on $\mathcal{A}$.

Theorem 8.1 *Let m be defined by (8.1). Then m satisfies conditions $(i),\ (ii)$ and (iii) in Definition 8.2, that is,*

(i) $m(\phi) = 0$.
(ii) *If $E, F \in \mathcal{R}(\mathbb{R}),\ E \cap F = \phi$, then $m(E \cup F) = m(E) + m(F)$.*

(iii) If $E_j \in \mathcal{R}(\mathbb{R})$ $(j = 1, 2, \cdots)$ $E_i \cap E_j = \phi$ $(i \neq j)$. Then

$$m\left(\bigcup_{j=1}^{\infty} E_j\right) = \sum_{j=1}^{\infty} m(E_j).$$

To prove this theorem we need some lemmas.

We recall the definition of the finite intersection property which is stated in the Sect. ??.

Definition 8.3 A collection $\mathcal{F}$ of sets is said to have the finite intersection property if $\bigcap_{j=1}^{N} F_j \neq \phi$ holds for any finite sets $F_1, F_2, \cdots, F_N \in \mathcal{F}$.

From Theorem **??** we have the following lemma.

Lemma 8.1 *The set K is compact if and only if for a class $\mathcal{F}$ consisting of closed sets, the class $\mathcal{F}_K = \{K \cap F : F \in \mathcal{F}\}$ has the finite intersection property, then $\bigcap_{F \in \mathcal{F}} K \cap F \neq \phi$.*

Lemma 8.2 *If $E, F \in \mathcal{R}(\mathbb{R})$, $E \subset F$, then $m(E) \leqslant m(F)$.*

Proof Since $F = (F - E) \cup E$, $F - E \in \mathcal{F}(\mathbb{R})$ and $(F - E) \cap E = \phi$, we have

$$m(F) = m(F - E) + m(E) \geqslant m(E).$$

■

Lemma 8.3 *If $E, E_j \in \mathcal{R}(\mathbb{R})$ $(j = 1, \cdots, k)$, $E \subset \bigcup_{j=1}^{k} E_j$, then*

$$m(E) \leqslant \sum_{j=1}^{k} m(E_j).$$

Proof Put $F_j = E \cap E_j$. Then we see

$$E = \bigcup_{j=1}^{k} F_j = F_1 \cup (F_2 - F_1) \cup (F_3 - F_1 - F_2) \cup \cdots \cup (F_k - F_1 - \cdots - F_{k-1}).$$

Thus

$$m(E) = m(F_1) + m(F_2 - F_1) + m(F_3 - F_1 - F_2) + \cdots + m(F_k - F_1 - \cdots - F_{k-1}).$$

Hence, by Lemma 8.2, we have

$$m(F_1) \leqslant m(E_1),\ m(F_2 - F_1) \leqslant m(E_2), \cdots, m(F_k - F_1 - \cdots - F_{k-1}) \leqslant m(E_k),$$

that is, we have the result. ■

Lemma 8.4 *Let $E_j \in \mathcal{R}(\mathbb{R})$ $(j = 1, 2, \cdots)$, $E_1 \supset E_2 \supset \cdots \to \phi$. Then*

$$m(E_j) \to 0\ (j \to \infty).$$

Proof Let $m(E_j) \not\to 0$. Then there exists a constant $\delta > 0$ such that $m(E_j) > \delta$, $j = 1, 2, \cdots$. We can take $F_j \in \mathcal{R}(\mathbb{R})$ such that

$$E_j \supset \overline{F_j} \supset F_j,\ \ m(E_j - F_j) < \delta/2^{j+1}. \tag{8.2}$$

In fact, let

$$E_j = \bigcup_{i=1}^{n}[a_{i,j}, b_{i,j}),\ a_{i,j}, b_{i,j} \in \mathbb{R}.$$

Then we take $a_{i,j}<b'_{i,j} < b_{i,j}$ as $\varepsilon_{i,j} = b_{i,j} - b'_{i,j} > 0$ with $\sum_{i=1}^{n} \varepsilon_{i,j} < \delta/2^{j+1}$. Let

$$F_j = \bigcup_{i=1}^{n}[a_{i,j}, b'_{i,j}).\ \text{Then}\ \overline{F}_j = \bigcup_{i=1}^{n}[a_{i,j}, b'_{i,j}] \subset E_j\ \text{and}\ m(E_j - F_j) < \delta/2^{j+1}.$$

Since every $E_j \in \mathcal{R}(\mathbb{R})$ is bounded, $\overline{F}_1$ is compact. Since

$$\phi = \bigcap_{j=1}^{\infty} E_j \supset \overline{F_1} \bigcap \bigcap_{j=2}^{\infty} \overline{F_j},$$

$\{\overline{F_1} \cap \overline{F_k}; k = 2, 3, \cdots\}$ does not have the finite intersection property by Lemma 8.1. Thus from the compactness of $\overline{F}_1$ there exists $N \in \mathbb{N}$ such that $\overline{F}_1 \bigcap \bigcap_{j=2}^{N} \overline{F}_j = \phi$. So

$$E_N = E_N - \bigcap_{j=1}^{N} F_j = \bigcup_{j=1}^{N}(E_N - F_j) \subset \bigcup_{j=1}^{N}(E_j - F_j).$$

Hence, from Lemma 8.3 and (8.2) we have

$$\delta \leqslant m(E_N) \leqslant \sum_{j=1}^{N} m(E_j - F_j) \leqslant \sum_{j=1}^{N} \delta/2^{j+1} < \delta/2.$$

But this contradicts. ■

Proof of Theorem 8.1 (i) By Lemma 8.4 we see $m(\phi) = 0$. (ii) From (8.1) it follows. (iii) Let E, E_j $(j = 1, 2, \cdots)$, $E_i \cap E_j = \phi$ $(i \neq j)$, and let $E = \bigcup_{j=1}^{\infty} E_j$. We set $F_k := E - \bigcup_{j=1}^{k} E_j$, and then we see $F_1 \supset F_2 \supset \cdots \to \phi$. Thus, by Lemma 8.4 we have

$$m\left(E - \bigcup_{j=1}^{k} E_j\right) \to 0 \ (k \to \infty).$$

We see

$$m(E) = m\left(\bigcup_{j=1}^{k} E_j\right) + m\left(E - \bigcup_{j=1}^{k} E_j\right) = \sum_{j=1}^{k} m(E_j) + m\left(E - \bigcup_{j=1}^{k} E_j\right).$$

Therefore, if we take $k \to \infty$, then we have

$$m(E) = \sum_{j=1}^{\infty} m(E_j). \ \blacksquare$$

8.1.2 Outer Measure and Measurable Sets

For $A \in \mathcal{S}(\mathbb{R})$ we define

$$m^*A = \inf\left\{\sum_{j=1}^{\infty} m(E_j);\ E_j \in \mathcal{R}(\mathbb{R}),\ A \subset \bigcup_{j=1}^{\infty} E_j\right\}.$$

From the definition of m^* we easily see that $m^*\phi = 0$ and if $A \subset B$, then $m^*A \leqslant m^*B$. Each set consisting of a single point has an outer measure zero, and $m^*[a, b] = m^*(a, b] = m^*[a, b) = b - a$.

Proposition 8.1 *Let* $\{A_n\}$ *be a countable collection of sets. Then we have*

$$m^*(\cup A_n) \leqslant \sum m^*A_n.$$

Proof We may suppose $m^*A_n < \infty$ for each $n \in \mathbb{N}$. For a given $\varepsilon > 0$ there exists a countable collection $\{I_{n,j}\}_{j=1}^{\infty}$ of open intervals such that $A_n \subset \bigcup_{j=1}^{\infty} I_{n,j}$ and

$$\sum_{j=1}^{\infty} m(I_{n,j}) < m^*A_n + \varepsilon/2^n.$$

Now we take a countable collection $\{I_{n,j}\}_{j,n}^{\infty}$ such as $\cup A_n \subset \bigcup_{n,j} \{I_{n,j}\}$, then by the definition m^* we have

$$m^*(\cup A_n) \leqslant \sum_{n,j} m(I_{n,j}) = \sum_n \sum_j m(I_{n,j}) < \sum_n (m^* A_n + \varepsilon/2^n)$$
$$= \sum m^* A_n + \varepsilon.$$

Since ε is an arbitrary positive number,

$$m^*(\cup A_n) \leqslant \sum m^* A_n.$$

■

Corollary 8.1 *If A is countable, then $m^*A = 0$. Especially, $m^*(\mathbb{Q}) = 0$ for the set of all rational numbers $\mathbb{Q}$.*

Definition 8.4 If $m^*(A) = 0$, then we say that A is a null set.

One point set $\{x\}$ and the set $\mathbb{Q}$ consisting of all rational numbers are null sets. And the Cantor set C is also a null set.

Definition 8.5 A set E is said to be measurable if for each $A \in \mathcal{S}(\mathbb{R})$ we have

$$m^*A = m^*(A \cap E) + m^*(A \cap E^c) \tag{8.3}$$

(we call (8.3) Carathéodory's Condition). We denote the class of all measurable sets by $\mathcal{M}(\mathbb{R})$.

Theorem 8.2 *The null set is measurable.*

Proof Let $m^*(E) = 0$. For any $A \in \mathcal{S}(\mathbb{R})$ we see $A = (A \cap E) \cup (A \cap E^c)$. By Proposition 8.1 we have

$$m^*A \leqslant m^*(A \cap E) + m^*(A \cap E^c).$$

On the other hand, by $m^*(A \cap E) \leqslant m^*(E) = 0$ we have

$$m^*A \geqslant m^*(A \cap E^c) = m^*(A \cap E) + m^*(A \cap E^c).$$

Hence, E is measurable. ■

Lemma 8.5 *If E_1 and E_2 are measurable, then $E_1 \cup E_2$ is also measurable.*

Proof Let A be any set. Since E_2 is measurable, we have

$$m^*(A \cap E_1^c) = m^*(A \cap E_1^c \cap E_2) + m^*(A \cap E_1^c \cap E_2^c).$$

On the other hand, since $A \cap (E_1 \cup E_2) = (A \cap E_1) \cup (A \cap E_2 \cap E_1^c)$, we see

$$m^*(A \cap [E_1 \cup E_2]) \leqslant m^*(A \cap E_1) + m^*(A \cap E_2 \cap E_1^c).$$

Hence, by Proposition 8.1

$$\begin{aligned} m^*A &\leqslant m^*(A \cap [E_1 \cup E_2]) + m^*(A \cap [E_1 \cup E_2]^c) \\ &\leqslant m^*(A \cap E_1) + m^*(A \cap E_2 \cap E_1^c) + m^*(A \cap E_1^c \cap E_2^c) \\ &= m^*(A \cap E_1) + m^*(A \cap E_1^c) = m^*A, \end{aligned}$$

by measurabilites of E_1 and E_2. Thus, we see that $E_1 \cup E_2$ is measurable. ■

Remark 8.1 For (8.3) we see that

$$m^*A \leqslant m^*(A \cap E) + m^*(A \cap E^c)$$

holds by Proposition 8.1. Thus, to prove (8.3) it is sufficient to show

$$m^*A \geqslant m^*(A \cap E) + m^*(A \cap E^c). \tag{8.4}$$

Hereafter, to show the measurability of E, we show only (8.4) for (8.3).

Theorem 8.3 *For $E_1, E_2 \in \mathcal{M}(\mathbb{R}),\ E_1 \cap E_2 = \phi$ we have*

$$m^*(E_1 \cup E_2) = m^*(E_1) + m^*(E_2). \tag{8.5}$$

Furthermore, for $E_1, E_2, \cdots \in \mathcal{M}(\mathbb{R}),\ E_i \cap E_j = \phi\ (i \neq j)$ we have

$$m^*\left(\bigcup_{j=1}^{\infty} E_j\right) = \sum_{j=1}^{\infty} m^*\left(E_j\right). \tag{8.6}$$

Proof If we take $A = E_1 \cup E_2$ in (8.3),

$$\begin{aligned} m^*(E_1 \cup E_2) &= m^*((E_1 \cup E_2) \cap E_1) + m^*((E_1 \cup E_2) \cap E_1^c) \\ &= m^*(E_1) + m^*(E_2). \end{aligned}$$

Thus we have (8.5). We will show (8.6). First we see for any $n \in \mathbb{N}$

$$m^*\left(\bigcup_{j=1}^{n} E_j\right) = \sum_{j=1}^{n} m^*(E_j). \tag{8.7}$$

In fact, if we suppose (8.7), then

$$m^*\left(\bigcup_{j=1}^{n+1} E_j\right) = m^*\left(\bigcup_{j=1}^{n} E_j\right) + m^*(E_{n+1})$$
$$= \sum_{j=1}^{n+1} m^*(E_j).$$

Thus, we see (8.7), and so

$$m^*\left(\bigcup_{j=1}^{n} E_j\right) \leqslant \sum_{j=1}^{\infty} m^*(E_j),$$

and as $n \to \infty$

$$m^*\left(\bigcup_{j=1}^{\infty} E_j\right) \leqslant \sum_{j=1}^{\infty} m^*(E_j).$$

Similarly, we have

$$m^*\left(\bigcup_{j=1}^{\infty} E_j\right) \geqslant \sum_{j=1}^{\infty} m^*(E_j).$$

Thus, we have (8.6). ■

Definition 8.6 Let

$$\tilde{m} = m^*|_{\mathcal{M}(\mathbb{R})}.$$

Then we call $\tilde{m}$ the Lebesgue measure.

Theorem 8.4 *(i) $\mathcal{M}(\mathbb{R})$ is a σ-algebra which contains $\mathcal{R}(\mathbb{R})$,*
(ii) $\tilde{m}$ is a measure on $\mathcal{R}(\mathbb{R})$.

Proof Step 1.

$$\phi \in \mathcal{M}(\mathbb{R}), \tag{8.8}$$

$$\text{if } E \in \mathcal{M}(\mathbb{R}), \text{ then } E^c \in \mathcal{M}(\mathbb{R}), \tag{8.9}$$

$$\text{if } E, F \in \mathcal{M}(\mathbb{R}), \text{ then } E - F \in \mathcal{M}(\mathbb{R}). \tag{8.10}$$

Since

$$m^*(A \cap \phi) + m^*(A \cap \mathbb{R}) = m^*(A),$$

we have (8.8). From (8.3) we have (8.9). We will show (8.10). Then we may show for $A \in \mathcal{S}(\mathbb{R})$

$$m^*(A) = m^*(A \cap (E - F)) + m^*\left(A \cap (E - F)^c\right). \tag{8.11}$$

Let $E, F \in \mathcal{M}(\mathbb{R})$. We have

$$\begin{aligned}
[A - (E - F)] \cap F &= A \cap \left(E \cap F^c\right)^c \cap F = A \cap \left(E^c \cup F\right) \cap F = A \cap F, \\
[A - (E - F)] - F &= A \cap \left(E \cap F^c\right)^c \cap F^c = A \cap \left(E^c \cup F\right) \cap F^c \\
&= \left(A \cap F^c\right) - E.
\end{aligned}$$

Thus, since F is measurable, we see

$$\begin{aligned}
m^*(A - (E - F)) &= m^*([A - (E - F)] \cap F) + m^*([A - (E - F)] - F) \\
&= m^*(A \cap F) + m^*\left(\left(A \cap F^c\right) - E\right).
\end{aligned} \tag{8.12}$$

And, we have

$$A \cap (E - F) = A \cap E \cap F^c = (A \cap F^c) \cap E. \tag{8.13}$$

Thus, from (8.12), (8.13) and the measurability of E, we have

$$\begin{aligned}
&m^*(A \cap (E - F)) + m^*(A - (E - F)) \\
&= m^*((A \cap F^c) \cap E) + m^*(A \cap F) + m^*((A \cap F^c) - E) \\
&= m^*(A \cap F) + m^*(A \cap F^c) = m^*(A).
\end{aligned}$$

Thus, we conclude (8.10).

Step 2.

Let $E_j \in \mathcal{M}(\mathbb{R})$ $(j = 1, 2, \cdots)$, $E_i \cap E_j = \phi$ $(i \neq j)$. Then we will show

$$\begin{aligned}
m^*A &\geqslant m^*\left(A \cap \bigcup_{j=1}^{\infty} E_j\right) + m^*\left(A - \bigcup_{j=1}^{\infty} E_j\right) \\
&= \sum_{j=1}^{\infty} m^*(A \cap E_j) + m^*\left(A - \bigcup_{j=1}^{\infty} E_j\right)
\end{aligned} \tag{8.14}$$

(see (8.6)). First, for any $n = 1, 2, \cdots$ we will show

$$m^* A \geqslant \sum_{j=1}^{n} m^*(A \cap E_j) + m^* \left(A - \bigcup_{j=1}^{n} E_j \right). \tag{8.15}$$

When $n = 1$, (8.15) holds. Let (8.15) hold for n. Then

$$\begin{aligned} m^*(A) &= m^*(A \cap E_{n+1}) + m^*(A - E_{n+1}) \\ &\geqslant m^*(A \cap E_{n+1}) \\ &+ \left\{ \sum_{j=1}^{n} m^* \left((A - E_{n+1}) \cap E_j \right) + m^* \left(A - E_{n+1} - \bigcup_{j=1}^{n} E_j \right) \right\}. \end{aligned} \tag{8.16}$$

Since $E_i \cap E_j = \phi \; (i \neq j)$, we see

$$(A - E_{n+1}) \cap E_j = A \cap (E_{n+1}^c \cap E_j) = A \cap E_j,$$

similarly

$$(A - E_{n+1}) - \bigcup_{j=1}^{n} E_j = A - \bigcup_{j=1}^{n+1} E_j.$$

Thus, substituting these formulas for (8.16), we have

$$\begin{aligned} & m^*(A \cap E_{n+1}) + m^*(A - E_{n+1}) \\ & \geqslant m^* \left(A \cap E_{n+1} \right) + \left\{ \sum_{j=1}^{n} m^* \left(A \cap E_j \right) + m^* \left(A - \bigcup_{j=1}^{n+1} E_j \right) \right\} \\ & = \sum_{j=1}^{n+1} m^*(A \cap E_j) + m^* \left(A - \bigcup_{j=1}^{n+1} E_j \right). \end{aligned}$$

Thus, we have (8.15). When $n \to \infty$, we have (8.14).

Step 3.

Let $E_j \in \mathcal{M}(\mathbb{R}) \; (j = 1, 2, \cdots)$, $E_i \cap E_j = \phi \, (i \neq j)$. Then we see $\bigcup_{j=1}^{\infty} E_j \in \mathcal{M}(\mathbb{R})$. From the property of an outer measure m^* we have

$$m^*\left(A\bigcap\bigcup_{j=1}^{\infty}E_j\right)\leqslant\sum_{j=1}^{\infty}m^*\left(A\cap E_j\right).$$

Thus, substituting this formula for (8.14), we have

$$m^*(A)\geqslant m^*\left(A\cap\bigcup_{j=1}^{\infty}E_j\right)+m^*\left(A-\bigcup_{j=1}^{\infty}E_j\right),$$

that is, $\bigcup_{j=1}^{\infty}E_j$ is measurable.

Step 4.

$\mathcal{M}(\mathbb{R})$ is a σ-algebra. In fact, for $E_j\in\mathcal{M}(\mathbb{R}),\ j=1,2,\cdots$ we see

$$\bigcup_{j=1}^{\infty}E_j=E_1\cup(E_2-E_1)\cup(E_3-E_1-E_2)\cup\cdots,$$

and then the right-hand side belongs to $\mathcal{M}(\mathbb{R})$ (see Step 3).

Step 5.

We show that $\tilde{m}$ is a measure on $\mathcal{M}(\mathbb{R})$. It is sufficient to show (iii) in Definition 8.2. Let

$$E_j\in\mathcal{M}(\mathbb{R})\ (i=1,2,\cdots),\ \ \bigcup_{i=1}^{\infty}E_i\in\mathcal{M}(\mathbb{R})\ \text{and}\ \ E_i\cap E_j=\phi\ (i\neq j).$$

In (8.14) we put $A=\bigcup_{j=1}^{\infty}E_j$. Then we see

$$\tilde{m}\left(\bigcup_{j=1}^{\infty}E_j\right)\geqslant\sum_{j=1}^{\infty}\tilde{m}\left(E_j\right).$$

Since

$$\tilde{m}\left(\bigcup_{j=1}^{\infty}E_j\right)\leqslant\sum_{j=1}^{\infty}\tilde{m}(E_j)$$

holds, we conclude

$$\tilde{m}\left(\bigcup_{j=1}^{\infty} E_j\right) = \sum_{j=1}^{\infty} \tilde{m}(E_j).$$

■

Theorem 8.5 *The system $(\mathcal{M}(\mathbb{R}), \tilde{m})$ is an extension of the system $(\mathcal{R}(\mathbb{R}), m)$, that is,*

(i) $\mathcal{R}(\mathbb{R}) \subset \mathcal{M}(\mathbb{R})$,
(ii) $\tilde{m}|_{\mathcal{R}(\mathbb{R})} = m$.

Proof (i) Let $A \in \mathcal{S}(\mathbb{R})$, $E, E_j \in \mathcal{R}(\mathbb{R})$ $(j = 1, 2, \cdots)$, and let

$$A \subset \bigcup_{j=1}^{\infty} E_j.$$

Then, since

$$A \cap E \subset \bigcup_{j=1}^{\infty} E_j \cap E, \quad A - E \subset \bigcup_{j=1}^{\infty} (E_j - E),$$

we have

$$\begin{aligned}\sum_{j=1}^{\infty} m(E_j) &= \sum_{j=1}^{\infty} \{m(E_j \cap E) + m(E_j - E)\} \\ &\geqslant m^*(A \cap E) + m^*(A - E).\end{aligned}$$

Here we see

$$m^* A = \inf \left\{ \sum_{j=1}^{\infty} m\left(E_j\right) \right\},$$

so we have

$$m^* A \geqslant m^*(A \cap E) + m^*(A - E).$$

Thus, we have (i).

(ii) Since $m^* = m$ on $\mathcal{R}(\mathbb{R})$, we have $\tilde{m}|_{\mathcal{R}(\mathbb{R})} = m$. ■

Hereafter, we write m instead of $\tilde{m}$ unless there is any confusion.

8.1.3 *Fundamental Properties for the Measure*

We consider the measure on a general set X. For $\{E_j\}$ we define the followings;

Let $\mathcal{S}(X)$ be the class of all subsets of X. Let $\mathcal{F} \subset \mathcal{S}(X)$ be a $\sigma-$algebra.

When $\{E_j\}$ is a monotonic increasing sequence, we denote $\lim_{j\to\infty} E_j = \bigcup_{j=1}^{\infty} E_j$.

When $\{E_j\}$ is a monotonic decreasing sequence, we denote $\lim_{j\to\infty} E_j = \bigcap_{j=1}^{\infty} E_j$.

In general, we define

$$\limsup_{j\to\infty} = \bigcap_{k=1}^{\infty}\bigcup_{j=k}^{\infty} E_j, \quad \liminf_{j\to\infty} E_j = \bigcup_{k=1}^{\infty}\bigcap_{j=k}^{\infty} E_j.$$

Since $\mathcal{F}$ is a σ-algebra, if $E_j \in \mathcal{F}$, we have

$$\limsup_{j\to\infty} = \bigcap_{k=1}^{\infty}\bigcup_{j=k}^{\infty} E_j, \quad \liminf_{j\to\infty} E_j = \bigcup_{k=1}^{\infty}\bigcap_{j=k}^{\infty} E_j \in \mathcal{F}.$$

In fact, we have $\bigcup_{j=k}^{\infty} E_j \in \mathcal{F}$ and $\bigcap_{j=i}^{\infty} E_j = \left(\bigcup_{j=i}^{\infty} E_j^c\right)^c \in \mathcal{F}$. Thus we see

$$\limsup_{j\to\infty} E_j, \quad \liminf_{j\to\infty} E_j \in F.$$

We consider μ which is non-negative measure (see Definition 8.2), for example, the Lebesgue measure m on $\mathcal{M}(\mathbb{R})$.

Theorem 8.6 *Let $\mathcal{F}$, and let μ be a non-negative measure. Let $E_j \in \mathcal{F}$ $(j = 1, 2, \cdots)$. Then we have the following:*

(i) $\mu\left(\bigcup_{j=1}^{\infty} E_j\right) \leqslant \sum_{j=1}^{\infty} \mu(E_j)$,

(ii) if $E_1 \subset E_2 \subset \cdots$, then

$$\lim_{j\to\infty} \mu(E_j) = \mu(\lim_{j\to\infty} E_j),$$

(iii) if $E_1 \supset E_2 \supset \cdots$ and $\mu(E_j) < \infty$, then

$$\lim_{j\to\infty} \mu(E_j) = \mu(\lim_{j\to\infty} E_j).$$

Proof (i) Since

$$\bigcup_{j=1}^{\infty} E_j = E_1 \cup (E_2 - E_1) \cup (E_3 - E_1 - E_2) \cup \cdots,$$

and

$$(E_i - E_1 - E_2 - \cdots - E_{i-1}) \cap (E_j - E_1 - E_2 - \cdots - E_{j-1}) = \phi \quad (i \neq j),$$

we have

$$\begin{aligned}\mu\left(\bigcup_{j=1}^{\infty} E_j\right) &= \mu(E_1) + \mu(E_2 - E_1) + \mu(E_3 - E_1 - E_2) + \cdots \\ &\leqslant \mu(E_1) + \mu(E_2) + \mu(E_3) + \cdots.\end{aligned}$$

(ii) If $\mu(E_k) = \infty$ for some k, then by $E_k \subset \lim\limits_{j\to\infty} E_j$ we see

$$\mu\left(\lim_{j\to\infty} E_j\right) = \infty.$$

On the other hand, we see $\mu(E_j) \geqslant \mu(E_k) = \infty$ $(j \geqslant k)$. Hence we have (ii). Let $\mu(E_k) < \infty$ for all k. As (i) we have

$$\mu\left(\lim_{j\to\infty} E_j\right) = \mu(E_1) + \mu(E_2 - E_1) + \mu(E_3 - E_2) + \cdots.$$

Since we see $\mu(E_{j+1}) = \mu(E_{j+1} - E_j) + \mu(E_j)$, we have

$$\begin{aligned}&\mu(E_1) + \mu(E_2 - E_1) + \mu(E_3 - E_2) + \cdots \\ &= \mu(E_1) + [\mu(E_2) - \mu(E_1)] + [\mu(E_3) - \mu(E_2)] + \cdots = \lim_{j\to\infty} \mu(E_j).\end{aligned}$$

(iii) If we put $F_j = E_1 - E_j$, then we see $F_1 \subset F_2 \subset F_3 \subset \cdots$. Hence, by (ii) we have

$$\mu\left(\lim_{j\to\infty} F_j\right) = \lim_{j\to\infty} \mu\left(F_j\right).$$

Here

$$\mu\left(\lim_{j\to\infty} F_j\right) = \mu\left(\lim_{j\to\infty} (E_1 - E_j)\right) = \mu(E_1) - \mu\left(\lim_{j\to\infty} E_j\right),$$

and

$$\lim_{j\to\infty} \mu(F_j) = \lim_{j\to\infty} \left[\mu(E_1) - \mu(E_j)\right] = \mu(E_1) - \lim_{j\to\infty} \mu(E_j).$$

Thus, we have (iii). ∎

For a general set X we consider a class $\mathcal{A}$ consisting subsets of X. Then we denote the smallest σ-algebra containing $\mathcal{A}$ by $\sigma(\mathcal{A})$.

Definition 8.7 We denote the class of all compact subsets in $\mathbb{R}$ by $\mathcal{C}(\mathbb{R})$, and we write $\mathcal{B}(\mathbb{R})$ the smallest σ-algebra which contains $\mathcal{C}(\mathbb{R})$. Then we call the element of $\mathcal{B}(\mathbb{R})$ the Borel set.

Let $\mathcal{F}$ be a σ-algebra containing $\mathcal{C}(\mathbb{R})$ (for example, $\mathcal{S}(\mathbb{R})$). We see

$$\mathcal{B}(\mathbb{R}) = \sigma(\mathcal{C}(\mathbb{R})) = \cap\{\mathcal{F}; \mathcal{C}(\mathbb{R}) \subset \mathcal{F} \subset \mathcal{S}(\mathbb{R})\}.$$

We denote the class of all open subsets in $\mathbb{R}$ by $\mathbb{O}(\mathbb{R})$. Let $B(0, n) = (-n, n)$. We have

$$O^c = \bigcup_{n=1}^{\infty} O^c \cap \overline{B(0, n)},$$

where $O \in \mathbb{O}(\mathbb{R})$. Thus we have $O \in \mathcal{B}(\mathbb{R})$, so we have $\sigma(\mathbb{O}(\mathbb{R})) \subset \mathcal{B}(\mathbb{R})$. And since the complement of a compact set is open, we see $\mathcal{C}(\mathbb{R}) \subset \sigma(\mathbb{O}(\mathbb{R}))$. Thus we have $\mathcal{B}(\mathbb{R}) = \sigma(\mathbb{O}(\mathbb{R}))$. Consequently, we have

$$\mathcal{B}(\mathbb{R}) = \sigma(\mathbb{O}(\mathbb{R})) = \cap\{\mathcal{F}; \mathbb{O}(\mathbb{R}) \subset \mathcal{F} \subset \mathcal{S}(\mathbb{R})\}.$$

When $O \in \mathbb{O}(\mathbb{R})$, we have an expression

$$O = \bigcup_{n=1}^{\infty} I_j, \quad I_j = (a_j, b_j) \ (j = 1, 2, \cdots).$$

Because of $I_j \in \mathcal{M}(\mathbb{R})$ we see

$$\mathcal{B}(\mathbb{R}) \subset \mathcal{M}(\mathbb{R}).$$

Let $N(\mathbb{R})$ denote the class of all null sets. Then by Theorem 8.2 we also have

$$\mathcal{N}(\mathbb{R}) \subset \mathcal{M}(\mathbb{R}).$$

Definition 8.8 A set is called a G_δ-set if it is a countable intersection of open sets;

$$\bigcap_{j=1}^{\infty} O_j, \quad O_j; \text{ open set}, \ j = 1, 2, \cdots,$$

and a set is called a F_σ-set if it is a countable union of closed sets;

$$\bigcup_{j=1}^{\infty} F_j, \quad F_j; \text{ closed set}, \ j = 1, 2, \cdots.$$

Theorem 8.7 *Let $E \in \mathfrak{M}(\mathbb{R})$. Then*

(*i*) *there exist a G_δ-set G and a null set N such that*

$$E = G - N, \quad m(E) = m(G),$$

(*ii*) *there exist a F_σ-set F and a null set N' such that*

$$E = F \cup N', \quad m(E) = m(F).$$

To prove the theorem we need some lemmas.

Lemma 8.6 *For any $A \subset \mathbb{R}$ we have*

$$m^*(A) = \inf\{m(O); A \subset O, \ O \in \mathfrak{O}(\mathbb{R})\}.$$

Proof We may suppose $m^*(A) < \infty$. By the definition of an outer measure, for $n = 1, 2, \cdots$ there exists $E_j^n \in \mathcal{R}(\mathbb{R})$ such that

$$A \subset \bigcup_{j=1}^{\infty} E_j^n, \quad m^*(A) + \frac{1}{n} > \sum_{j=1}^{\infty} m\left(E_j^n\right). \tag{8.17}$$

Since E_j^n is a finite sum of semi-closed intervals, there exists $O_j^n \in \mathfrak{O}(\mathbb{R})$ such that

$$E_j^n \subset O_j^n, \quad m^*(E_j^n) + \frac{1}{n2^j} > m(O_j^n). \tag{8.18}$$

Therefore, if we put $O^n = \bigcup_{j=1}^{\infty} O_j^n$, then $O^n \in \mathfrak{O}(\mathbb{R}), \ A \subset O^n$. Hence, by (8.17) and (8.18) we have

$$m(O^n) \leqslant \sum_{j=1}^{\infty} m(O_j^n) \leqslant \sum_{j=1}^{\infty} \left[m(E_j^n) + \frac{1}{n2^j} \right] < m^*(A) + \frac{1}{n} + \frac{1}{n}.$$

Since $O^n \in \mathfrak{O}(\mathbb{R})$, we conclude

$$m^*(A) \geqslant \inf\{m(O); A \subset O, \ O \in \mathfrak{O}(\mathbb{R})\}.$$

The inverse inequality is trivial. ∎

Corollary 8.2 *Let E be a measurable set with finite measure. For $\varepsilon > 0$ there exist intervals $I_1, I_2, \cdots, I_N, \ I_i \cap I_j = \phi$ such that*

$$m\left(E-\bigcup_{j=1}^{N} I_j\right)<\varepsilon \text{ and } m\left(\bigcup_{j=1}^{N} I_j-E\right)<\varepsilon.$$

Proof Let $\varepsilon>0$. By Lemma 8.6 there exists an open set $O\supset E$ such that

$$m(E)>m(O)-\varepsilon,$$

that is,

$$m(O-E)=m(O)-m(E)<\varepsilon.$$

Since there exist intervals $I_1, I_2, \cdots$ such that

$$O=\bigcup_{j=1}^{\infty} I_j,\ I_i\cap I_j=\phi\quad(i\neq j),$$

we have for N large enough

$$m\left(O-\bigcup_{j=1}^{N} I_j\right)=m\left(\bigcup_{j=N+1}^{\infty} I_j\right)\leqslant\sum_{j=N}^{\infty} m\left(I_j\right)<\varepsilon.$$

Thus we have

$$m\left(E-\bigcup_{j=1}^{N} I_j\right)\leqslant m\left(O-\bigcup_{j=1}^{N} I_j\right)<\varepsilon.$$

And

$$m\left(\bigcup_{j=1}^{N} I_j-E\right)\leqslant m\left(O-E\right)<\varepsilon.$$

■

Proof of Theorem 8.7. Let $E\in\mathcal{M}(\mathbb{R})$. If we put $E_k=E\cap B(0,k),\ k=1,2,\cdots,$ then there exists an open set O_k^n such that

$$E_k\subset O_k^n,\ \ m\left(O_k^n-E_k\right)\leqslant\frac{1}{n2^k}.$$

If we set $O^n=\bigcup_{k=1}^{\infty} O_k^n$, then we see that O^n is an open set and $E=\bigcup_{k=1}^{\infty} E_k\subset O^n$.

Here $G=\bigcap_{n=1}^{\infty} O^n$ is a G_δ-set which contains E. Since

$$G-E\subset O^n-E=\bigcup_{k=1}^{\infty}(O_k^n-E)\subset\bigcup_{k=1}^{\infty}(O_k^n-E_k),$$

we have

$$m(G - E) \leqslant \sum_{k=1}^{\infty} m(O_k^n - E_k) \leqslant \sum_{k=1}^{\infty} \frac{1}{n2^k} = \frac{1}{n}.$$

When $n \to \infty$, we see that $N = G - E$ is a null set. From $E = G - N$ we have (i). (ii) follows by applying (i) for E^c.

8.2 Lebesgue Integral on $\mathbb{R}$

8.2.1 Measurable Functions

Definition 8.9 Let f be a function on $\mathbb{R}$ which has its values in $\mathbb{R} \cup \{\pm\infty\}$. If for all $a \in \mathbb{R}$ the set

$$\{x;\ f(x) > a\}$$

is measurable, then we say that f is a measurable function.

Example 8.1 A continuous function f is measurable.

Example 8.2 Let $\{E_j,\ j = 1, 2, \cdots\}$ be a sequence of sets with $E_i \cap E_j = \phi\ (i \neq j)$, and let $\{a_j,\ j = 1, 2, \cdots\}$ be a real sequence with $a_i \neq a_j\ (i \neq j)$. Then the function

$$f(x) = \sum_{j=1}^{k} a_j \chi_{E_j}(x),$$

where

$$\chi_{E_j}(x) = \begin{cases} 1, & x \in E_j; \\ 0, & x \notin E_j, \end{cases}$$

is measurable if and only if all $E_j,\ j = 1, 2, \cdots$ are measurable.

Definition 8.10 If a function f on $\mathbb{R}$ satisfies

(i) f has its values in $\mathbb{R} \cup \{+\infty\}$,
(ii) $\{x;\ f(x) > a\}$ is open for any $a \in \mathbb{R}$,

then f is semi-continuous below. Similarly, if a function f on $\mathbb{R}$ has its values in $\mathbb{R} \cup \{-\infty\}$, and $\{x;\ f(x) < a\}$ is open for any $a \in \mathbb{R}$, then f is semi-continuous above.
In this chapter we suppose that f has its values in $\mathbb{R} \cup \{\pm\infty\}$.

Lemma 8.7 *For a function f the following $(i) - (iv)$ are equivalent.*

(i) $\{x;\, f(x) > a\} \in \mathcal{M}(\mathbb{R})$ $(a \in \mathbb{R})$.
(ii) $\{x;\, f(x) \geqslant a\} \in \mathcal{M}(\mathbb{R})$ $(a \in \mathbb{R})$.
(iii) $\{x;\, f(x) < a\} \in \mathcal{M}(\mathbb{R})$ $(a \in \mathbb{R})$.
(iv) $\{x;\, f(x) \leqslant a\} \in \mathcal{M}(\mathbb{R})$ $(a \in \mathbb{R})$.

Proof $(i) \Rightarrow (ii)$.

$$\{x;\, f(x) \geqslant a\} = \bigcap_{n=1}^{\infty} \left\{x;\, f(x) > a - \frac{1}{n}\right\} \in \mathcal{M}(\mathbb{R}).$$

$(ii) \Rightarrow (iii)$.

$$\{x;\, f(x) < a\} = \{x;\, f(x) \geqslant a\}^c \in \mathcal{M}(\mathbb{R}).$$

$(iii) \Rightarrow (iv)$.

$$\{x;\, f(x) \leqslant a\} = \left\{x;\, f(x) < a + \frac{1}{n}\right\} \in \mathcal{M}(\mathbb{R}).$$

$(iv) \Rightarrow (i)$.

$$\{x;\, f(x) > a\} = \{x;\, f(x) \leqslant a\}^c \in \mathcal{M}(\mathbb{R}).$$

■

Remark 8.2 By Lemma 8.7 we see

$$\{x;\, f(x) = a\} \in \mathcal{M}(\mathbb{R}) \ (a \in \mathbb{R}).$$

Theorem 8.8 *A function f is measurable if and only if*

$$f^{-1}(O) \in \mathcal{M}(\mathbb{R}), \quad O \in \mathcal{O}(\mathbb{R}).$$

Proof Especially, we take $O = (a, \infty)$, $a \in \mathbb{R}$. Then

$$\{x;\, f(x) > a\} = f^{-1}(O) \in \mathcal{M}(\mathbb{R}).$$

Similarly, when $O = (a, b)$, we see

$$\{x;\, f(x) > a\} \cap \{x;\, f(x) < b\} = f^{-1}(O) \in \mathcal{M}(\mathbb{R}).$$

Let O be an open set. Since O has a representation

$$O = \bigcup_{j=1}^{\infty} (a_j, b_j),$$

we have

$$f^{-1}(O) = f^{-1}\left(\bigcup_{j=1}^{\infty}(a_j, b_j)\right) = \bigcup_{j=1}^{\infty} f^{-1}(a_j, b_j) \in \mathcal{M}(\mathbb{R}).$$

■

Definition 8.11 When a proposition $\mathbf{P}(x)$ holds without points of a null set N, that is,

$$\mathbf{P}(x) \quad (x \notin N),$$

we say that the proposition $\mathbf{P}(x)$ holds almost everywhere. Then we write $\mathbf{P}(x)$ *a.e.* x or $\mathbf{P}$ *a.e.*.

Lemma 8.8 *If f is measurable, and $f = g$ a.e., then g is also measurable.*

Proof Since $N = \{x;\ f(x) \neq g(x)\}$ is a null set,

$$\{x;\ g(x) > a\} = [\{x;\ f(x) > a\} - N] \cup [\{x;\ g(x) > a\} \cap N] \in \mathcal{M}(\mathbb{R}).$$

■

Theorem 8.9 (i) *$af + bg$ $(a, b \in \mathbb{R})$ is measurable.*
(ii) *$f \cdot g$ is measurable, and if $g \neq 0$ a.e., then f/g is measurable.*
(iii) *$|f|^p$ $(p > 0)$ is measurable.*
(iv) *$f^+ = \max(f, 0)$ is measurable.*

Proof (i) Since af and bg are measurable, we may show that $f + g$ is measurable. If $f(x) + g(x) < \alpha$, then $f(x) < -g(x) + \alpha$, and then there exists a rational number r such that

$$f(x) < r < -g(x) + \alpha.$$

Thus

$$\{x;\ f(x) + g(x) < \alpha\} = \bigcup_{r \in \mathbb{Q}} (\{x;\ f(x) < r\} \cap \{x;\ g(x) < \alpha - r\}) \in \mathcal{M}(\mathbb{R}).$$

(ii) Let $\alpha \geqslant 0$. We see

$$\left\{x;\ f^2(x) > \alpha\right\} = \left\{x;\ f(x) > \sqrt{\alpha}\right\} \cup \left\{x;\ f(x) < -\sqrt{\alpha}\right\}.$$

And if $\alpha < 0$, then we have the domain of f; $D = \{x;\ f^2(x) > \alpha\} \in \mathcal{M}(\mathbb{R})$. Hence f^2 is measurable. Thus we see

$$fg = \frac{1}{2}\left[(f+g)^2 - f^2 - g^2\right]$$

is measurable, and we can show that $1/g$ is measurable, so f/g is also measurable.
(iii) Let $\alpha \geqslant 0,\ p > 0$. Thus we have

$$\{x;\ |f|^p(x) > \alpha\} = \{x;\ f(x) > \alpha^{1/p}\} \cup \{x;\ f(x) < -\alpha^{1/p}\} \in \mathcal{M}(\mathbb{R}).$$

For $\alpha < 0$ we have the domain of f; $D = \{x;\ |f|^p(x) > \alpha\} \in \mathcal{M}(\mathbb{R})$.
(iv) By (i), (iii) we have that $f^+ = (|f| + f)/2$ is measurable. ■

Theorem 8.10 *If f_j $(j = 1, 2, \cdots)$ are measurable, then*

$$\sup_j f_j(x),\ \inf_j f_j(x),\ \limsup_{j\to\infty} f_j(x),\ \liminf_{j\to\infty} f_j(x)$$

are measurable. Especially, if $\lim f_j$ *converges almost everywhere, then* $\lim f_j$ *is measurable. Furthermore, if the sum* $\sum_{j=1}^{\infty} u_j$ *for measurable functions* u_j $(j = 1, 2, \cdots)$ *converges almost everywhere, then* $\sum_{j=1}^{\infty} u_j$ *is measurable.*

Proof We see

$$\left\{x;\ \sup_j f_j(x) > a\right\} = \bigcup_{j=1}^{\infty} \{x;\ f_j(x) > a\} \in \mathcal{M}(\mathbb{R}),$$

$$\left\{x;\ \inf_j f_j(x) \geqslant a\right\} = \bigcap_{j=1}^{\infty} \{x;\ f_j(x) \geqslant a\} \in \mathcal{M}(\mathbb{R}).$$

Since $\left\{x;\ \sup_{j\geqslant k} f_j(x) > a\right\} = \bigcup_{j\geqslant k}^{\infty} \{x;\ f_j(x) > a\} \in \mathcal{M}(\mathbb{R})$, we have

$$\limsup_{j\to\infty} f_j(x) = \inf_{k\geqslant 1} \sup_{j\geqslant k} f_j(x) \in \mathcal{M}(\mathbb{R}).$$

Similarly, we see

$$\liminf_{j\to\infty} f_j(x) \in \mathcal{M}(\mathbb{R}).$$

■

8.2.2 Simple Functions and Integrals

Definition 8.12 A real-valued function s is called simple if it is measurable and assumes only a finite number of values.

If s is a simple function and $\{a_1, a_2, \cdots, a_n\}$ are a set of non-zero distinct values of s, then we have

$$s = \sum_{j=1}^{n} a_j \chi_{A_j},$$

where

$$\chi_{A_j}(x) = \begin{cases} 1, & x \in A_j; \\ 0, & x \notin A_j. \end{cases}$$

This representation for s is called the canonical representation and it is characterized by the fact that the $\{A_j\}$ are disjoint and the a_j is distinct and non-zero. The sum, product, and difference of two simple functions are simple.

If s vanishes outside a set of finite measures, we define the integral of s by

$$\int s(x)dx = \sum_{j=1}^{n} a_j m(A_j), \tag{8.19}$$

where s has the canonical representation $s = \sum_{j=1}^{n} a_j \chi_{A_j}$. We sometimes abbreviate the expression for this integral to $\int s$. If E is any measurable set, we define

$$\int_E s = \int s \cdot \chi_E.$$

Theorem 8.11 *Let f be measurable, and let $0 \leqslant f(x) \leqslant \infty$. Then there exist simple functions $\{s_j\}$ such that*

(i) for all x, $0 \leqslant s_1(x) \leqslant s_2(x) \leqslant \cdots \to f(x)$,
(ii) each s_j is bounded, and $m\left(\{x; s_j(x) \neq 0\}\right) < \infty$.

Proof For $j = 1, 2, \cdots$ we define $s_j(x)$ as follows:

$$s_j(x) := \begin{cases} 2^{-j}(k-1), \ (|x| \leqslant j \text{ and } 2^{-j}(k-1) \leqslant f(x) < k \cdot 2^{-j}), \ k = 1, 2, \cdots, j \cdot 2^j, \\ j, \qquad (|x| \leqslant j \text{ and} f(x) \geqslant j), \\ 0, \qquad (|x| > j). \end{cases}$$

Proof of (ii) is trivial. (i) We see for all x and j

$$0 \leqslant s_j(x) \leqslant s_{j+1}(x) \leqslant f(x),$$

and for $|x| \leqslant j$ and $f(x) < j$

$$0 \leqslant f(x) - s_j(x) \leqslant 2^{-j}.$$

Thus when $f(x) < \infty$, we have $s_j(x) \to f(x)$. If $f(x) = \infty$, then $s_j(x) = j \to \infty$, that is, we have (i). ∎

Hereafter, we consider measurable sets and measurable functions. It is often convenient to use representations that are not canonical, and then the following lemma is useful.

Lemma 8.9 *Let* $s = \sum_{j=1}^{n} a_j \chi_{E_j}$, $E_i \cap E_j = \phi$ $(i \neq j)$. *Suppose each set* E_j *is a measurable set of finite measure. Then*

$$\int s(x)dm(x) = \sum_{j=1}^{n} a_j m(E_j).$$

Proof We set $A_a = \{x; s(x) = a\} = \bigcup_{a_j = a} E_j$. Hence $a \cdot m(A_a) = \sum_{a_j = a} a_j m\left(E_j\right)$, by the additivity of m and so

$$\int s(x)dx = \sum a\, m(A_a) = \sum a_j m(E_j).$$

∎

We consider the case for a simple function $s = \sum_{j=1}^{k} a_j \chi_{E_j}$ which is not always $E_i \cap E_j = \phi$ $(i \neq j)$.

Definition 8.13 Let $f(x) \geqslant 0$, and let us take a sequence $\{s_j\}$ of simple functions such as

$$0 \leqslant s_1(x) \leqslant s_2(x) \leqslant \cdots \to f(x). \tag{8.20}$$

Then we define

$$\int_{\mathbb{R}} f(x)dm(x) = \lim_{p \to \infty} \int_{\mathbb{R}} s_p(x)dm(x). \tag{8.21}$$

For $E \in \mathcal{M}(\mathbb{R})$ we write

$$\int_E f dm = \int_{\mathbb{R}} f \chi_E dm.$$

Let $s = \sum_{j=1}^{k} a_j \chi_{E_j}$, and we define

$$\int s dm = \sum_{j=1}^{k} a_j m(E_j). \tag{8.22}$$

Then we note that s may have another representation, furthermore (8.20) may have another sequence. So we need to show the following lemma.

Lemma 8.10 *We have that*

(i) *the integral in (8.22) is independent from the representation of $s(x)$,*
(ii) *the right* lim *of (8.21) is independent from the choice of $\{s_p(x)\}$.*

Proof [Proof of (i)] Let

$$s(x) = \sum_{j=1}^{k} a_j \chi_{E_j}(x) = \sum_{p=1}^{q} b_p \chi_{F_p}(x), \quad a_j, b_p > 0.$$

Then we will show

$$\int s dm = \sum_{j=1}^{k} a_j m\left(E_j\right) = \sum_{p=1}^{q} b_p m\left(F_p\right). \tag{8.23}$$

We note $\cup E_j = \cup F_p$. Now, we consider the class Δ:

$$\Delta = \{D; D = A_1 \cap \cdots \cap A_k \cap B_1 \cap \cdots \cap B_q\},$$

where

$$A_1 = E_1 \text{ or } E_1^c, \cdots, A_k = E_k \text{ or } E_k^c, \quad B_1 = F_1 \text{ or } F_1^c \cdots, \quad B_q = E_q \text{ or } E_q^c.$$

We see that $D \cap D' = \phi$ if $D \neq D'$. In fact, when

$$D = A_1 \cap \cdots \cap A_k \cap B_1 \cap \cdots \cap B_q,$$
$$D' = A_1' \cap \cdots \cap A_k' \cap B_1' \cap \cdots \cap B_q',$$

there exists j or p such that $A_j \neq A_j'$ or $B_p \neq B_p'$, that is, $A_j \cap A_j' = \phi$ or $B_p \cap B_p' = \phi$. Hence we see $D \cap D' = \phi$. Next we will show $E_j = \cup\{D \in \Delta; D \subset$

$E_j\}$. $E_j \supset \cup\{D \in \Delta;\ D \subset E_j\}$ is trivial. Let $x \in E_j$. For each $l = 1, 2, ..., k$, if $x \in E_l$ then we put $A_l = E_l$, and if $x \notin E_l$, then we put $A_l = E_l^c$. Similarly, for each $p = 1, 2, ..., q$, if $x \in F_p$ then we put $B_p = F_p$, and if $x \notin F_p$, then we put $B_p = F_p^c$. Then we see $E_j \subset \cup\{D \in \Delta;\ D \subset E_j\}$, that is, we have $E_j = \cup\{D \in \Delta;\ D \subset E_j\}$.

Now, if $x \in D \in \Delta$, then we see

$$s(x) = \sum_{\{j;D\subset E_j\}} a_j = \sum_{\{p;D\subset E_p\}} b_p.$$

Thus we see

$$\begin{aligned}\sum_{j=1}^{k} a_j m\left(E_j\right) &= \sum_{j=1}^{k} a_j \left(\sum \{m(D);\ D \in \Delta,\ D \subset E_j\}\right)\\ &= \sum_{\{D\in\Delta\}} \left(\sum_{\{j;D\subset E_j\}} a_j \right) m(D)\\ &= \sum_{\{D\in\Delta\}} \left(\sum_{\{p;D\subset F_p\}} b_p \right) m(D)\\ &= \sum_{p=1}^{q} b_p m\left(F_p\right).\end{aligned}$$

Therefore we have (8.23), and so we conclude (i).
To prove (ii) in Lemma 8.10 we need some lemmas. ■

Lemma 8.11 *Let $s(x) \geqslant 0$ and $t(x) \geqslant 0$. Then we have*

$$(i) \quad \int (s+t)dm = \int sdm + \int tdm,$$

and if $s \leqslant t$, then

$$(ii) \quad \int sdm \leqslant \int tdm.$$

Proof (i) Let $s(x) = \sum_i a_i \chi_{E_i}(x)$, $t(x) = \sum_j b_j \chi_{F_j}(x)$. Here, we assume

$$a_i,\ b_j \geqslant 0,\quad E_i \cap E_{i'} = \phi,\quad F_j \cap F_{j'} = \phi,\ \text{and}\ \cup E_i = \cup F_j = \mathbb{R}.$$

Then we see

$$(E_i \cap F_j) \cap (E_{i'} \cap F_{j'}) = \phi \quad ((i, j) \neq (i', j'))$$

and

$$E_i = \bigcup_j E_i \cap F_j, \quad F_j = \bigcup_i E_i \cap F_j.$$

Thus we see

$$s(x) + t(x) = \sum_{i,j} (a_i + b_j)\chi_{E_i \cap F_j}(x),$$

so we have

$$\begin{aligned}
\int (s+t)dm &= \sum_{i,j} (a_i + b_j) m(E_i \cap F_j) \\
&= \sum_{i,j} a_i m(E_i \cap F_j) + \sum_{i,j} b_j m(E_i \cap F_j) \\
&= \sum_i a_i m(E_i) + \sum_j b_j m(F_j) \\
&= \int sdm + \int tdm.
\end{aligned}$$

(ii) If $s \leqslant t$, then $u = t - s \geqslant 0$. Thus, by (i) we see

$$\int tdm = \int sdm + \int udm \geqslant \int sdm.$$

■

Lemma 8.12 *Let* $\{s_p\}$ *and* $\{t_q\}$ *be sequences of positive and monotone increasing simple functions. If*

$$\lim_{p\to\infty} t_q(x) \leqslant \lim_{p\to\infty} s_p(x) \ \ (x \in \mathbb{R}),$$

then we have

$$\lim_{q\to\infty} \int t_q dm \leqslant \lim_{p\to\infty} \int s_p dm.$$

Proof For a fixed q we write

$$t_q(x) = \sum_{j=1}^{k} b_j \chi_{F_j}(x), \tag{8.24}$$

where $b_j > 0$ and $F_i \cap F_j = \phi$. Let $\int t_q dm < \infty$. Then we see $m(\cup F_j) < \infty$. We put $F = \cup F_j$, and for a fixed $\varepsilon > 0$, we set

$$E_p = \{x \in F; s_p(x) + \varepsilon > t_q(x)\}.$$

Then we see $E_p \nearrow F (p \to \infty)$. Thus, noting $m(F) < \infty$, we have $m(F - E_p) \to 0\ (p \to \infty)$ (see Theorem 8.6 (iii)). Hence we have

$$\begin{aligned}\int t_q dm &= \int t_q \chi_{F-E_p} + \int t_q \chi_{E_p} dm \\ &\leqslant \sup_x t_q(x) \cdot m(F - E_p) + \int [s_p(x) + \varepsilon] \chi_{E_p}(x) dm \\ &\leqslant \varepsilon \cdot \sup_x t_q(x) + \varepsilon \cdot m(E_p) + \int s_p dm.\end{aligned}$$

Thus, when $p \to \infty$, we see

$$\int t_q dm \leqslant \varepsilon \left[\sup_x t_q(x) + m(F)\right] + \lim_{p\to\infty} \int s_p dm.$$

Let $\varepsilon \to 0$, and let $q \to \infty$. Then we have

$$\lim_{q\to\infty} \int t_q dm \leqslant \lim_{p\to\infty} \int s_p dm.$$

If there exists q such that $\int t_q dm = \infty$, then for any $N > 0$ we can find $a > 0$ and a set F such that

$$t_q(x) > a\chi_F(x)\ (x \in \mathbb{R}) \text{ and } \infty > \int a\chi_F dm > N. \tag{8.25}$$

In fact, for the representation (8.24) we can find a certain j such that $b_j m(F_j) = \infty$. Thus we see $m(F_j) = \infty$. If we set $F_j^n = \{|x| < n\} \cap F_j$, then we see $F_j^n \nearrow F_j\ (n \to \infty)$. Thus $m(F_j^n) \to \infty\ (n \to \infty)$, and then $b_j m(F_j^n) > N$ for n large enough.

Now we take $a = b_j$ and $F = F_j^n\ (\subset F_j)$. Then by Lemma 8.12

$$t_q(x) > a\chi_F(x)\ (x \in \mathbb{R}) \quad \text{and} \quad N < a \cdot m(F) = \int b_j \chi_{F_j^n} dm < \infty.$$

Thus we have (8.25). Now, using the method in the case of $\int t_q dm < \infty$, we see

$$\lim_{p\to\infty} \int s_p dm \geqslant \int a\chi_F dm > N,$$

that is, we have

$$\lim_{q\to\infty}\int t_q dm = \infty = \lim_{p\to\infty}\int s_p dm.$$

■

Now we show the proof of (ii) in Lemma 8.10.

[Proof of (ii) in Lemma 8.10]. Let simple functions $\{s_p\}$, $\{t_q\}$ be non-negative and monotonic, furthermore

$$\lim s_p = \lim t_q = f$$

holds. By Lemma 8.12 we have

$$\lim \int s_p dm \leqslant \lim \int t_q dm \text{ and } \lim \int s_p dm \geqslant \lim \int t_q dm.$$

Thus we conclude

$$\lim \int s_p dm = \lim \int t_q dm.$$

■

8.2.3 Fundamental Properties of the Lebesgue Integral

Definition 8.14 For a real-valued measurable function $f(x)$ on a set E we put

$$f_+(x) := \max(f(x), 0), \quad f_-(x) := f_+(x) - f(x).$$

When

$$\int f_+ dm < \infty \text{ and } \int f_- dm < \infty,$$

we say that f is Lebesgue integrable, and then we define the value of the integral by

$$\int f dm = \int f_+ dm - \int f_- dm.$$

Then we write $f \in \mathcal{L}(E, dm)$, $f \in \mathcal{L}(E)$ or $f \in \mathcal{L}$, simply.

Lemma 8.13 *Let f and g be measurable functions. If $0 \leqslant f \leqslant g$, and $g \in \mathcal{L}$, then we have $f \in \mathcal{L}$, and*

$$0 \leqslant \int f dm \leqslant \int g dm.$$

Proof It follows that from Theorem 8.11 there exist simple functions $\{s_p\}$ and $\{t_q\}$ such that

$$0 \leqslant s_1(x) \leqslant s_2(x) \leqslant \cdots \to g(x),$$
$$0 \leqslant t_1(x) \leqslant t_2(x) \leqslant \cdots \to f(x).$$

Then, $f \leqslant g$ means

$$\lim_{q\to\infty} t_q(x) \leqslant \lim_{p\to\infty} s_p(x) \ \ (x \in \mathbb{R}).$$

Thus from Lemma 8.12 we see

$$\lim_{q\to\infty} \int t_q dm = \int f dm \leqslant \lim_{p\to\infty} \int s_p dm = \int g dm.$$

■

Theorem 8.12 *Let f be measurable. Then*

$$f \in \mathcal{L} \Longleftrightarrow |f| \in \mathcal{L},$$

and

$$\left| \int f dm \right| \leqslant \int |f| dm. \tag{8.26}$$

Proof Let $f = f_+ - f_-$. Then we see $|f| = f_+ + f_-$. If $f \in \mathcal{L}$, then we have $f_+, \ f_- \in \mathcal{L}$. Thus

$$|f| = f_+ + f_- \in \mathcal{L}.$$

Conversely, if $|f| \in \mathcal{L}$, then we have $f_+, \ f_- \in \mathcal{L}$. Thus we see

$$f = f_+ - f_- \in \mathcal{L}.$$

We will show (8.26). Let

$$0 \leqslant s_1(x) \leqslant s_2(x) \leqslant \cdots \to f_+(x),$$
$$0 \leqslant t_1(x) \leqslant t_2(x) \leqslant \cdots \to f_-(x),$$

and $s_j = \sum\limits_{j=1}^{k} a_j \chi_{E_j}, \ t_p = \sum\limits_{p=1}^{q} \chi_{F_p}$. Here we set

$$a_j, \ b_p > 0 \text{ and } E_{j_1} \cap E_{j_2} \cap F_{p_1} \cap F_{p_2} = \phi \ \ (J_1 \neq j_2 \text{ or } p_1 \neq p_2).$$

Then we have

$$\left|\int fdm\right| = \left|\int (f_+ - f_-)dm\right| = \lim_{j,p\to\infty}\left|\sum_j a_j m(E_j) - \sum_p b_p m(F_p)\right|$$
$$\leqslant \lim_{j,p\to\infty}\left(\sum_j a_j m\left(E_j\right) + \sum_p b_p m\left(F_p\right)\right)$$
$$= \int (f_+ + f_-) = \int |f|dm.$$

Hence we have (8.26). ■

Lemma 8.14 *(i) If $f \in \mathcal{L}(\mathbb{R})$, then for any measurable set E we have $f \in \mathcal{L}(E)$. (ii) If $E \cap F = \phi$, then we have*

$$\int_{E\cup F} fdm = \int_E fdm + \int_F fdm.$$

Proof (i) is trivial, and (ii) follows from the definition of integral. ■

Theorem 8.13 *If $f, g \in \mathcal{L}$ and $a, b \in \mathbb{R}$, then*

$$\int (af + bg)dm = a\int fdm + b\int gdm. \tag{8.27}$$

Proof For (8.27) we may show

$$(1)\ \int (af)dm = a\int fdm,\quad (2)\ \int (f+g)dm = \int fdm + \int gdm.$$

(1) Let $f = f_+ - f_-$, and let $a \geqslant 0$. Then we have

$$\int (af_+)dm = a\int f_+dm,\quad \int (af_-)dm = a\int f_-dm.$$

Thus we have

$$\int (af)dm = \int (af_+)dm - \int (af_-)dm = a\int (f_+ - f_-)dm = a\int fdm.$$

We consider the case of $a < 0$. First we see

$$\int (-f)dm = \int (f_- - f_+)dm = \int f_-dm - \int f_+dm = -\int fdm. \tag{8.28}$$

Thus, by (8.28)

$$\int (-af)dm = -a\int fdm,$$

we conclude

$$\int (af)dm = -\int (-af)dm = a\int fdm.$$

(2) We consider sets $E_j,\ j = 1, 2, \cdots, 6$ as follows:

$$\begin{aligned}
E_1 &= \{x;\ f \geqslant 0 \text{ and } g \geqslant 0\},\\
E_2 &= \{x;\ f < 0 \text{ and } g < 0\},\\
E_3 &= \{x;\ f \geqslant 0,\ g < 0 \ \text{ and } f + g \geqslant 0\},\\
E_4 &= \{x;\ f < 0,\ g \geqslant 0 \ \text{ and } f + g \geqslant 0\},\\
E_5 &= \{x;\ f \geqslant 0,\ g < 0 \ \text{ and } f + g < 0\},\\
E_6 &= \{x;\ f < 0,\ g \geqslant 0 \ \text{ and } f + g < 0\}.
\end{aligned}$$

Here we see

$$\bigcup_{j=1}^{6} E_j = \mathbb{R},\quad E_i \cap E_j = \phi\ (i \neq j).$$

Thus, by Lemma 8.14 we have

$$\int (f+g)dm = \sum_{j=1}^{6}\int_{E_j}(f+g)dm.$$

Now we may show

$$\int_{E_j}(f+g)dm = \int_{E_j} fdm + \int_{E_j} gdm \quad (j = 1, 2, \cdots, 6). \tag{8.29}$$

For $j = 1$ we have (8.29) by Lemma 8.12. For $j = 2$, we apply the case of $j = 1$ to $-(f+g)$, then we have (8.29). Let $j = 3$. Since $f + g \geqslant 0,\ \ -g \geqslant 0$ on E_3, we have

$$\begin{aligned}
\int_{E_3} fdm &= \int_{E_3} [(f+g) + (-g)]dm = \int_{E_3}(f+g)dm + \int_{E_3}(-g)dm\\
&= \int_{E_3}(f+g)dm - \int_{E_3} gdm.
\end{aligned}$$

Thus, (8.29) holds. Similarly, for the cases of $j = 4, 5, 6$ we have (8.29). ■

Theorem 8.14 *Let $f \in \mathcal{L}(\mathbb{R})$. For any $\varepsilon > 0$ there exist $a_1, \cdots, a_k$ and intervals $I_1, \cdots, I_k$ such that*

$$\int \left| f - \sum_{j=1}^{k} a_j \chi_{I_j} \right| dm < \varepsilon.$$

Proof We may suppose $f \geqslant 0$. Then there exists a simple function $s = \sum_{j=1}^{k} a_j \chi_{E_j}$, $E_i \cap E_j = \phi (i \neq j)$ such that

$$\int |f - s| \, dm < \varepsilon/2.$$

From Corollary 8.2 for each E_j there exist intervals $I_1^j, I_2^j, \cdots, I_{N_j}^j$ such that

$$m\left(E_j - \bigcup_{n=1}^{N_j} I_n^j\right) + m\left(\bigcup_{n=1}^{N_j} I_n^j - E_j\right) < \varepsilon' \quad (j = 1, 2, \cdots, k).$$

Then we take $\varepsilon' > 0$ satisfying

$$(|a_1| + \cdots + |a_k|)\varepsilon' < \varepsilon/2.$$

Then

$$\begin{aligned}
&\int \left| f - \sum_{j=1}^{k} a_j \sum_{n=1}^{N_j} \chi_{I_n^j} \right| dm \\
&\leqslant \int \left| f - \sum_{j=1}^{k} a_j \chi_{E_j} \right| dm + \int \left| \sum_{j=1}^{k} a_j \chi_{E_j} - \sum_{j=1}^{k} a_j \sum_{n=1}^{N_j} \chi_{I_n^j} \right| dm \\
&\leqslant \varepsilon/2 + \int \sum_{j=1}^{k} |a_j| \left[m\left(E_j - \bigcup_{n=1}^{N_j} I_n^j\right) + m\left(\bigcup_{n=1}^{N_j} I_n^j - E_j\right)\right] \\
&\leqslant \varepsilon/2 + \varepsilon' \sum_{j=1}^{k} |a_j| < \varepsilon/2 + \varepsilon/2 = \varepsilon.
\end{aligned}$$

■

Theorem 8.15 *Let $f \in \mathcal{L}$. Then for any $\lambda > 0$ we have*

$$m(\{x; |f(x)| > \lambda\}) \leqslant \frac{1}{\lambda} \int |f| dm. \quad \textit{(the Chebyshev's inequality)} \tag{8.30}$$

Proof Let $E_\lambda = \{x; |f(x)| > \lambda\}$. Then

$$\int |f|dm \geqslant \int |f|\chi_{E_\lambda}dm \geqslant \int \lambda\chi_{E_\lambda}dm = \lambda m(E_\lambda).$$

Thus, we have (8.30). ■

Theorem 8.16 *Let $f \in \mathcal{L}$. If $|f(x)| < \infty$ a.e., then f takes finite values at almost all points of x.*

Proof Let $N = \{x; |f(x)| = \infty\}$. Since $N \subset \{x; |f(x)| > \lambda\}$, $\lambda > 0$, by the Chebyshev's inequality we see

$$m(N) \leqslant m(\{x; |f(x)| > \lambda\}) \leqslant \frac{1}{\lambda}\int |f|dm.$$

As $\lambda \to \infty$, we have $m(N) = 0$. ■

Theorem 8.17 *Let $f \in \mathcal{L}$. If $f = g$ a.e., then $g \in \mathcal{L}$, and*

$$\int fdm = \int gdm.$$

Proof By Lemma 8.8, g is measurable. Let $N = \{x; f(x) \neq g(x)\}$, and let $u(x) = 0\ (x \notin N),\ = \infty\ (x \in N)$. Since N is a null set, we see $\int udm = 0$ (consider simple functions $u(x) = 0\ (x \notin N),\ = M\ (x \in N)$, and let $M \to \infty$). Thus we have

$$\left|\int fdm - \int gdm\right| = \left|\int (f-g)dm\right| \leqslant \int |f-g|dm \leqslant \int udm = 0.$$

■

8.2.4 The Convergence of Sequences of Functions

Definition 8.15 Let $\{f_n\}$ be a sequence of functions defined on a set E.

(*i*) If there exists a function $f(x)$ such that for all $x \in E$

$$f_n(x) \to f(x)\ \ (n \to \infty),$$

then we say that $\{f_n\}$ is pointwise convergence on E.

(*ii*) If there exists a null set N such that $\{f_n\}$ converges pointwise on $E - N$, then we say that the sequence $\{f_n\}$ converges to a function f almost all in E, and then we write

$$f_n \to f \text{ a.e. in } E.$$

(iii) For any $\varepsilon > 0$ there exists a number $N > 0$ such that

$$|f_n(x) - f(x)| < \varepsilon \ (x \in E, \ n \geqslant N),$$

then we say that $\{f_n\}$ converges uniformly on E.

(iv) For any $\varepsilon > 0$

$$m(\{x \in E; |f_n(x) - f(x)| > \varepsilon\}) \to 0 \ (n \to \infty),$$

then we say that the sequence $\{f_n\}$ converges to a function f in measure, and we have

$$f_n \to f \text{ in measure} \quad \text{on } E.$$

(v) When

$$\int_E |g| dm < \infty,$$

we say that g is L_1-integrable, and then we write $g \in L^1(E)$. Let $f_n, f \in L^1(E)$. If

$$\int_E |f_n - f| dm \to 0 \ (n \to \infty),$$

then we say that $\{f_n\}$ converges to f in L^1-norm, and we write

$$\|f_n - f\|_1 \to 0 \ (n \to \infty) \ \text{ or } \ f_n \to f \text{ in } L^1.$$

Theorem 8.18 *Let $\{f_n\}$ be a sequence of functions on a set E which converges to a function f in L^1-norm. Then $\{f_n\}$ converges to f in measure.*

Proof Let $\varepsilon > 0$. By Chebyshev's inequality we see

$$m\left(\{x \in E; |f_n(x) - f(x)| > \varepsilon\}\right) \leqslant \frac{1}{\varepsilon} \int_E |f_n(x) - f(x)| dm \to 0 \ (n \to \infty)$$

■

Definition 8.16 Let $f, f_n \in \mathcal{L}(E), \ n = 1, 2, \cdots$, where $E (\subseteq \mathbb{R})$ is a measurable set, and let $0 < p < \infty$. Then if

$$\|f_n - f\|_{L^p(E)} := \left\{ \int_E |f_n - f|^p dm \right\}^{1/p} \to 0 \ (n \to \infty)$$

holds, then we say that $\{f_n\}$ converges to f in L^p-norm. Furthermore, for $f \in C(\mathbb{E})$ (the class of all continuous functions on E) we define

$$\|f\|_\infty := \sup_{x\in E} |f(x)|.$$

If for $f,\ f_n \in C(\mathbb{R}),\ n = 1, 2, \cdots$ we have

$$\|f - f_n\|_\infty \to 0 \ (n \to \infty), \tag{8.31}$$

then we say that $\{f_n\}$ converges to f uniformly (or in sup-norm). The formula (8.31) is equivalent to

$$\forall \varepsilon > 0 \ \exists N; \ \|f_j - f\|_\infty < \varepsilon \ (j \geqslant N). \tag{8.32}$$

Theorem 8.19 *If* $\{f_n\}$ $(f_n \in C(E))$ *converges to* f *uniformly, then* $f \in C(E)$.

Proof For $\varepsilon > 0$ we suppose that for N (8.32) holds. Since f_N is continuous, for any x there exists $\delta > 0$ such that

$$|f_N(x) - f_N(y)| < \varepsilon \ (y \in E,\ \ |x - y| < \delta). \tag{8.33}$$

Thus, from (8.32) and (8.33) we have

$$\begin{aligned} |f(x) - f(y)| &\leqslant |f(x) - f_N(x)| + |f_N(x) - f_N(y)| + |f_N(y) - f(y)| \\ &< \varepsilon + \varepsilon + \varepsilon = 3\varepsilon \quad (y \in E,\ |x - y| < \delta). \end{aligned}$$

Thus f is continuous at x. ■

Theorem 8.20 (Dini) *Let* $K \subset \mathbb{R}$ *be a compact set. If*

(*i*) $f_i, f \in C(K)$,
(*ii*) $f_1(x) \leqslant f_2(x) \leqslant f_3(x) \leqslant \cdots \to f(x)$,

then $\{f_j\}$ *converges to* f *uniformly.*

Proof Let $g_j(x) = f(x) - f_j(x)$. Then we may show $g_j \to 0$ uniformly. Since $\{g_j\}$ is monotonic decreasing, and converges to 0, for a given $\varepsilon > 0$ and any $x \in K$ there exists

$$0 \leqslant g_{j_x}(x) < \varepsilon/2.$$

Since $g_{j_x}(y)$ is a continuous function with respect to y, there exists $\delta_x > 0$ such that

$$|g_{j_x}(y) - g_{j_x}(x)| < \varepsilon/2, \quad y \in K \cap B(x, \delta_x).$$

Thus, if $y \in K \cap B(x, \delta_x)$, then

$$0 \leqslant g_{j_x}(y) \leqslant |g_{j_x}(y) - g_{j_x}(x)| + g_{j_x}(x) < \varepsilon. \tag{8.34}$$

Since $\{B(x, \delta_x); x \in K\}$ is an open covering of K, noting the compactness of K, there exist $B(x_n, \delta_{x_n}), \ n = 1, 2, \cdots, N$

$$K \subset \bigcup_{n=1}^{N} B(x_n, \delta_{x_n}).$$

Now let $j_{x_1}, j_{x_2}, \cdots, J_{x_N} \leqslant J$. When $y \in K$, we can find an open set $B(x_n, \delta_{x_n})$ such that $y \in B(x_n, \delta_{x_n})$. Since $\{g_{j_x}\}$ is a monotonic decreasing sequence, by (8.34) we have

$$0 \leqslant g_J(y) \leqslant g_{j_{xn}}(y) < \varepsilon.$$

Thus, for all $y \in K$ we have $0 \leqslant g_j(y) < \varepsilon$ $(j \geqslant J)$, that is, $g_j \to 0$ uniformly. ■

Theorem 8.21 (Egorov) *Let E be a measurable set with $m(E) < \infty$, and let $\{f_n\}$ be a sequence of measurable functions which converges to f, $|f(x)| < \infty$ almost all $x \in E$. Then for $\varepsilon > 0$ there exists a set E_0 such that*

$$(i) \ E_0 \subset E, \ \ m(E - E_0) < \varepsilon,$$
$$(ii) \ f_n \to f \ \ on \ \ E_0, \ \ uniformly.$$

Proof For $p, q = 1, 2, \cdots$ we set

$$E_p(q) = \{x \in E; |f(x) - f_j(x)| < 1/q, \ \ \forall j \geqslant p\}.$$

Then we see

$$E_p(q) = \bigcap_{j=p}^{\infty} \left\{x \in E; \left|f(x) - f_j(x)\right| < 1/q\right\}.$$

Thus, $E_p(q) \in \mathcal{M}(\mathbb{R})$, and we see $E_1(q) \subset E_2(q) \subset \cdots$. Since for $x \in E$,

$$f_j(x) \to f(x) \text{ (as } j \to \infty),$$

we have $\lim_{p\to\infty} E_p(q) = E$. Now by Theorem 8.6 (ii) we have

$$m\left(E - E_p(q)\right) \to 0 \ \ (p \to \infty). \tag{8.35}$$

Let us take $\varepsilon > 0$. From (8.35) for each q we take p_q such that

$$m\left(E - E_{p_q}(q)\right) < \varepsilon/2^q.$$

We set $E_0 = \bigcap_{q=1}^{\infty} E_{p_q}(q)$. Then

$$m(E - E_0) = m\left(\bigcup_{q=1}^{\infty} \{E - E_{p_q}(q)\}\right) \leqslant \sum_{q=1}^{\infty} m\left(E - E_{p_q}(q)\right) < \sum_{q=1}^{\infty} \frac{\varepsilon}{2^q} = \varepsilon.$$

Since for all $q > 1$, $E_0 \subset E_{p_q}(q)$, we see

$$|f(x) - f_j(x)| \leqslant 1/q \quad (x \in E_0,\ j \geqslant p_q).$$

When $q \to \infty$, we have (ii). ∎

Theorem 8.22 (Lusin) *Let E be a measurable set with $m(E) < \infty$, and let f be a measurable function with $|f| < \infty$ a.e. on E. Then for any $\varepsilon > 0$ there exists a compact set K_0 such that*

$$(i)\ K_0 \subset E,\ m(E - K_0) < \varepsilon,$$
$$(ii)\ f \text{ is continuous on } K_0.$$

Proof We may suppose $f \geqslant 0$. First, we suppose that f is a simple function, and we write

$$f(x) = \sum_{j=1}^{k} a_j \chi_{E_j},\quad E_i \cap E_j = \phi\ (i \neq j).$$

For $\varepsilon > 0$ we take some compact sets C_j as

$$C_j \subset E_j,\ m(E_j - C_j) < \varepsilon/k \quad (j = 1, 2, \cdots, k).$$

Since f is constant on each C_j, it is continuous on C_j. We may put $C = \sum_{j=1}^{k} C_j$. Let us consider the case of a general function. Then we set a sequence $\{s_n\}$ which is a non-negative monotonic simple function as

$$s_n \to f\ (n \to \infty).$$

As we see the above, for each $n = 1, 2, \cdots$ there exists a compact set $K_n \subset E$ such that

$$m(E - K_n) < \varepsilon/2^{n+1} \text{ and } s_n \text{ is continuous on } K_n.$$

Now we put

$$K = \bigcap_{n=1}^{\infty} K_n.$$

Then K is a compact set, and

$$m(E-K)=m\left(\bigcup_{n=1}^{\infty}(E-K_n)\right)\leqslant\sum_{n=1}^{\infty}m\left(E-K_n\right)<\varepsilon/2.$$

By Egorov's Theorem there exists a compact set K_0 such that

$$K_0\subset K,\ m(K-K_0)<\varepsilon/2,\ \text{ and }\ s_n\to f\text{ on }K_0,\ \text{uniformly}.$$

Here, since s_n is continuous on K_n, and is continuous on K_0, f is also continuous. Thus, by Theorem 8.19 we see that $\lim s_n=f$ is continuous on K_0. Since

$$m(E-K_0)=m(E-K)+m(K-K_0)<\varepsilon,$$

we see that K_0 satisfies the conditions (i) and (ii). ■

Theorem 8.23 *Let $m(E)<\infty$, and let a sequence $\{f_j\}$ converge a.e. on E to a function f with finite values. Then*

$$f_j\to f\ \textit{ in measure.}$$

Proof Let us give $\varepsilon>0$. By Theorem 8.21 (Egorov) there exist a set $E_0\subset E$ and N such that

$$m(E-E_0)<\varepsilon,\ \ |f_j(x)-f(x)|<\varepsilon\ \ (x\in E_0,\ \ j>N).$$

Thus

$$\left\{x\in E;\left|f_j(x)-f(x)\right|>\varepsilon\right\}\subset E-E_0\ \ (j>N).$$

Therefore, we have

$$m\left(\left\{x\in E;\left|f_j(x)-f(x)\right|>\varepsilon\right\}\right)\leqslant m(E-E_0)<\varepsilon\ (j>N).$$

This means that $f_j\to f$ in measure. ■

Theorem 8.24 *If $\{f_j\}$ converges to f in measure on E, then there exists a subsequence $\{f_{n_k}\}$ such that*

$$f_{n_k}\to f\ \ a.e.\ (k\to\infty).$$

Proof We see that for any k there exists n_k such that

$$m\left(\left\{x\in E;\left|f_j(x)-f(x)\right|>2^{-k}\right\}\right)<2^{-k}\ \ (j\geqslant n_k).$$

Now we put

$$N_k = \left\{x \in E; \left|f_{n_k}(x) - f(x)\right| > 2^{-k}\right\}, \quad N = \limsup_{k\to\infty} N_k = \bigcap_{l=1}^{\infty}\bigcup_{k=l}^{\infty} N_k.$$

Then we have $m(N) = 0$. In fact, for any l we have

$$m(N) \leqslant m\left(\bigcup_{k=l}^{\infty} N_k\right) \leqslant \sum_{k=l}^{\infty} m(N_k) \leqslant \sum_{k=l}^{\infty} 2^{-k} = 2^{-l+1},$$

so we see that N is a null set.

If $x \notin N$, we have $x \in \bigcup_{l=1}^{\infty}\bigcap_{k=l}^{\infty} N_k^c$. Thus there exists l such that $x \in \bigcap_{k=l}^{\infty} N_k^c$, that is, $x \notin N_k$ $(k \geqslant l)$. Hence, when $k \geqslant l$, we have $|f_{n_k}(x) - f(x)| \leqslant 2^{-k}$. Consequently,

$$f_{n_k}(x) \to f(x) \ \ (x \notin N,\ k \to \infty).$$

■

Theorem 8.25 (B. Levi) *Let* $\{u_j\}_{j=1}^{\infty}$, $u_j \geqslant 0$ *be a sequence of measurable functions. Then we have*

$$\int \left(\sum_{j=1}^{\infty} u_j\right) dm = \sum_{j=1}^{\infty} \int u_j dm. \tag{8.36}$$

Proof Put $f = \sum_{j=1}^{\infty} u_j$. Since $f \geqslant \sum_{j=1}^{n} u_j$, we see

$$\int f dm \geqslant \int \left(\sum_{j=1}^{n} u_j\right) dm = \sum_{j=1}^{n} \int u_j dm.$$

Thus

$$\int f dm \geqslant \sum_{j=1}^{\infty} \int u_j dm. \tag{8.37}$$

We will show the inverse inequality. Let us define a sequence of simple functions for each j and x

$$0 \leqslant s_j^1(x) \leqslant s_j^2(x) \leqslant \cdots \to u_j(x),$$

and let

$$S^p(x) = \sum_{j=1}^{p} s_j^p(x).$$

Then we see

$$S^p(x) = \sum_{j=1}^{p} s_j^p(x) \leqslant \sum_{j=1}^{p} s_j^{p+1}(x) \leqslant \sum_{j=1}^{p+1} s_j^{p+1}(x) = S^{p+1}(x) \leqslant f(x). \tag{8.38}$$

Thus, $\{S^p(x)\}$ is a sequence of monotonic simple functions, and for a fixed p we have

$$\sum_{j=1}^{p} u_j(x) = \sum_{j=1}^{p} \lim_{q\to\infty} s_j^q(x) \leqslant \lim_{q\to\infty} \sum_{j=1}^{p} s_j^q(x) \leqslant \lim_{q\to\infty} \sum_{j=1}^{q} s_j^q(x) = \lim_{q\to\infty} S^q(x).$$

When $p \to \infty$, we have

$$f(x) = \sum_{j=1}^{\infty} u_j(x) \leqslant \lim_{q\to\infty} S^q(x).$$

With (8.38) we have $f(x) = \lim_{q\to\infty} S^q(x)$. Thus, from the definition of the integral we have

$$\int f dm = \lim_{q\to\infty} \int S^q(x) dm.$$

On the other hand,

$$\begin{aligned} \lim_{q\to\infty} \int S^q(x) dm &= \lim_{q\to\infty} \sum_{j=1}^{q} \int s_j^q(x) dm \\ &\leqslant \lim_{q\to\infty} \sum_{j=1}^{q} \int u_j(x) dm = \sum_{j=1}^{\infty} \int u_j(x) dm. \end{aligned}$$

Hence, we have

$$\int f dm \leqslant \sum_{j=1}^{\infty} \int u_j(x) dm.$$

From (8.37) we conclude

$$\int f dm = \sum_{j=1}^{\infty} \int u_j(x) dm,$$

that is, we have (8.36) ■

Theorem 8.26 *Let* $0 \leqslant f_1(x) \leqslant f_2(x) \leqslant \cdots$. *Then we have*

$$\int \left(\lim_{j\to\infty} f_j \right) dm = \lim_{j\to\infty} \int f_j dm.$$

Proof We put $u_1 = f_1,\ u_j = f_{j+1} - f_j\ (j \geqslant 2)$. Then we see $\sum_{j=1}^{\infty} u_j = \lim_{k\to\infty} f_k$. Thus, by Theorem 8.25 we have the result. ■

Theorem 8.27 *Let* $f \in L^1(E)$, *and let* $E = \cup E_j,\ E_i \cap E_j = \phi\ (i \neq j)$. *Then we have*

$$\int_E f dm = \sum_{j=1}^{\infty} \int_{E_j} f dm. \tag{8.39}$$

Proof We suppose $f \geqslant 0$. By Theorem 8.25 (B. Levi), if we put $u_j(x) = f(x)\chi_{E_j}(x)$, then we have (8.39). ■

Theorem 8.28 (Fatou's lemma) *If* $f_j \geqslant 0,\ j = 1, 2, \cdots$, *then*

$$\int \left(\liminf_{j\to\infty} f_j \right) dm \leqslant \liminf_{j\to\infty} \int f_j dm.$$

Proof Let $g_k = \inf_{j \geqslant k} f_j$. Then we see

$$0 \leqslant g_1(x) \leqslant g_2(x) \leqslant \cdots \to \lim_{k\to\infty} g_k(x) = \liminf_{j\to\infty} f_j(x).$$

Thus, by Theorem 8.26 we have

$$\lim_{k\to\infty} \int g_k dm = \int \left(\lim_{k\to\infty} g_k \right) dm = \int \left(\liminf_{j\to\infty} f_j(x) \right) dm. \tag{8.40}$$

Since $g_k \leqslant f_j\ (j \geqslant k)$, we see $\int g_k dm \leqslant \inf_{j \geqslant k} \int f_j dm$, and we have

$$\lim_{k\to\infty} \int g_k dm \leqslant \liminf_{j\to\infty} \int f_j(x) dm. \tag{8.41}$$

From (8.40) and (8.41) we have the result. ■

Theorem 8.29 (The Lebesgue Convergence Theorem) *Let* $\{f_j\}$ *be a sequence satisfying the following conditions:*

(i) f_j *is convergent a.e.,*
(ii) there exists $g(x) \in L^1$ *such that* $|f_j(x)| \leqslant g(x)$ *a.e.* $(j = 1, 2, \cdots)$.

Then we have

$$\lim_{j\to\infty} \int f_j(x)dm = \int \lim_{j\to\infty} f_j(x)dm. \tag{8.42}$$

Proof We may assume $f \geqslant 0$. Since $g(x) + f_j(x) \geqslant 0$, we have by Theorem 8.28 (Fatou)

$$\begin{aligned}\int gdm + \int \lim f_j dm &= \int (g + \lim f_j)dm = \int \liminf (g + f_j)dm \\ &\leqslant \liminf \int (g + f_j)dm = \int gdm + \liminf \int f_j dm.\end{aligned}$$

Thus,

$$\int \lim f_j dm \leqslant \liminf \int f_j dm. \tag{8.43}$$

Similarly, from $g(x) - f_j(x) \geqslant 0$, we have

$$\begin{aligned}\int gdm - \int \lim f_j dm &= \int (g - \lim f_j)dm = \int \liminf (g - f_j)dm \\ &\leqslant \liminf \int (g - f_j)dm = \int gdm - \limsup \int f_j dm.\end{aligned}$$

Thus,

$$\limsup \int f_j dm \leqslant \int \lim f_j dm. \tag{8.44}$$

Hence, by (8.43) and (8.44) we have

$$\int \lim f_j dm \leqslant \liminf \int f_j dm \leqslant \limsup \int f_j dm \leqslant \int \lim f_j dm.$$

This means

$$\lim_{j\to\infty} \int f_j dm = \int \lim_{j\to\infty} f_j dm,$$

that is, we have (8.42). ■

Corollary 8.3 (Bounded Convergence Theorem) *Let $m(E) < \infty$, and let $\{f_j\}$ be a sequence satisfying the following conditions:*

(i) f_j is convergent a.e. on E,
(ii) there exists $M < \infty$ such that $|f_j(x)| \leqslant M$ a.e. $(j = 1, 2, \cdots)$.

Then we have

$$\lim_{j\to\infty} \int f_j(x)dm = \int \lim_{j\to\infty} f_j(x)dm. \tag{8.45}$$

Proof We apply Theorem 8.29 for $M\chi_E(x) \in L^1$. Then we have (8.45). ■

8.2.5 The Riemann Integral and the Lebesgue Integral

In Chap. 6 we investigate the Riemann integral. Here we consider the relation of the Riemann integral and the Lebesgue integral. For simplicity, we suppose that the domain is $[a, b]$, and the function f is bounded. We do not always suppose the continuity or the measurability.

Theorem 8.30 *Let $f(x)$ be a function bounded on a finite interval I. $f(x)$ is Riemann-integrable if and only if the set of discontinuous points of $f(x)$ is a null set. Then the Riemann-integral is equal to the Lebesgue integral.*

We recall the definition of the Riemann-integral. For an interval $[a, b]$ we consider a partition :

$$\Delta; a = a_0 < a_1 < \cdots < a_k = b \tag{8.46}$$

and we put

$$I_j = [a_{j-1}, a_j] \quad j = 1, 2, \cdots, k.$$

Now we define the upper integral and the lower integral respectively by

$$\overline{\int} f dx = \inf_{\Delta} \overline{s}(f, \Delta), \quad \underline{\int} f dx = \sup_{\Delta} \underline{s}(f, \Delta),$$

where

$$\overline{s}(f, \Delta) = \sum_{j=1}^{k} \sup_{x\in I_j} f(x)m(I_j), \quad \underline{s}(f, \Delta) = \sum_{j=1}^{k} \inf_{x\in I_j} f(x)m(I_j).$$

Here we put

$$\overline{f}(x) = \inf_{\delta>0} \sup_{|x-y|<\delta,\ y\in[a,b]} f(y), \quad \underline{f}(x) = \sup_{\delta>0} \inf_{|x-y|<\delta,\ y\in[a,b]} f(y).$$

To prove the Theorem 8.30 we need some lemmas.

Lemma 8.15 *Both the functions $\overline{f}$ and $\underline{f}$ are measurable.*

Proof We show that $\underline{f}$ is measurable (similarly, for $\overline{f}$ we can show the measurability). We may show that $\left\{x;\ \underline{f}(x) > \lambda\right\}$ is an open set. Let $\underline{f}(c) > \lambda$. Then there exists $\delta > 0$ such that

$$\inf_{|c-y|<2\delta,\ y\in[a,b]} f(y) > \lambda.$$

Thus if $|c - x| < \delta$, then we see

$$\inf_{|x-y|<\delta,\ y\in[a,b]} f(y) > \lambda,$$

that is,

$$(c - \delta, c + \delta) \subset \{x;\ \underline{f}(x) > \lambda\}.$$

Hence we have the result. ∎

We lay stress on the Lebesgue integral by $\mathcal{L} \int$.

Lemma 8.16 *Let f be a bounded function on $[a, b]$. Then we have*

$$\overline{\int} f dx = \mathcal{L} \int \overline{f} dm, \quad \underline{\int} f dx = \mathcal{L} \int \underline{f} dm.$$

Proof From the definition of the upper integral of f there exists a partition Δ^n; $a = a_0^n < a_1^n < \cdots < a_{k_n}^n = b$ such that

$$\overline{\int} f dx = \lim_{n\to\infty} \overline{s}(f, \Delta^n).$$

Now we define a simple function f^n as

$$f^n(x) = \sup_{y\in[a_{j-1}^n, a_j^n]} f(y), \quad x \in [a_{j-1}^n, a_j^n), \ j = 1, 2, \cdots, k_n.$$

Here we see f^n is measurable and

$$\overline{s}\left(f, \Delta^n\right) = \mathcal{L} \int f^n dm.$$

Thus we have

$$\overline{\int} f dx = \lim_{n\to\infty} \mathcal{L}\int f^n dm.$$

If we set $M = \sup_y |f(y)|$, then

$(i)\ |f^n(x)| \leqslant M < \infty,\ \ x \in [a, b],$
$(ii)\ x \notin N = \{a_j^n;\ j = 1, 2, \cdots, k_n,\ n = 1, 2, \cdots\} \Longrightarrow f^n(x) \to \overline{f}(x).$

Since $m(N) = 0$, applying Corollary 8.3, we have

$$\overline{\int} f dx = \mathcal{L}\int \overline{f} dm.$$

Similarly, we see

$$\underline{\int} f dx = \mathcal{L}\int \underline{f} dm.$$

■

Proof of Theorem 8.30. The necessary and sufficient conditions which f is Riemann integrable

$$\overline{\int} f dx = \underline{\int} f dx.$$

Hence by Lemma 8.16 we see

$$\mathcal{L}\int \left(\overline{f} - \underline{f}\right) dm = 0,$$

thus we have $\overline{f} = \underline{f}$ *a.e.* This means that f is continuous *a.e.*, on $[a, b]$. ■

Remark 8.3 A bounded monotonic function is Riemann-integrable.

8.2.6 Riemann-Stieltjes Integral

(a) Function of Bounded Variation

Definition 8.17 Let f be a real-valued function on ℝ. We consider a partition

$$\Delta;\ a = a_0 < a_1 < \cdots < a_k = x.$$

Then we define

$$V_f([a,x]) = \sup_{\Delta} \sum_{j=1}^{k} |f(a_j) - f(a_{j-1})|,$$

and for $V_f([a,x]) < \infty$ we say that f is a function of bounded variation. Here we permit the case of $a = -\infty$ or $x = +\infty$. Then we write $f \in \mathcal{V}([a,b])$, and for simplifying we suppose $f(x) = f(x+0)$ (see Theorem 8.31 below).

Theorem 8.31 *A function of bounded variation expresses a difference of two monotonic functions.*

Proof Let f be of bounded variation, and put

$$2P_f([a,x]) = V_f([a,x]) + f(x) - f(a), \quad 2N_f([a,x]) = V_f([a,x]) - f(x) + f(a).$$

Then we have

$$f(x) = f(a) + P_f([a,x]) - N_f([a,x]).$$

Here, we see that $P_f([a,x])$ is a monotonic increasing function. In fact, for $h > 0$

$$\begin{aligned}
2[P_f([a,x+h]) - P_f([a,x])] &= [V_f([a,x+h]) - V_f([a,x])] + [f(x+h) - f(x)] \\
&= [V_f([a,x]) + V_f([x,x+h]) - V_f([a,x])] \\
&\quad + [f(x+h) - f(x)] \\
&= V_f([x,x+h]) + [f(x+h) - f(x)] \\
&\geqslant |f(x+h) - f(x)| + [f(x+h) - f(x)] \geqslant 0.
\end{aligned}$$

Similarly, we can show that $N_f([a,x])$ is a monotonic increasing function. ∎

We say that $P_f([a,x])$ is a positive bounded variation, and $N_f([a,x])$ is a negative bounded variation.

Remark 8.4 If f is continuous, then $P_f([a,x])$ and $N_f([a,x])$ are also continuous.

For a monotonic increasing function f the values

$$\lim_{h \to +0} f(x+h) = f(x+0), \quad \text{and} \quad \lim_{h \to -0} f(x+h) = f(x-0)$$

exist, therefore for the function f of bounded variation there exist $f(x+0)$, $f(x-0)$. Furthermore, since the monotone function has at most countable discontinuous points, the function of bounded variation has also at most countable discontinuous points.

Example 8.3 Let $f \in L^1([a,b])$ and let

$$\mu_f(x) = \int_a^x f(t)dt.$$

Then we see

$$V_{\mu f}([a,b]) = \int_a^b |f(t)|dt, \tag{8.47}$$

so we have

$$P_{\mu f}([a,x]) = \frac{1}{2}\left(\int_a^x |f(t)|dt + \int_a^x f(t)dt\right) = \int_a^x f_+(t)dt.$$

Similarly we see

$$N_{\mu f}([a,x]) = \frac{1}{2}\left(\int_a^x |f(t)|dt - \int_a^x f(t)dt\right) = \int_a^x f_-(t)dt.$$

In fact, when we consider a partition

$$\Delta; a = a_0 < a_1 < \cdots < a_N = b,$$

we see

$$\sum_{j=1}^{N} \left|\mu_f(a_j) - \mu_f(a_{j-1})\right| = \sum_{j=1}^{N} \left|\int_{a_{j-1}}^{a_j} f(t)dt\right|. \tag{8.48}$$

Hence we have

$$V_{\mu f}([a,b]) \leqslant \int_a^b |f(t)|dt.$$

We show the inverse inequality. For $\varepsilon > 0$ we take a simple function $s(x) = \sum_j c_j \chi_{I_j}(x)$ as

$$\int |f - s| < \varepsilon,$$

where $\{I_j\}$ is a sequence of finite intervals with $I_j \cap I_k = \phi$ $(j \neq k)$. The right side in (8.48) is estimated as

$$\geqslant \sum_{j=1}^{N} \left| \int_{a_{j-1}}^{a_j} s(t)dt \right| - \sum_{j=1}^{N} \left| \int_{a_{j-1}}^{a_j} (f-s)dt \right|$$

$$\geqslant \int |s|dt - \int_a^b |f-s|dt.$$

Hence we have

$$V_{\mu f}([a,b]) \geqslant \sum_{j=1}^{N} |\mu_f(a_j) - \mu_f(a_{j-1})| \geqslant \int |s|dt - \int_a^b |f-s|dt$$

$$\geqslant \int_a^b |f|dt - 2\varepsilon.$$

Thus we have (8.47).

(b) **Riemann-Stieltjes Integral**

For a partition

$$\Delta; a = a_0 < a_1 < \cdots < a_k = b$$

we put

$$I_j = [a_{j-1}, a_j] \quad j = 1, 2, \cdots, k.$$

Let $I = [a, b]$ be a finite interval, and let μ be a monotonic increasing function with $\mu(x) = \mu(x+0),\ x \in [a, b)$.
We set

$$\Delta_j \mu = \mu(a_j) - \mu(a_{j-1}), \quad j = 1, 2, \cdots, k,$$

and for a real-valued function f we define

$$\overline{s}_\mu(f, \Delta) = \sum_{j=1}^{k} \overline{f_j} \Delta_j \mu, \quad \underline{s}_\mu(f, \Delta) = \sum_{j=1}^{k} \underline{f_j} \Delta_j \mu,$$

where

$$\overline{f_j} = \sup_{x \in I_j} f(x), \quad \underline{f_j} = \inf_{x \in I_j} f(x).$$

Then, we define

$$\overline{\int} f d\mu = \inf_{\Delta} \overline{s}_\mu(f, \Delta), \quad \underline{\int} f d\mu = \sup_{\Delta} \underline{s}_\mu(f, \Delta).$$

If these values are equal, then we say that f is Riemann-Stieltjes integrable with respect to μ, and we write

$$\int_I f d\mu = \overline{\int} f d\mu = \inf_{\Delta} \overline{s}_\mu(f, \Delta) = \underline{\int} f d\mu = \sup_{\Delta} \underline{s}_\mu(f, \Delta).$$

If μ is a function of bounded variation on $[a, b]$, then we define

$$\int f d\mu = \int f d\nu_1 - \int f d\nu_2, \quad \mu = \nu_1 - \nu_2,$$

$$(\nu_1 \text{ and } \nu_2 \text{ are monotonic increasing functions}).$$

We say that $\int f d\mu$ is the Riemann-Stieltjes integral with respect to μ. If $\mu(x) = x$, then $\int f d\mu$ equals a Riemann-integral.

Theorem 8.32 *Let μ be a monotonic increasing function on a finite interval $I = [a, b]$.*

(i) If f is a continuous function on I, then f is Riemann-Stieltjes integrable with respect to μ.
(ii) Let f be a function of bounded variation, and let μ be continuous. Then f is Riemann-Stieltjes integrable with respect to μ.

Proof (i) Let Δ be a partition

$$a = a_0 < a_1 < \cdots < a_k = b,$$

and let $\varepsilon > 0$. Since f is continuous on I, if Δ is small enough, then

$$|f(x) - f(y)| < \frac{\varepsilon}{\mu(b) - \mu(a) + 1}; \; x, y \in I_j = [a_{j-1} - a_j] \; (j = 1, 2, \cdots).$$

Thus,

$$\overline{s}_\mu(f, \Delta) - \underline{s}_\mu(f, \Delta) = \sum_{j=1}^{k} [\overline{f}_j - \underline{f}_j] \Delta_j \mu \leqslant \sum_{j=1}^{k} \frac{\varepsilon}{\mu(b) - \mu(a) + 1} \Delta_j \mu$$

$$= \frac{\varepsilon[\mu(b) - \mu(a)]}{\mu(b) - \mu(a) + 1} < \varepsilon.$$

Thus, we see that f is Riemann-Stieltjes integrable.
(ii) We may suppose that f is a monotonic increasing function. Let $\varepsilon > 0$, and let Δ be small enough. Then, since $\Delta_j \mu < \varepsilon / 2[f(b) - f(a)]$, we have

$$\overline{s}_\mu(f, \Delta) - \underline{s}_\mu(f, \Delta) = \sum_{j=1}^{k} [\overline{f}_j - \underline{f}_j] \Delta_j \mu$$

$$\leqslant \sum_{j=1}^{k} \frac{\varepsilon\left[f(a_j) - f(a_{j-1})\right]}{2[f(b) - f(a)]}$$

$$\leqslant \frac{2[f(b) - f(a)]\varepsilon}{2[f(b) - f(a)]} = \varepsilon.$$

■

Theorem 8.33 *Let w be an integrable function, and let $\mu(x) = \int_{-\infty}^{x} w(t)dt$. Then for a bounded Riemann-integrable function f on a finite interval, we have*

$$\int_I f(t)d\mu(t) = \int_I f(t)w(t)dt. \tag{8.49}$$

Then we say that w is a weight and the right side of (8.49) is a weighted integral.

Proof We may suppose $w \geqslant 0$. By the definition $\Delta_j \mu$

$$\overline{s}_\mu(f, \Delta) = \sum_{j=1}^{k} \overline{f}_j \Delta_j \mu = \sum_{j=1}^{k} \overline{f}_j \int_{a_{j-1}}^{a_j} w(t)dt. \tag{8.50}$$

We put

$$\overline{f}_j^{\Delta}(t) = \overline{f}_j, \quad t \in [a_{j-1}, a_j].$$

Then we see that $\overline{f}_j^{\Delta}(t)$ is a simple function. From (8.50) and the bounded convergence theorem of Lebesgue

$$\overline{s}_\mu(f, \Delta) = \sum_{j=1}^{k} \overline{f}_j \Delta_j \mu = \sum_{j=1}^{k} \overline{f}_j \int_{a_{j-1}}^{a_j} w(t)dt = \int_a^b \overline{f}_j^{\Delta}(t)w(t)dt$$

$$\to \int_a^b f(t)w(t)dt.$$

Thus, we conclude

$$\int_I f(t)d\mu(t) = \int_a^b f(t)w(t)dt,$$

that is, we have (8.49). ■

Theorem 8.34 (Integration by Parts) *Let f and μ be functions of bounded variation on $[a, b]$. Then we have*

$$\int_a^b f d\mu = f(b)\mu(b) - f(a)\mu(a) - \int_a^b \mu df.$$

Proof We may suppose that f and μ are monotonic increasing functions. Let

$$\Delta;\ a = a_0 < a_1 < \cdots < a_k = b.$$

Since $\sup\limits_{a_{j-1} \leqslant t \leqslant a_j} f(t) = f(a_j)$, we have

$$\begin{aligned}\overline{s}_\mu(f, \Delta) &= \sum_{j=1}^{k} f(a_j)\left[\mu(a_j) - \mu(a_{j-1})\right] \\ &= f(b)\mu(b) - f(a)\mu(a) - \sum_{j=1}^{k} \mu(a_{j-1})\left[f(a_j) - f(a_{j-1})\right].\end{aligned}$$

Then, we see

$$\overline{s}_\mu(f, \Delta) = \sum_{j=1}^{k} f(a_j)[\mu(a_j) - \mu(a_{j-1})] \to \int_a^b f d\mu,$$

and

$$\sum_{j=1}^{k} \mu(a_{j-1})\left[f(a_j) - f(a_{j-1})\right] \to \int_a^b \mu df.$$

■

8.3 Exercises

1. Define the Simple as well as Step functions and let s and t be non-negative measurable simple functions then show that

$$\int_X (s+t)d\mu = \int_X s d\mu + \int_X t d\mu.$$

2. Define the Lebesgue outer measure and show that a countable union of measurable sets is measurable.
3. Compare Lebesgue and Riemann integrals.
4. Prove that a function $f : [a, b] \to \mathbb{R}$ is measurable if and only if there is a sequence of simple functions $\{\varphi_n\}$ converging to f almost everywhere.
5. Define types of convergence. If $\{f_n\}$ is a uniformly bounded sequence of measurable functions to f a.e. on $[a, b]$, then show that f is measurable and

$$\lim_{n\to\infty}\int_a^b f_n d\mu = \int_a^b \lim_{n\to\infty} f_n d\mu = \int_a^b f d\mu.$$

6. Prove that the constant functions are measurable.
7. Prove that the continuous functions are measurable.
8. If

$$f(x) = \begin{cases} 1/x, & \text{if } 0 < x \leqslant 1 \\ 9, & \text{if } x = 0. \end{cases}$$

   then show that f is not Lebesgue integrable.
9. If

$$f(x) = \begin{cases} 1/x^{2/3}, & \text{if } 0 < x < 1 \\ 0, & \text{if } x = 0. \end{cases}$$

   Show that f is Lebesgue integrable on [0, 1] and

$$\int_0^1 1/x^{2/3} dx = 3.$$

   Find also $F(x, 2)$.

Selected Answers

Answer 1

3. $0, 4$ and 3^4.
5. $\sin x^2$, $(\sin x)^2$.
8. (i), (iii), (iv) are countable and (ii), (v) are uncountable.
10. $l.u.b. = 1/2,\ g.l.b. = 4/3$.
11. Yes.
12. $A = [0, 1)$ with $\sup A = 1$, but max does not exist.
13. $A = \{x \in Q : 0 < x^2 < 2\}$.
14. Singleton set.
24. 1.
25. It is not one to one but it is onto.

Answer 2

1. (i) (b) (ii) (a) (iii) (c).
2. (i) $3n - 1$.
   (ii) $x_n = 2$ when n is odd, and $x_n = 3$ when n is even.
   (iii) $x_n = x_{n-1} + n,\ n > 2$ with $x_1 = 1$.
   (iv) $x_1 = 1,\ x_2 = 1$ and $x_n = x_{n-1} + x_{n-2},\ n > 2$.
3. (i), (ii), (iv) are convergent and (iii) is divergent.
4. (i) $1/2$. (ii) $\{0\} \cup \mathbb{R}^+$. (iii) $-1, 1$. (iv) $A \cup \{0\}$.
5. $(-1)^n$.
6. n.
8. Yes.
12. $(-1)^n$.
17. $n, \frac{1}{n} - n$.
18. (i) $\limsup = 1,\ \liminf = -2$.
    (ii) $\limsup = \sqrt{2},\ \liminf = -\sqrt{2}$.
    (iii) $\limsup = 2,\ \liminf = -2$.

N. Deo and R. Sakai, *Introduction to Mathematical Analysis*,
https://doi.org/10.1007/978-981-97-6568-3

Answers 3

1. (b), (c), (d), (f), (e) are convergent. (a), (g), (h), (i), (j) are divergent, and (k) is convergent if $x < 1$, and divergent if $x \geqslant 1$.
2. (a) Convergent for $x > 1$ and divergent for $x \geqslant 1$.
   (b) Convergent for $x > 1$ and divergent for $x \geqslant 1$.
   (c) Convergent if $x < 1/4$ and divergent if $x \geqslant 1/4$.
8. 50: due to an equal quantity of milk in A and water in B, the answer is obvious.
9. Convergent for $x < 1$ and divergent for $x \geqslant 1$.
10. Convergent for $x \geqslant 1$ and divergent for $x > 1$.

Answer 4

1. 0, 0, 1.
3. 1, 1/3.
4. $\alpha = \pm 1$.
12. $a^n f'(a) - f(a)na^{n-1}$.

Answer 5

6. The discrete metric space.
7. (0, 1).
11 $f(x) = x^2,\ x \in \mathbb{R}$.

Answers 7

1. (i) Divergent when $x < 0$ and convergent when $x \geqslant 0$.
   (ii) Convergent for all x.
5. Non-uniformly convergent.
9. Uniform convergence when $x \neq 0$.
12. The series is not term-by-term integrable.

Bibliography

1. M. Tom Apostol, *Mathematical Analysis*, 2nd edn. (Addison-Wesley Publishing Company, Inc., 1974). ISBN: 0-201-00288-4
2. R.R. Goldberg, *Methods of Real Analysis*, 2nd edn. (Wiley Inc, New York, 1976)
3. K.A. Ross, *Elementary Analysis: The Theory of Calculus* (Springer, New York, 2013)
4. R.B. Reisel, *Elementary Theory of Metric Spaces* (Springer, New York, 1982), p. 13. 978-0-387-90706-2
5. J. R. Giles, *Introduction to the Analysis of Metric Spaces*, Australian Mathematical Society Lecture Series, vol. 3 (Cambridge University Press, Melbourne, 1987), ISBN: 0521350514
6. E.T. Copson, *Metric Spaces* (Cambridge University Press, UK); February 1988; ISBN: 9780511863455
7. C.G.C. Pitts, *Introduction to Metric Spaces* (Oliver & Boyd, Edinburgh, UK, 1972). 0050024531
8. C.G. Denlinger, *Elements of Real Analysis* (Jones & Bartlett, New Delhi, 2011). 978-9380853154
9. S. Kumaresan, *Topology of Metric Spaces* (Alpha Science International Ltd., Harrow, U.K., 2005). 1-84265-250-8
10. S.C. Malik, S. Arora, *Mathematical Analysis, New Age International Private Limited*, 1st edn. (2017). ISBN-13: 978-9385923869
11. E.D. Gaughan, *Introduction to Analysis*, 5th edn. (American Mathematical Society, 2009)
12. P.K. Jain, K. Ahmad, *Metric Space*, 2nd edn. (Narosa Publishing House, New Delhi, 2015), Ninth Reprint
13. A. Kumar, S. Kumaresan, *A Basic Course in Real Analysis* (CRC Press and Taylor & Francis Group, LLC, 2014); ISBN:13: 978-1-4822-1638-7 (eBook-PDF)
14. M.H. Protter, *Basic Elements of Real Analysis* (Springer, New York, 1998). ISBN: 978-0-387-98479-7
15. E. Kreyszig, *Introductory Functional Analysis, with Applications* (Wiley Inc, Toronto, 1978). 0-471-50731-8 (1978)
16. V.A. Zorich, *Mathematical Analysis II* (Springer, Berlin, 2016). ISBN: 978-3-662-48991-8
17. W.F. Trench, *Introduction to Real Analysis* (Trinity University San Antonio, Texas, 2013); ISBN: 0-13-045786-8

N. Deo and R. Sakai, *Introduction to Mathematical Analysis*,
https://doi.org/10.1007/978-981-97-6568-3

18. H.L. Royden, *Real Analysis*, 2nd edn. (Macmillan, New York, 1968)
19. V. Gupta, R.P. Agarwal, *Convergence Estimates in Approximation Theory* (Springer, Cham, 2014). ISBN: 978-3-319-02764-7
20. Naokant Deo, Vijay Gupta, Ana-Maria Acu, Purshottam Narain Agrawal, Mathematical Analysis I - Approximation Theory, Springer-Verlag, Print ISBN: 978-981-15-1152-3; Electronic ISBN: 978-981-15-1153-0

Printed in the United States
by Baker & Taylor Publisher Services